Engineering Materials 1

An Introduction to Properties, Applications and Design

Engineering Materials 1

An Introduction to Properties, Applications and Design

Third Edition

by

Michael F. Ashby

and

David R. H. Jones

Department of Engineering, University of Cambridge, UK

ELSEVIER
BUTTERWORTH
HEINEMANN

Amsterdam • Boston • Heidelberg • London • New York • Oxford
Paris • San Diego • San Francisco • Singapore • Sydney • Tokyo

Elsevier Butterworth-Heinemann
Linacre House, Jordan Hill, Oxford OX2 8DP
200 Wheeler Road, Burlington, MA 01803

First published 1980
Second edition 1996
Reprinted 1998 (twice), 2000, 2001, 2002, 2003
Third edition 2005

British Library Cataloguing in Publication Data
A catalogue record for this book is available from the British Library

Library of Congress Cataloging in Publication Data
A catalog record for this book is available from the Library of Congress

ISBN 0 7506 63804

For information on all Elsevier Butterworth-Heinemann publications visit our website at http://www.books.elsevier.com

Typeset by Newgen Imaging Systems (P) Ltd, Chennai, India
Printed and bound in Great Britain

Contents

General introduction

To the student

Innovation in engineering often means the clever use of a new material — new to a particular application, but not necessarily (although sometimes) new in the sense of recently developed. Plastic paper clips and ceramic turbine-blades both represent attempts to do better with polymers and ceramics what had previously been done well with metals. And engineering disasters are frequently caused by the misuse of materials. When the plastic teaspoon buckles as you stir your tea, and when a fleet of aircraft is grounded because cracks have appeared in the tailplane, it is because the engineer who designed them used the wrong materials or did not understand the properties of those used. So it is vital that the professional engineer should know how to select materials which best fit the demands of the design — economic and aesthetic demands, as well as demands of strength and durability. The designer must understand the properties of materials, and their limitations.

This book gives a broad introduction to these properties and limitations. It cannot make you a materials expert, but it can teach you how to make a sensible choice of material, how to avoid the mistakes that have led to embarrassment or tragedy in the past, and where to turn for further, more detailed, help.

You will notice from the Contents list that the chapters are arranged in *groups*, each group describing a particular class of properties: elastic modulus; fracture toughness; resistance to corrosion; and so forth. Each group of chapters starts by *defining the property*, describing how it is *measured*, and giving *data* that we use to solve problems involving design with materials. We then move on to the *basic science* that underlies each property, and show how we can use this fundamental knowledge to choose materials with better properties. Each group ends with a chapter of *case studies* in which the basic understanding and the data for each property are applied to practical engineering problems involving materials.

At the end of each chapter you will find a set of examples; each example is meant to consolidate or develop a particular point covered in the text. Try to do the examples from a particular chapter while this is still fresh in your mind. In this way you will gain confidence that you are on top of the subject.

No engineer attempts to learn or remember tables or lists of data for material properties. But you *should* try to remember the broad orders of magnitude of these quantities. All foodstores know that "a kg of apples is about 10 apples" — they still weigh them, but their knowledge prevents them making silly mistakes which might cost them money. In the same way an engineer should know that "most elastic moduli lie between 1 and $10^3\,\mathrm{GN\,m^{-2}}$; and are around $10^2\,\mathrm{GN\,m^{-2}}$ for metals" — in any real design you need an accurate value, which you can get from suppliers' specifications; but an order of magnitude knowledge prevents you getting the units wrong, or making other silly, and possibly expensive, mistakes. To help you in this, we have added at the end of the book a list of the important definitions and formulae that you should know, or should be able to derive, and a summary of the orders of magnitude of materials properties.

To the lecturer

This book is a course in Engineering Materials for engineering students with no previous background in the subject. It is designed to link up with the teaching of Design, Mechanics, and Structures, and to meet the needs of engineering students for a first materials course, emphasizing design applications.

The text is deliberately concise. Each chapter is designed to cover the content of one 50-minute lecture, thirty-one in all, and allows time for demonstrations and graphics. The text contains sets of worked case studies which apply the material of the preceding block of lectures. There are examples for the student at the end of the each chapter.

We have made every effort to keep the mathematical analysis as simple as possible while still retaining the essential physical understanding, and still arriving at results which, although approximate, are useful. But we have avoided mere description: most of the case studies and examples involve analysis, and the use of data, to arrive at numerical solutions to real or postulated problems. This level of analysis, and these data, are of the type that would be used in a preliminary study for the selection of a material or the analysis of a design (or design-failure). It is worth emphasizing to students that the next step would be a detailed analysis, using *more precise mechanics* and *data from the supplier of the material or from in-house testing*. Materials data are notoriously variable. Approximate tabulations like those given here, though useful, should never be used for final designs.

Acknowledgements

The authors and publishers are grateful to the following copyright holders for permission to reproduce their photographs in the following figures: 1.3, Rolls—Royce Ltd; 1.5, Catalina Yachts Inc; 7.1, Photo Labs, Royal Observatory, Edinburgh; 9.11, Dr Peter Southwick; 31.7, Group Lotus Ltd; 31.2 Photo credit to Brian Garland © 2004, Courtesy of Volkswagen.

Accompanying Resources

The following accompanying web-based resources are available to teachers and lecturers who adopt or recommend this text for class use. For further details and access to these resources please go to http://books.elsevier.com/manuals

Instructor's Manual

A full Solutions Manual with worked answers to the exercises in the main text is available for downloading.

Image bank

An image bank of downloadable PDF versions of the figures from the book is available for use in lecture slides and class presentations.

Online Materials Science Tutorials

A series of online materials science tutorials accompanies *Engineering Materials 1* and 2. These were developed by Alan Crosky, Mark Hoffman, Paul Munroe and Belinda Allen at the University of New South Wales (UNSW) Australia, based upon earlier editions of the books. The group is particularly interested in the effective and innovative use of technology in teaching. They realised the potential of the material for the teaching of Materials Engineering to their students in an online environment and have developed and then used these very popular tutorials for a number of years at UNSW. The results of this work have also been published and presented extensively.

The tutorials are designed for students of materials science as well as for those studying materials as a related or elective subject, for example mechanical or civil engineering students. They are ideal for use as ancillaries to formal teaching programs, and may also be used as the basis for quick refresher courses for more advanced materials science students. By picking selectively from the range of tutorials available they will also make ideal subject primers for students from related faculties.

The software has been developed as a self-paced learning tool, separated into learning modules based around key materials science concepts. For further information on accessing the tutorials, and the conditions for their use, please go to http://books.elsevier.com/manuals

About the authors of the Tutorials

Alan Crosky is an Associate Professor in the School of Materials Science and Engineering, UNSW. His teaching specialties include metallurgy, composites and fractography.

Belinda Allen is an Educational Graphics Manager and Educational Designer at the Educational Development and Technology Centre, UNSW. She provides consultation and production support for the academic community and designs and presents workshops and online resources on image production and web design.

Mark Hoffman is an Associate Professor in the School of Materials Science and Engineering, UNSW. His teaching specialties include fracture, numerical modelling, mechanical behaviour of materials and engineering management.

Paul Munroe has a joint appointment as Professor in the School of Materials Science and Engineering and Director of the Electron Microscope Unit, UNSW. His teaching specialties are the deformation and strengthening mechanisms of materials and crystallographic and microstructural characterisation.

Chapter 1

Engineering materials and their properties

Chapter contents

1.1 Introduction

There are, it is said, more than 50,000 materials available to the engineer. In designing a structure or device, how is the engineer to choose from this vast menu the material which best suits the purpose? Mistakes can cause disasters. During the Second World War, one class of welded merchant ship suffered heavy losses, not by enemy attack, but by breaking in half at sea: the *fracture toughness* of the steel—and, particularly, of the welds 1-1 was too low. More recently, three Comet aircraft were lost before it was realized that the design called for a *fatigue strength* that—given the design of the window frames—was greater than that possessed by the material. You yourself will be familiar with poorly designed appliances made of plastic: their excessive "give" is because the designer did not allow for the low *modulus* of the polymer. These bulk properties are listed in Table 1.1, along with other common classes of property that the designer must consider when choosing a material. Many of these

Table 1.1 Classes of property

Economic	Price and availability Recyclability
General Physical	Density
Mechanical	Modulus Yield and tensile strength Hardness Fracture toughness Fatigue strength Creep strength Damping
Thermal	Thermal conductivity Specific heat Thermal expansion coefficient
Electrical and Magnetic	Resistivity Dielectric constant Magnetic permeability
Environmental Interaction	Oxidation Corrosion Wear
Production	Ease of manufacture Joining Finishing
Aesthetic	Colour Texture Feel

properties will be unfamiliar to you—we will introduce them through examples in this chapter. They form the basis of this first course on materials.

In this first course, we shall also encounter the *classes of materials* shown in Table 1.2 and Figure 1.1. More engineering components are made of *metals and alloys* than of any other class of solid. But increasingly, *polymers* are replacing metals because they offer a combination of properties which are more attractive to the designer. And if you've been reading the newspaper, you will know that the new *ceramics*, at present under development world wide, are an emerging class of engineering material which may permit more efficient heat engines, sharper knives, and bearings with lower friction. The engineer can combine the best properties of these materials to make *composites* (the most familiar is fiberglass) which offer specially attractive packages of

Table 1.2 Classes of materials

Metals and alloys	Iron and steels Aluminium and its alloys Copper and its alloys Nickel and its alloys Titanium and its alloys
Polymers	Polyethylene (PE) Polymethylmethacrylate (acrylic and PMMA) Nylon, alias polyamide (PA) Polystyrene (PS) Polyurethane (PU) Polyvinylchloride (PVC) Polyethylene terephthalate (PET) Polyethylether ketone (PEEK) Epoxies (EP) Elastomers, such as natural rubber (NR)
Ceramics and glasses*	Alumina (Al_2O_3, emery, sapphire) Magnesia (MgO) Silica (SiO_2) glasses and silicates Silicon carbide (SiC) Silicon nitride (Si_3N_4) Cement and concrete
Composites	Fiberglass (GFRP) Carbon-fiber reinforced polymers (CFRP) Filled polymers Cermets
Natural materials	Wood Leather Cotton/wool/silk Bone

* Ceramics are crystalline, inorganic, nonmetals. Glasses are noncrystalline (or *amorphous*) solids. Most engineering glasses are nonmetals, but a range of *metallic glasses* with useful properties is now available.

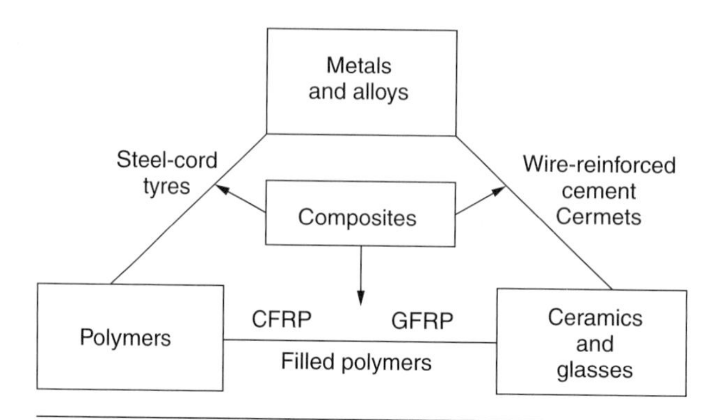

Figure 1.1 The classes of engineering materials from which articles are made.

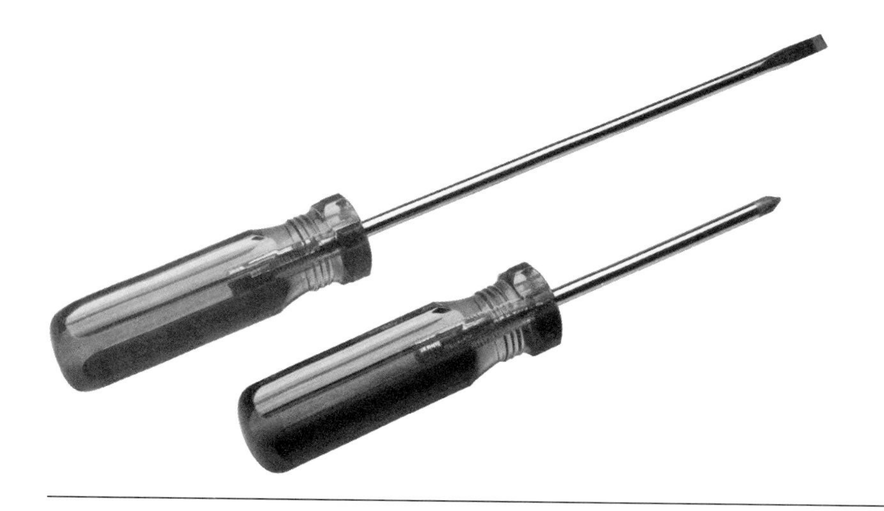

Figure 1.2 Typical screwdrivers, with steel shaft and polymer (plastic) handle.

properties. And—finally—one should not ignore *natural materials* like wood and leather which have properties which—even with the innovations of today's materials scientists—are hard to beat.

In this chapter we illustrate, using a variety of examples, how the designer selects materials so that they provide him or her with the properties needed.

1.2 Examples of materials selection

A typical screwdriver (Figure 1.2) has a shaft and blade made of carbon steel, a metal. Steel is chosen because its *modulus* is high. The modulus measures the resistance of the material to elastic deflection or bending. If you made the shaft out of a polymer like polyethylene instead, it would twist far too much. A high modulus is one criterion in the selection of a material for this application. But it is not the only one. The shaft must have a high *yield strength.* If it does not, it will bend or twist if you turn it hard (bad screwdrivers do). And the blade must have a high *hardness,* otherwise it will be damaged by the head of the screw. Finally, the material of the shaft and blade must not only do all these things, it must also resist fracture—glass, for instance, has a high modulus, yield strength, and hardness, but it would not be a good choice for this application because it is so brittle. More precisely, it has a very low *fracture toughness.* That of the steel is high, meaning that it gives a bit before it breaks.

The handle of the screwdriver is made of a polymer or plastic, in this instance polymethylmethacrylate, otherwise known as PMMA, plexiglass or perspex. The handle has a much larger section than the shaft, so its twisting, and thus its modulus, is less important. You could not make it satisfactorily out of a soft rubber (another polymer) because its modulus is much too low, although a thin skin of rubber might be useful because its *friction coefficient* is high, making it easy to grip. Traditionally, of course, tool handles were made of another natural polymer—wood—and, if you measure importance by the volume consumed per year, wood is still by far the most important polymer available to the engineer. Wood has been replaced by PMMA because PMMA becomes soft when hot and can be molded quickly and easily to its final shape. Its *ease of fabrication* for this application is high. It is also chosen for aesthetic reasons: its *appearance,* and feel or *texture,* are right; and its *density* is low, so that the screwdriver is not unnecessarily heavy. Finally, PMMA is cheap, and this allows the product to be made at a reasonable *price.*

Now a second example (Figure 1.3), taking us from low technology to the advanced materials design involved in the turbofan aeroengines which power large planes. Air is propelled past (and into) the engine by the turbofan, providing aerodynamic thrust. The air is further compressed by the compressor blades, and is then mixed with fuel and burnt in the combustion chamber. The expanding gases drive the turbine blades, which provide power to the turbofan and the compressor blades, and finally pass out of the rear of the engine, adding to the thrust.

The *turbofan blades* are made from a titanium alloy, a metal. This has a sufficiently good modulus, yield strength, and fracture toughness. But the metal must also resist *fatigue* (due to rapidly fluctuating loads), *surface wear* (from striking everything from water droplets to large birds) and *corrosion* (important when taking off over the sea because salt spray enters the engine). Finally, *density* is extremely important for obvious reasons: the heavier the engine, the less the payload the plane can carry. In an effort to reduce weight even further, composite blades made of carbon-fiber reinforced polymers (CFRP) with density less than one-half of that of titanium, have been tried. But CFRP, by itself is simply not tough enough for turbofan blades—a "bird strike" demolishes a CFRP blade. The problem can be overcome by *cladding,* giving the CFRP a metallic leading edge.

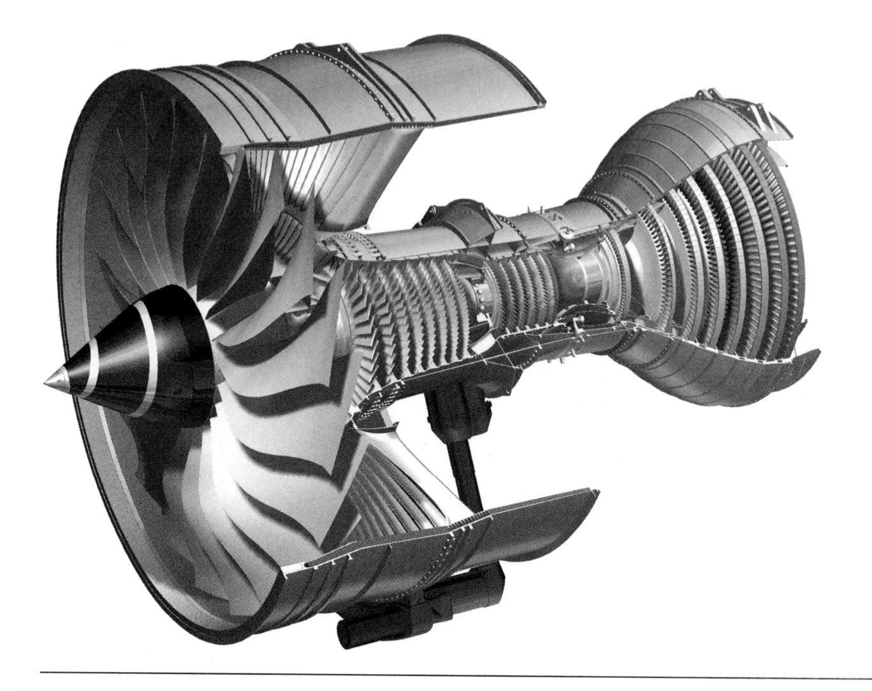

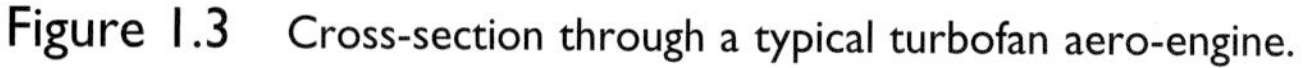

Figure 1.3 Cross-section through a typical turbofan aero-engine.

Turning to the *turbine blades* (those in the hottest part of the engine) even more material requirements must be satisfied. For economy the fuel must be burnt at as high a temperature as possible. The first row of engine blades (the "HP1" blades) runs at metal temperatures of about 950°C, requiring resistance to *creep* and to *oxidation*. Nickel-based alloys of complicated chemistry and structure are used for this exceedingly stringent application; they are one pinnacle of advanced materials technology.

An example which brings in somewhat different requirements is the *spark plug* of an internal combustion engine (Figure 1.4). The *spark electrodes* must resist *thermal fatigue* (from rapidly fluctuating temperatures), *wear* (caused by spark erosion), and *oxidation and corrosion* from hot upper-cylinder gases containing nasty compounds of sulphur. Tungsten alloys are used for the electrodes because they have the desired properties.

The *insulation* around the central electrode is an example of a nonmetallic material—in this case, alumina, a ceramic. This is chosen because of its electrical insulating properties and because it also has good thermal fatigue resistance and resistance to corrosion and oxidation (it is an oxide already).

The use of nonmetallic materials has grown most rapidly in the consumer industry. Our next example, a sailing cruiser (Figure 1.5), shows just how extensively polymers and manmade composites and fibers have replaced the "traditional" materials of steel, wood, and cotton. A typical cruiser has a *hull*

Figure 1.4 A petrol engine spark plug, with tungsten electrodes and ceramic body.

made from GFRP, manufactured as a single molding; GFRP has good *appearance* and, unlike steel or wood, does not rust or become eaten away by Terido worm. The *mast* is made from aluminum alloy, which is lighter for a given strength than wood; advanced masts are now being made by reinforcing the alloy with carbon or boron fibers (man-made composites). The sails, formerly of the natural material cotton, are now made from the polymers nylon, Terylene or Kevlar, and, in the running rigging, cotton ropes have been replaced by polymers also. Finally, polymers like PVC are extensively used for things like fenders, anoraks, buoyancy bags, and boat covers.

Three man-made composite materials have appeared in the items we have considered so far: GFRP; the much more expensive CFRP; and the still *more* expensive boron-fiber reinforced alloys (BFRP). The range of composites is a large and growing one (Figure 1.1); during the next decade composites will, increasingly, compete with steel and aluminium in many traditional uses of these metals.

So far we have introduced the mechanical and physical properties of engineering materials, but we have yet to discuss a consideration which is often of overriding importance: that of *price and availability*.

Table 1.3 shows a rough breakdown of material prices. Materials for large-scale structural use—wood, cement and concrete, and structural steel—cost between UK£50 and UK£500 (US$90 and US$900) per tonne. There are many materials which have all the other properties required of a structural material—nickel or titanium, for example—but their use in this application is eliminated by their price.

Figure 1.5 A sailing cruiser, with composite (GFRP) hull, aluminum alloy mast and sails made from synthetic polymer fibers.

Table 1.3 Breakdown of material prices

Class of use	*Material*	*Price per tonne*	
Basic construction	Wood, concrete, structural steel	UK£50–500	US$90–900
Medium and light engineering	Metals, alloys, and polymers for aircraft, automobiles, appliances, etc.	UK£500–5000	US$900–9000
Special materials	Turbine-blade alloys, advanced composites (CFRP, BFRP), etc.	UK£5000–50,000	US$9000–90,000
Precious metals, etc.	Sapphire bearings, silver contacts, gold microcircuits	UK£50,000–10m	US$90,000–18m
Industrial diamond	Cutting and polishing tools	>UK£100m	>US$180m

The value that is added during light- and medium-engineering work is larger, and this usually means that the economic constraint on the choice of materials is less severe — a far greater proportion of the cost of the structure is that associated with labor or with production and fabrication. Stainless steels, most aluminum alloys and most polymers cost between UK£500 and UK£5000 (US$900 and US$9000) per tonne. It is in this sector of the market that the competition between materials is most intense, and the greatest scope for imaginative design exists. Here polymers and composites compete directly with metals, and new structural ceramics (silicon carbide and silicon nitride) may compete with both in certain applications.

Next there are the materials developed for high-performance applications, some of which we have mentioned already: nickel alloys (for turbine blades), tungsten (for spark-plug electrodes), and special composite materials such as CFRP. The price of these materials ranges between UK£5000 and UK£50,000 (US$9000 and US$90,000) per tonne. This the régime of high materials technology, actively under research, and in which major new advances are continuing to be made. Here, too, there is intense competition from new materials.

Finally, there are the so-called precious metals and gemstones, widely used in engineering: gold for microcircuits, platinum for catalysts, sapphire for

Figure 1.6 The wooden bridge at Queens' College, Cambridge, a 1902 reconstruction of the original bridge built in 1749 to William Etheridge's design.

Figure 1.7 Clare Bridge, built in 1640, is Cambridge's oldest surviving bridge; it is reputed to have been an escape route from the college in times of plague.

Figure 1.8 Magdalene Bridge built in 1823 on the site of the ancient Saxon bridge over the Cam. The present cast-iron arches carried, until recently, loads far in excess of those envisaged by the designers. Fortunately, the bridge has now undergone a well-earned restoration.

Figure 1.9 A typical twentieth-century mild-steel bridge; a convenient crossing to the Fort St George inn!

Figure 1.10 The reinforced concrete footbridge in Garret Hostel Lane. An inscription carved nearby reads: "This bridge was given in 1960 by the Trusted family members of Trinity Hall. It was designed by Timothy Guy Morgan an undergraduate of Jesus College who died in that year."

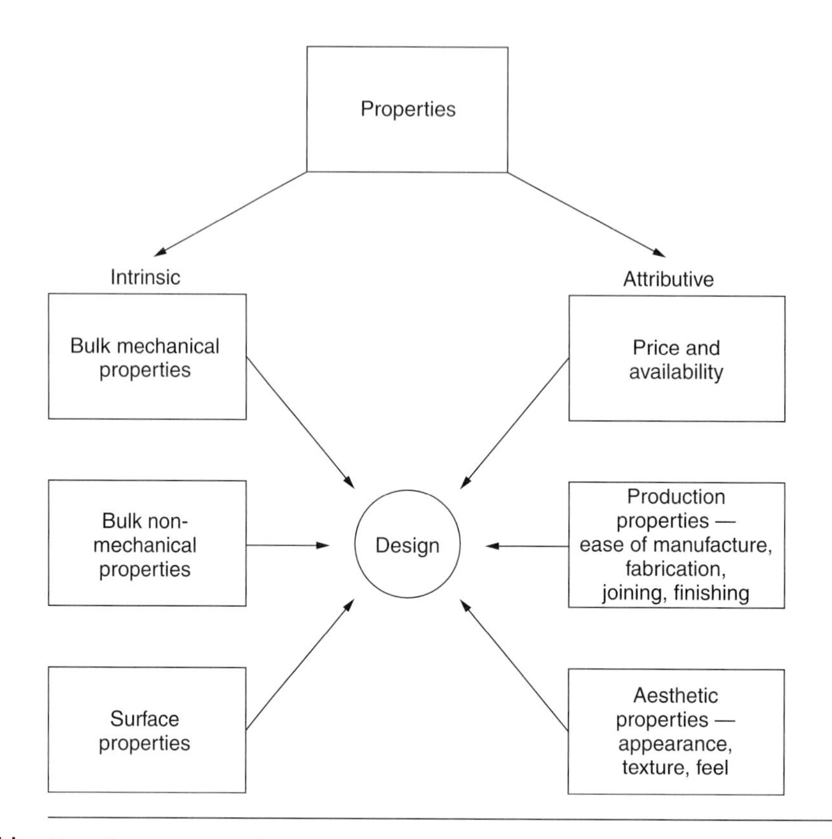

Figure 1.11 How the properties of engineering materials affect the way in which products are designed.

bearings, diamond for cutting tools. They range in price from UK£50,000 (US$90,000) to well over UK£100m (US$180m) per tonne.

As an example of how price and availability affect the choice of material for a particular job, consider how the materials used for building bridges in Cambridge have changed over the centuries. As our photograph of Queens' Bridge (Figure 1.6) suggests, until 150 years or so ago wood was commonly used for bridge building. It was cheap, and high-quality timber was still available in large sections from natural forests. Stone, too, as the picture of Clare Bridge (Figure 1.7) shows, was widely used. In the eighteenth century the ready availability of cast iron, with its relatively low assembly costs, led to many cast-iron bridges of the type exemplified by Magdalene Bridge (Figure 1.8). Metallurgical developments of the later nineteenth century allowed large mild-steel structures to be built (the Fort St George footbridge, Figure 1.9). Finally, the advent of cheap reinforced concrete led to graceful and durable structures like that of the Garret Hostel Lane bridge (Figure 1.10). This evolution clearly illustrates how availability influences the choice of materials.

Nowadays, wood, steel, and reinforced concrete are often used interchangeably in structures, reflecting the relatively small *price* differences between them. The choice of which of the three materials to use is mainly dictated by the kind of structure the architect wishes to build: chunky and solid (stone), structurally efficient (steel), slender, and graceful (pre-stressed concrete).

Engineering design, then, involves many considerations (Figure 1.11). The choice of a material must meet certain criteria on bulk and surface properties (e.g. strength and corrosion resistance). But it must also be easy to fabricate; it must appeal to potential consumers; and it must compete economically with other alternative materials. In the next chapter we consider the economic aspects of this choice, returning in later chapters to a discussion of the other properties.

Part A

Price and availability

Chapter 2

The price and availability of materials

Chapter contents

2.1 Introduction

In the first chapter we introduced the range of properties required of engineering materials by the design engineer, and the range of materials available to provide these properties. We ended by showing that the *price* and *availability* of materials were important and often overriding factors in selecting the materials for a particular job. In this chapter we examine these economic properties of materials in more detail.

2.2 Data for material prices

Table 2.1 ranks materials by their relative cost per unit weight. The most expensive materials—diamond, platinum, gold—are at the top. The cheapest—cast iron, wood, cement—are at the bottom. Such data are obviously important in choosing a material. How do we keep informed about materials price changes and what controls them?.

The Financial Times and the *Wall Street Journal* give some, on a daily basis. Trade supply journals give more extensive lists of current prices. A typical such journal is *Procurement Weekly*, listing current prices of basic materials, together with prices 6 months and a year ago. All manufacturing industries take this or something equivalent—the workshop in your engineering department will have it—and it gives a guide to prices and their trends.

Figure 2.1 shows the variation in price of two materials—copper and rubber. These short-term price fluctuations have little to do with the real scarcity or abundance of materials. They are caused by small differences between the rate of supply and demand, much magnified by speculation in commodity futures. The volatile nature of the commodity market can result in large changes over a period of a few days—that is one reason speculators are attracted to it—and there is little that an engineer can do to foresee these changes. Political factors are also extremely important—a scarcity of cobalt in 1978 was due to the guerilla attacks on mineworkers in Zaire, the world's principal producer of cobalt; the low price of aluminum and diamonds in 1995 was partly caused by a flood of both from Russia at the end of the Cold War.

The long-term changes are of a different kind. They reflect, in part, the real cost (in capital investment, labor, and energy) of extracting and transporting the ore or feedstock and processing it to give the engineering material. Inflation and increased energy costs obviously drive the price up; so, too, does the necessity to extract materials, like copper, from increasingly lean ores; the leaner the ore, the more machinery and energy are required to crush the rock containing it, and to concentrate it to the level that the metal can be extracted.

In the long term, then, it is important to know which materials are basically plentiful, and which are likely to become scarce. It is also important to know the extent of our dependence on materials.

Table 2.1 Approximate relative price per tonne (mild steel = 100)

Material	*Relative price $*
Diamonds, industrial	200m
Platinum	5m
Gold	2m
Silver	150,000
CFRP (mats. 70% of cost; fabr. 30% of cost)	20,000
Cobalt/tungsten carbide cermets	15,000
Tungsten	5000
Cobalt alloys	7000
Titanium alloys	10,000
Nickel alloys	20,000
Polyimides	8000
Silicon carbide (fine ceramic)	7000
Magnesium alloys	1000
Nylon 66	1500
Polycarbonate	1000
PMMA	700
Magnesia, MgO (fine ceramic)	3000
Alumina, Al_2O_3 (fine ceramic)	3000
Tool steel	500
GFRP (mats. 60% of cost; fabr. 40% of cost)	1000
Stainless steels	600
Copper, worked (sheets, tubes, bars)	400
Copper, ingots	400
Aluminum alloys, worked (sheet, bars)	400
Aluminum ingots	300
Brass, worked (sheet, tubes, bars)	400
Brass, ingots	400
Epoxy	1000
Polyester	500
Glass	400
Foamed polymers	1000
Zinc, worked (sheet, tubes, bars)	400
Zinc, ingots	350
Lead, worked (bars, sheet, tube)	250
Lead, ingots	200
Natural rubber	300
Polypropylene	200
Polyethylene, high density	200
Polystyrene	250
Hard woods	250
Polyethylene, low density	200
Polyvinyl chloride	300
Plywood	200
Low-alloy steels	130

Table 2.1 (*Continued*)

Material	*Relative price $*
Mild steel, worked (angles, sheet, bars)	100
Cast iron	90
Iron, ingots	70
Soft woods	70
Concrete, reinforced (beams, columns, slabs)	50
Fuel oil	50
Cement	20
Coal	20

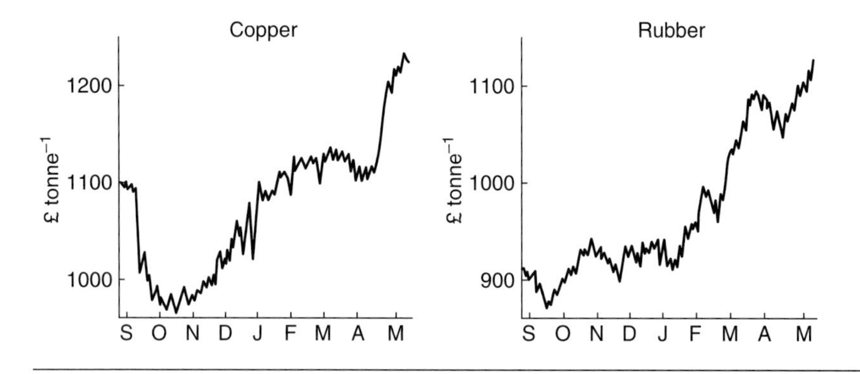

Figure 2.1 Recent fluctuations in the price of copper and rubber.

2.3 The use-pattern of materials

The way in which materials are used in an industrialized nation is fairly standard. It consumes steel, concrete, and wood in construction; steel and aluminum in general engineering; copper in electrical conductors; polymers in appliances, and so forth; and roughly in the same proportions. Among metals, steel is used in the greatest quantities by far: 90 percent of all the metal produced in the world is steel. But the nonmetals wood and concrete beat steel—they are used in even greater volume.

About 20 percent of the total import bill is spent on engineering materials. Table 2.2 shows how this spend is distributed. Iron and steel, and the raw materials used to make them, account for about a quarter of it. Next are wood and lumber—widely used in light construction. More than a quarter is spent on the metals copper, silver, aluminum, and nickel. All polymers taken together, including rubber, account for little more than 10 percent. If we include the further metals zinc, lead, tin, tungsten, and mercury, the list accounts for

Table 2.2 Imports of engineering materials, raw, and semis: percentage of total cost

Iron and steel	27
Wood and lumber	21
Copper	13
Plastics	9.7
Silver and platinum	6.5
Aluminum	5.4
Rubber	5.1
Nickel	2.7
Zinc	2.4
Lead	2.2
Tin	1.6
Pulp/paper	1.1
Glass	0.8
Tungsten	0.3
Mercury	0.2
Etc.	1.0

99 percent of all the money spent abroad on materials, and we can safely ignore the contribution of materials which do not appear on it.

2.4 Ubiquitous materials

The composition of the earth's crust

Let us now shift attention from what we *use* to what is widely *available*. A few engineering materials are synthesized from compounds found in the earth's oceans and atmosphere: magnesium is an example. Most, however, are won by mining their ore from the earth's crust, and concentrating it sufficiently to allow the material to be extracted or synthesized from it. How plentiful and widespread are these materials on which we depend so heavily? How much copper, silver, tungsten, tin, and mercury in useful concentrations does the crust contain? All five are rare: workable deposits of them are relatively small, and are so highly localized that many governments classify them as of strategic importance, and stockpile them.

Not all materials are so thinly spread. Table 2.3 shows the relative abundance of the commoner elements in the earth's crust. The crust is 47 percent oxygen by weight or — because oxygen is a big atom, it occupies 96 percent of the volume (geologists are fond of saying that the earth's crust is solid oxygen

Table 2.3 Abundance of elements

Crust		Oceans		Atmosphere	
Element	weight %	Element	weight %	Element	weight %
Oxygen	47	Oxygen	85	Nitrogen	79
Silicon	27	Hydrogen	10	Oxygen	19
Aluminum	8	Chlorine	2	Argon	2
Iron	5	Sodium	1	Carbon dioxide	0.04
Calcium	4	Magnesium	0.1		
Sodium	3	Sulphur	0.1		
Potassium	3	Calcium	0.04		
Magnesium	2	Potassium	0.04		
Titanium	0.4	Bromine	0.007		
Hydrogen	0.1	Carbon	0.002		
Phosphorus	0.1				
Manganese	0.1				
Fluorine	0.06				
Barium	0.04				
Strontium	0.04				
Sulphur	0.03				
Carbon	0.02				

* The total mass of the crust to a depth of 1 km is 3×10^{21} kg; the mass of the oceans is 10^{20} kg; that of the atmosphere is 5×10^{18} kg.

containing a few percent of impurities). Next in abundance are the elements silicon and aluminum; by far the most plentiful solid materials available to us are silicates and alumino-silicates. A few metals appear on the list, among them iron and aluminum both of which feature also in the list of widely used materials. The list extends as far as carbon because it is the backbone of virtually all polymers, including wood. Overall, then, oxygen and its compounds are overwhelmingly plentiful—on every hand we are surrounded by oxide-ceramics, or the raw materials to make them. Some materials are widespread, notably iron and aluminum; but even for these the local concentration is frequently small, usually too small to make it economic to extract them. In fact, the raw materials for making polymers are more readily available at present than those for most metals. There are huge deposits of carbon in the earth: on a world scale, we extract a greater tonnage of carbon every month than we extract iron in a year, but at present we simply burn it. And the second ingredient of most polymers—hydrogen—is also one of the most plentiful of elements. Some materials—iron, aluminum, silicon, the elements to make glass, and cement—are plentiful and widely available. But others—mercury, silver, tungsten are examples—are scarce and highly localized, and—if the current pattern of use continues—may not last very long.

2.5 Exponential growth and consumption doubling-time

How do we calculate the lifetime of a resource like mercury? Like almost all materials, mercury is being consumed at a rate which is growing exponentially with time (Figure 2.2), simply because both population and living standards grow exponentially. We analyze this in the following way. If the current rate of consumption in tonnes per year is C then exponential growth means that

$$\frac{dC}{dt} = \frac{r}{100}C \tag{2.1}$$

where, for the generally small growth rates we deal with here (1–5 percent per year), r can be thought of as the percentage fractional rate of growth per year. Integrating gives

$$C = C_0 \exp\left\{\frac{r(t - t_0)}{100}\right\} \tag{2.2}$$

where C_0 is the consumption rate at time $t = t_0$. The *doubling-time* t_D of consumption is given by setting $C/C_0 = 2$ to give

$$t_D = \frac{100}{r}\log_e 2 \approx \frac{70}{r} \tag{2.3}$$

Steel consumption is growing at less than 2 percent per year—it doubles about every 35 years. Polymer consumption is rising at about 5 percent per

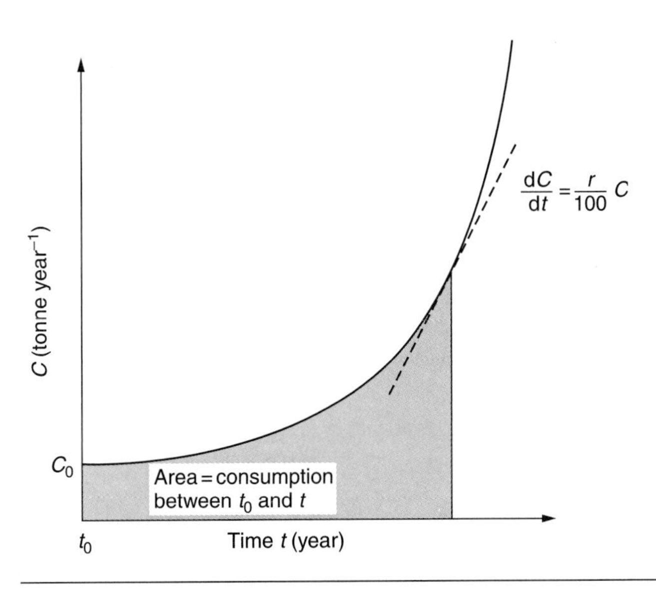

Figure 2.2 The exponentially rising consumption of materials.

year — it doubles every 14 years. During times of boom — the 1960s and 1970s for instance — polymer production increased much faster than this, peaking at 18 percent per year (it doubled every 4 years), but it has now fallen back to a more modest rate.

2.6 Resource availability

The availability of a resource depends on the degree to which it is *localized* in one or a few countries (making it susceptible to production controls or cartel action); on the *size* of the reserves, or, more accurately, the resource base (explained shortly); and on the *energy* required to mine and process it. The influence of the last two (size of reserves and energy content) can, within limits, be studied and their influence anticipated.

The calculation of resource life involves the important distinction between *reserves* and *resources*. The current reserve is the known deposits which can be extracted profitably at today's price using today's technology; it bears little relationship to the true magnitude of the resource base; in fact, the two are not even roughly proportional.

The resource base includes the current reserve. But it also includes all deposits that might become available given diligent prospecting and which, by various extrapolation techniques, can be estimated. And it includes, too, all known and unknown deposits that cannot be mined profitably now, but

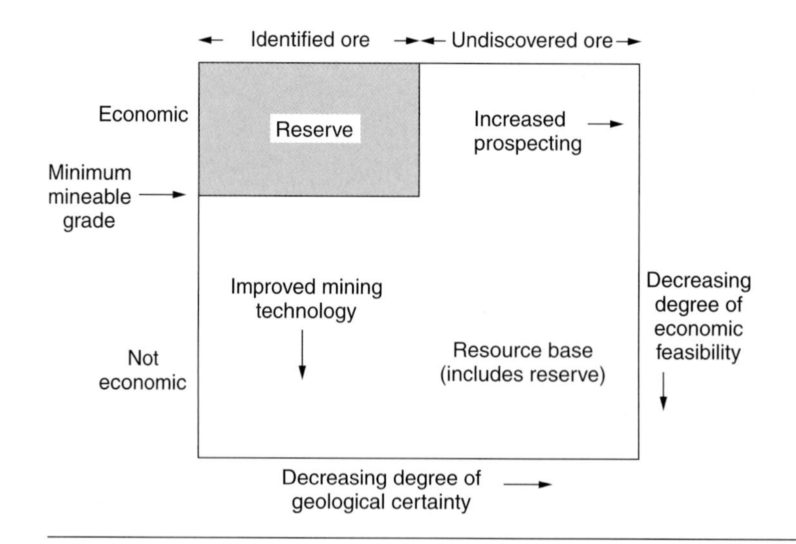

Figure 2.3 The distinction between the reserve and the resource base, illustrated by the McElvey diagram.

which—due to higher prices, better technology or improved transportation—might reasonably become available in the future (Figure 2.3). The reserve is like money in the bank—you know you have got it. The resource base is more like your total potential earnings over your lifetime—it is much larger than the reserve, but it is less certain, and you may have to work very hard to get it. The resource base is the realistic measure of the total available material. Resources are almost always much larger than reserves, but because the geophysical data and economic projections are poor, their evaluation is subject to vast uncertainty.

Although the resource base is uncertain, it obviously is important to have some estimate of how long it can last. Rough estimates do exist for the size of the resource base, and, using these, our exponential formula gives an estimate of how long it would take us to use up *half* of the resources. The half-life is an important measure: at this stage prices would begin to rise so steeply that supply would become a severe problem. For a number of important materials these half-lives lie within your lifetime: for silver, tin, tungsten, zinc, lead, mercury, and oil (the feed stock of polymers) they lie between 40 and 70 years. Others (most notably iron, aluminum, and the raw materials from which most ceramics and glasses are made) have enormous resource bases, adequate for hundreds of years, even allowing for continued exponential growth.

The cost of energy enters here. The extraction of materials requires energy (Table 2.4). As a material becomes scarcer—copper is a good example—it must be extracted from leaner and leaner ores. This expends more and more energy, per tonne of copper *metal* produced, in the operations of mining, crushing, and concentrating the ore; and these energy costs rapidly become prohibitive. The rising energy content of copper shown in Table 2.4 reflects the fact that the richer copper ores are, right now, being worked out.

Table 2.4 Approximate energy content of materials (GJ $tonne^{-1}$)

Aluminum	280
Plastics	85–180
Copper	140, rising to 300
Zinc	68
Steel	55
Glass	20
Cement	7
Brick	4
Timber	2.5–7
Gravel	0.2
Oil	44
Coal	29

2.7 The future

How are we going to cope with the shortages of engineering materials in the future? One way obviously is by

Material-efficient design

Many current designs use far more material than is necessary, or use potentially scarce materials where the more plentiful would serve. Often, for example, it is a surface property (e.g. low friction, or high corrosion resistance) which is wanted; then a thin surface film of the rare material bonded to a cheap plentiful substrate can replace the bulk use of a scarcer material. Another way of coping with shortages is by

Substitution

It is the property, not the material itself, that the designer wants. Sometimes a more readily available material can replace the scarce one, although this usually involves considerable outlay (new processing methods, new joining methods, etc.). Examples of substitution are the replacement of stone and wood by steel and concrete in construction; the replacement of copper by polymers in plumbing; the change from wood and metals to polymers in household goods; and from copper to aluminum in electrical wiring.

There are, however, technical limitations to substitution. Some materials are used in ways not easily filled by others. Platinum as a catalyst, liquid helium as a refrigerant, and silver on electrical contact areas cannot be replaced; they perform a unique function — they are, so to speak, the vitamins of engineering materials. Others — a replacement for tungsten for lamp filaments, for example — would require the development of a whole new technology, and this can take many years. Finally, substitution increases the demand for the replacement material, which may also be in limited supply. The massive trend to substitute plastics for other materials puts a heavier burden on petrochemicals, at present derived from oil. A third approach is that of

Recycling

Recycling is not new: old building materials have been recycled for millennia; scrap metal has been recycled for centuries; both are major industries. Recycling is labor intensive, and therein lies the problem in expanding its scope. Over the last 30 years, the rising cost of labor made most recycling less than economic.

2.8 Conclusion

Overall, the materials-resource problem is not as critical as that of energy. Some materials have an enormous base or (like wood) are renewable—and fortunately these include the major structural materials. For others, the resource base is small, but they are often used in small quantities so that the price could rise a lot without having a drastic effect on the price of the product in which they are incorporated; and for some, substitutes are available. But such adjustments can take time—up to 25 years if a new technology is needed; and they need capital too. Rising energy costs mean that the relative costs of materials will change in the next 20 years: designers must be aware of these changes, and continually on the look-out for opportunities to use materials as efficiently as possible. But increasingly, governments are imposing compulsory targets on recycling materials from a wide range of mass-produced consumer goods (such as cars, electronic equipment, and white goods). Manufacturers must now design for the whole life cycle of the product: it is no longer sufficient for one's mobile phone to work well for two years and then be thrown into the trash can—it must also be designed so that it can be dismantled easily and the materials recycled into the next generation of mobile phones.

Environmental impact

As well as simply consuming materials, the mass production of consumer goods places two burdens on the environment. The first is the volume of waste generated. Materials which are not recycled go eventually to landfill sites, which cause groundwater pollution and are physically unsustainable. Unless the percentage of materials recycled increases dramatically in the near future, a significant proportion of the countryside will end up as a rubbish dump. The second is the production of the energy necessary to extract and process materials, and manufacture and distribute the goods made from them. Fossil fuels, as we have seen, are a finite resource. And burning fossil fuels releases carbon dioxide into the atmosphere, with serious implications for global warming. Governments are already setting targets for carbon dioxide emissions—and also imposing carbon taxes—the overall effect being to drive up energy costs.

Examples

2.1 (a) Commodity A is currently consumed at the rate C_A tonnes per year, and commodity B at the rate C_B tonnes per year ($C_A > C_B$). If the two consumption rates are increasing exponentially to give growths in consumption after each year of r_A% and r_B%, respectively ($r_A < r_B$), derive an equation for the time, measured from the present day, before the annual consumption of B exceeds that of A.

(b) The table shows figures for consumption and growth rates of steel, aluminum and plastics. What are the doubling-times (in years) for consumption of these commodities?

(c) Calculate the number of years before the consumption of (a) aluminum and (b) polymers would exceed that of steel, *if exponential growth continued.*

Material	*Current consumption (tonnes year^{-1})*	*Projected growth rate in consumption (% year^{-1})*
Iron and steel	3×10^8	2
Aluminum	4×10^7	3
Polymers	1×10^8	4

Answers

(a) $t = \dfrac{100}{r_B - r_A} \ln\left(\dfrac{C_A}{C_B}\right)$

(b) Doubling-times: steel, 35 years; aluminum, 23 years; plastics, 18 years.

(c) If exponential growth continued, aluminum would overtake steel in 201 years; polymers would overtake steel in 55 years.

2.2 Discuss ways of conserving engineering materials, and the technical and social problems involved in implementing them.

2.3 (a) Explain what is meant by *exponential growth* in the consumption of a material.

(b) A material is consumed at C_0 tonne year^{-1} in 2005. Consumption in 2005 is increasing at r% year^{-1}. If the resource base of the material is Q tonnes, and consumption *continues* to increase at r% year^{-1}, show that the resource will be half exhausted after a time, $t_{\frac{1}{2}}$, given by

$$t_{1/2} = \frac{100}{r} \ln\left\{\frac{rQ}{200C_0} + 1\right\}$$

2.4 Discuss, giving specific examples, the factors that might cause a decrease in the rate of consumption of a potentially scarce material.

Part B

The elastic moduli

Chapter 3

The elastic moduli

Chapter contents

3.1 Introduction

The next material property that we shall examine is the *elastic modulus*. The modulus measures the resistance of a material to elastic (or "springy") deformation. If rods of identical cross section are laid on two widely spaced supports and then identical weights are hung at their centers, they bend elastically by very different amounts depending on the material of which they are made: wood or nylon deflect much more than steel or glass. Low modulus materials are floppy and deflect a lot when they are loaded. Sometimes this is desirable, of course: springs, cushions, vaulting poles—these structures are designed to deflect, and the right choice of modulus may be a low one. But in the great majority of mechanical applications, deflection is undesirable, and the engineer seeks a material with a high modulus. The modulus is reflected, too, in the natural frequency of vibration of a structure. A beam of low modulus has a lower natural frequency than one of higher modulus (although the density matters also) and this, as well as the deflection, is important in design calculations.

Before we look in detail at the modulus, we must first define stress and strain.

3.2 Definition of stress

Imagine a block of material to which we apply a force F, as in Figure 3.1(a). The force is transmitted through the block and is balanced by the equal, opposite force which the base exerts on the block (if this were not so, the block would move). We can replace the base by the equal and opposite force, F, which acts on all sections through the block parallel to the original surface; the whole of the block is said to be in a state of stress. The intensity of the stress, σ, is measured by the force F divided by the area, A, of the block face, giving

$$\sigma = \frac{F}{A} \tag{3.1}$$

This particular stress is caused by a force pulling at right angles to the face; we call it the *tensile* stress.

Suppose now that the force acted not normal to the face but at an angle to it, as shown in Figure 3.1(b). We can resolve the force into two components, one, F_t, normal to the face and the other, F_s, parallel to it. The normal component creates a tensile stress in the block. Its magnitude, as before, is F_t/A.

The other component, F_s, also loads the block, but it does so in *shear*. The shear stress, τ, in the block parallel to the direction of F_s, is given by

$$\tau = \frac{F_s}{A} \tag{3.2}$$

The important point is that the magnitude of a stress is always equal to the magnitude of a *force* divided by the *area* of the face on which it acts. Forces are measured in newtons, so stresses are measured in units of newtons per meter

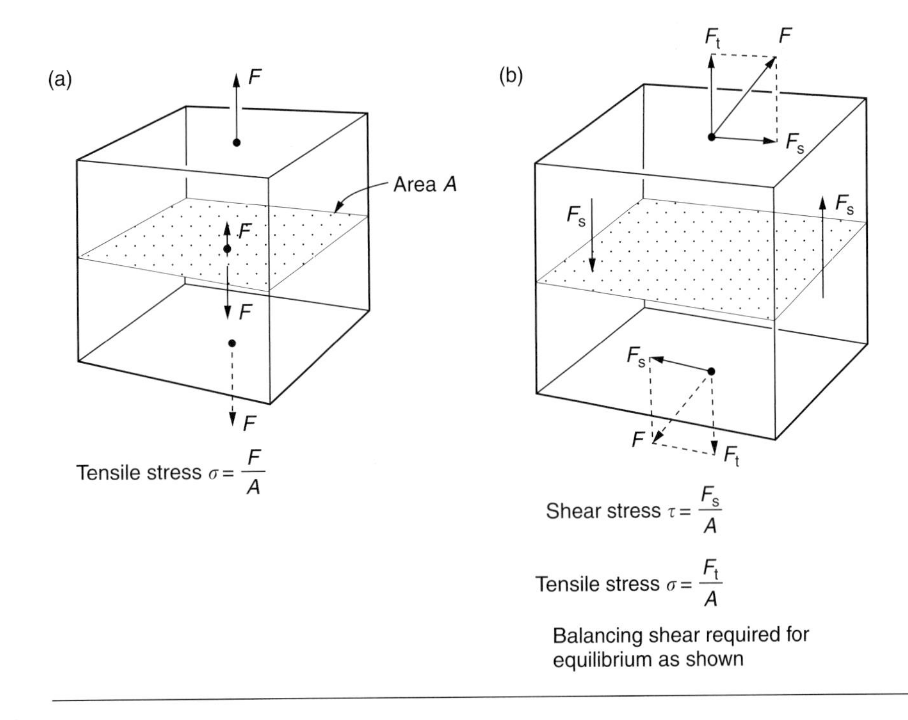

Figure 3.1 Definitions of tensile stress σ and shear stress τ.

squared ($\mathrm{N\,m^{-2}}$). For many engineering applications, this is inconveniently small, and the normal unit of stress is the mega newton per meter squared or mega (10^6) pascal ($\mathrm{MN\,m^{-2}}$ or MPa) or even the giga (10^9) newtons per meter squared or gigapascal ($\mathrm{GN\,m^{-2}}$ or GPa).

There are four commonly occurring states of stress, shown in Figure 3.2. The simplest is that of *simple tension* or *compression* (as in a tension member loaded by pin joints at its ends or in a pillar supporting a structure in compression). The stress is, of course, the force divided by the section area of the member or pillar. The second common state of stress is that of *biaxial tension*. If a spherical shell (like a balloon) contains an internal pressure, then the skin of the shell is loaded in two directions, not one, as shown in Figure 3.2. This state of stress is called biaxial tension (unequal biaxial tension is obviously the state in which the two tensile stresses are unequal). The third common state of stress is that of *hydrostatic pressure*. This occurs deep in the earth's crust, or deep in the ocean, when a solid is subjected to equal compression on all sides. There is a convention that stresses are *positive* when they *pull*, as we have drawn them in earlier figures. Pressure, however, is positive when it *pushes*, so that the magnitude of the pressure differs from the magnitude of the other stresses in its sign. Otherwise it is defined in exactly the same way as before: the force divided by the area on which it acts. The final common state of stress is

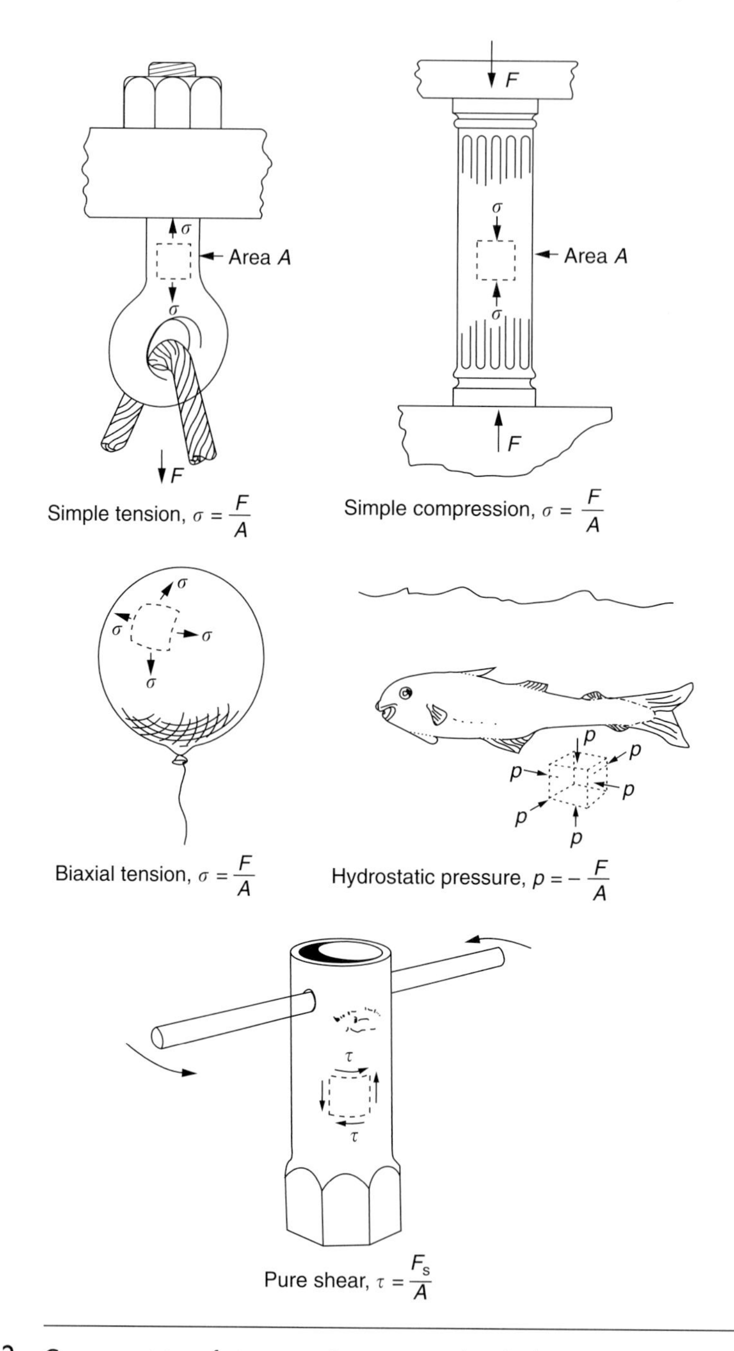

Figure 3.2 Common states of stress: tension, compression, hydrostatic pressure, and shear.

that of *pure shear*. If you try to twist a tube, then elements of it are subjected to pure shear, as shown. This shear stress is simply the shearing force divided by the area of the face on which it acts.

Remember one final thing; if you know the stress in a body, then the *force* acting across any face of it is the stress times the area.

3.3 Definition of strain

Materials respond to stress by *straining*. Under a given stress, a stiff material (like steel) strains only slightly; a floppy or compliant material (like polyethylene) strains much more. The modulus of the material describes this property, but before we can measure it, or even define it, we must define strain properly.

The kind of stress that we called a tensile stress induces a tensile strain. If the stressed cube of side l, shown in Figure 3.3(a) extends by an amount u parallel to the tensile stress, the *nominal tensile strain* is

$$\epsilon_n = \frac{u}{l} \tag{3.3}$$

When it strains in this way, the cube usually gets thinner. The amount by which it shrinks inwards is described by Poisson's ratio, ν, which is the negative of the ratio of the inward strain to the original tensile strain:

$$\nu = -\frac{\text{lateral strain}}{\text{tensile strain}}$$

A shear stress induces a shear strain. If a cube shears sideways by an amount ω then the *shear strain* is defined by

$$\gamma = \frac{\omega}{l} = \tan\theta \tag{3.4}$$

where θ is the angle of shear and l is the edge-length of the cube (Figure 3.3(b)). Since the elastic strains are almost always very small, we may write, to a good approximation,

$$\gamma = \theta$$

Finally, hydrostatic pressure induces a volume change called *dilatation* (Figure 3.3(c)). If the volume change is ΔV and the cube volume is V, we define the dilatation by

$$\Delta = \frac{\Delta V}{V} \tag{3.5}$$

Since strains are the ratios of two lengths or of two volumes, they are dimensionless.

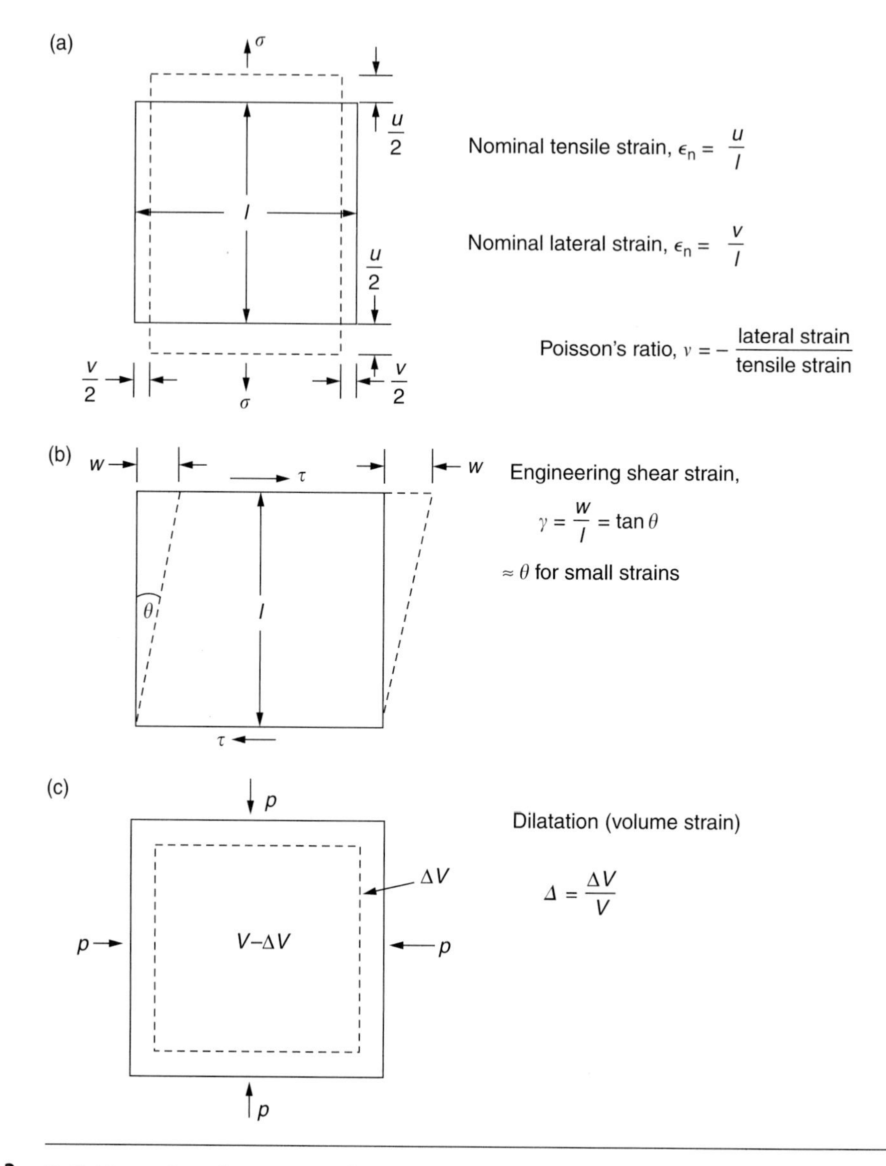

Figure 3.3 Definitions of tensile strain, ϵ_n, shear strain, γ, and dilatation, Δ.

3.4 Hooke's law

We can now define the elastic moduli. They are defined through Hooke's law, which is merely a description of the experimental observation that, when strains are small, the strain is very nearly proportional to the stress; that is, they are *linear elastic*. The nominal tensile strain, for example, is proportional to the tensile stress; for simple tension

$$\sigma = E\epsilon_n \tag{3.6}$$

where E is called *Young's modulus*. The same relationship also holds for stresses and strains in simple compression, of course.

In the same way, the shear strain is proportional to the shear stress, with

$$\tau = G\gamma \tag{3.7}$$

where G is the *shear modulus*.

Finally, the negative of the dilatation is proportional to the pressure (because positive pressure causes a shrinkage of volume) so that

$$p = -K\Delta \tag{3.8}$$

where K is called the *bulk modulus*.

Because strain is dimensionless, the moduli have the same dimensions as those of stress: force per unit area ($\mathrm{N\,m^{-2}}$). In those units, the moduli are enormous, so they are usually reported instead in units of GPa.

This linear relationship between stress and strain is a very handy one when calculating the response of a solid to stress, but it must be remembered that most solids are elastic only to *very small* strains: up to about 0.001. Beyond that some break and some become plastic—and this we will discuss in later chapters. A few solids like rubber are elastic up to very much larger strains of order 4 or 5, but they cease to be *linearly* elastic (that is the stress is no longer proportional to the strain) after a strain of about 0.01.

One final point. We earlier defined Poisson's ratio as the negative of the lateral shrinkage strain to the tensile strain. This quantity, Poisson's ratio, is also an elastic constant, so we have four elastic constants: E, G, K, and ν. In a moment when we give data for the elastic constants we list data only for E. For many materials it is useful to know that

$$K \approx E, \quad G \approx \frac{3}{8}E \quad \text{and} \quad \nu \approx 0.33 \tag{3.9}$$

although for some the relationship can be more complicated.

3.5 Measurement of Young's modulus

How is Young's modulus measured? One way is to compress the material with a known compressive force, and measure the strain. Young's modulus is then given by $E = \sigma/\epsilon_n$, each defined as described earlier. But this is not generally a good way to measure the modulus. For one thing, if the modulus is large, the extension u may be too small to measure with precision. And, for another, if anything else contributes to the strain, like creep (which we will discuss in

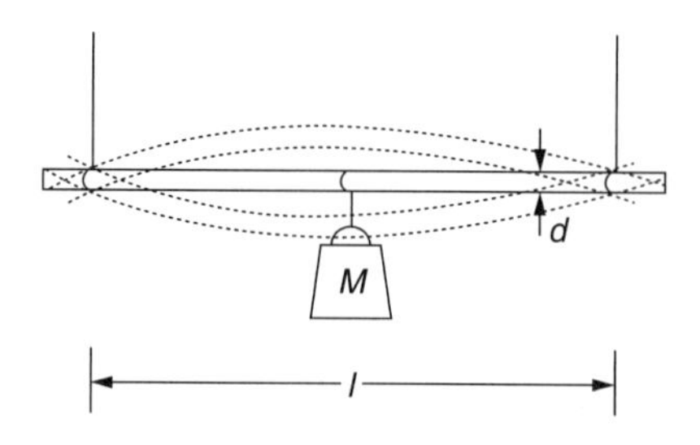

Figure 3.4 A vibrating bar with a central mass, M.

a later chapter), or deflection of the testing machine itself, then it will lead to an incorrect value for E—and these spurious strains can be serious.

A better way of measuring E is to measure the natural frequency of vibration of a round rod of the material, simply supported at its ends (Figure 3.4) and heavily loaded by a mass M at the middle (so that we may neglect the mass of the rod itself). The frequency of oscillation of the rod, f cycles per second (or hertz), is given by

$$f = \frac{1}{2\pi}\left\{\frac{3\pi E d^4}{4l^3 M}\right\}^{1/2} \tag{3.10}$$

where l is the distance between the supports and d is the diameter of the rod. From this,

$$E = \frac{16\pi M l^3 f^2}{3d^4} \tag{3.11}$$

Use of stroboscopic techniques and carefully designed apparatus can make this sort of method very accurate.

The best of all methods of measuring E is to measure the velocity of sound in the material. The velocity of longitudinal waves, ν_l, depends on Young's modulus and the density, ρ:

$$\nu_l = \left(\frac{E}{\rho}\right)^{1/2} \tag{3.12}$$

ν_l is measured by "striking" one end of a bar of the material (by glueing a piezo-electric crystal there and applying a charge-difference to the crystal surfaces) and measuring the time sound takes to reach the other end (by attaching a second piezo-electric crystal there). Most moduli are measured by one of these last two methods.

3.6 Data for Young's modulus

Now for some real numbers. Table 3.1 is a ranked list of Young's modulus of materials—we will use it later in solving problems and in selecting materials for particular applications. Diamond is at the top, with a modulus of 1000 GPa;

soft rubbers and foamed polymers are at the bottom with moduli as low as 0.001 GPa. You can, of course, make special materials with lower moduli—jelly, for instance, has a modulus of about 10^{-6} GPa. Practical engineering materials lie in the range 10^{-3} to 10^{+3} GPa—a range of 10^6. This is the range you have to choose from when selecting a material for a given application. A good perspective of the spread of moduli is given by the bar chart shown in Figure 3.5. Ceramics and metals—even the floppiest of them, like lead—lie near the top of this range. Polymers and elastomers are much more compliant, the common ones (polyethylene, PVC, and polypropylene) lying several decades lower. Composites span the range between polymers and ceramics.

To understand the origin of the modulus, why it has the values it does, why polymers are much less stiff than metals, and what we can do about it,

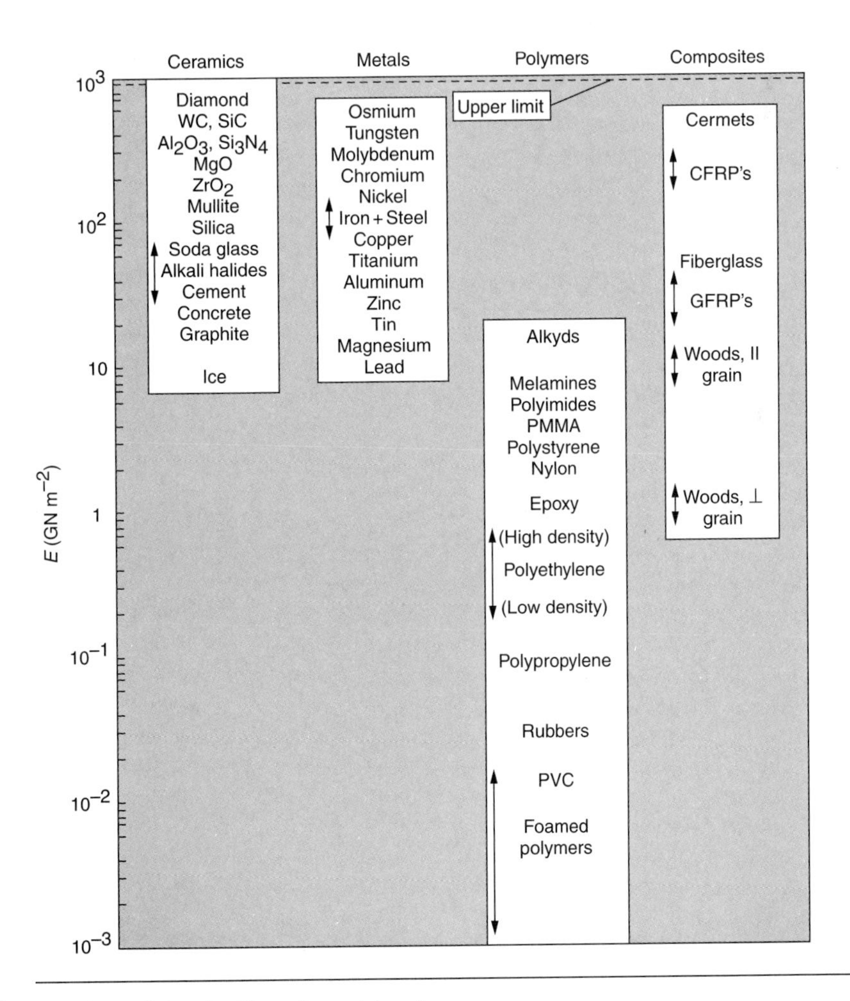

Figure 3.5 Bar chart of data for Young's modulus, E.

Table 3.1 Data for Young's modulus, E

Material	E ($GN\,m^{-2}$)
Diamond	1000
Tungsten carbide, WC	450–650
Osmium	551
Cobalt/tungsten carbide cermets	400–530
Borides of Ti, Zr, Hf	450–500
Silicon carbide, SiC	430–445
Boron	441
Tungsten and alloys	380–411
Alumina, Al_2O_3	385–392
Beryllia, BeO	375–385
Titanium carbide, TiC	370–380
Tantalum carbide, TaC	360–375
Molybdenum and alloys	320–365
Niobium carbide, NbC	320–340
Silicon nitride, Si_3N_4	280–310
Beryllium and alloys	290–318
Chromium	285–290
Magnesia, MgO	240–275
Cobalt and alloys	200–248
Zirconia, ZrO_2	160–241
Nickel	214
Nickel alloys	130–234
CFRP	70–200
Iron	196
Iron-based super-alloys	193–214
Ferritic steels, low-alloy steels	196–207
Stainless austenitic steels	190–200
Mild steel	200
Cast irons	170–190
Tantalum and alloys	150–186
Platinum	172
Uranium	172
Boron/epoxy composites	80–160
Copper	124
Copper alloys	120–150
Mullite	145
Vanadium	130
Titanium	116
Titanium alloys	80–130
Palladium	124
Brasses and bronzes	103–124
Niobium and alloys	80–110
Silicon	107
Zirconium and alloys	96

Table 3.1 *(Continued)*

Material	E ($GN\,m^{-2}$)
Silica glass, SiO_2 (quartz)	94
Zinc and alloys	43–96
Gold	82
Calcite (marble, limestone)	70–82
Aluminum	69
Aluminum and alloys	69–79
Silver	76
Soda glass	69
Alkali halides (NaCl, LiF, etc.)	15–68
Granite (Westerly granite)	62
Tin and alloys	41–53
Concrete, cement	30–50
Fiberglass (glass-fiber/epoxy)	35–45
Magnesium and alloys	41–45
GFRP	7–45
Calcite (marble, limestone)	31
Graphite	27
Shale (oil shale)	18
Common woods, ‖ to grain	9–16
Lead and alloys	16–18
Alkyds	14–17
Ice, H_2O	9.1
Melamines	6–7
Polyimides	3–5
Polyesters	1.8–3.5
Acrylics	1.6–3.4
Nylon	2–4
PMMA	3.4
Polystyrene	3–3.4
Epoxies	2.6–3
Polycarbonate	2.6
Common woods, ⊥ to grain	0.6–1.0
Polypropylene	0.9
PVC	0.2–0.8
Polyethylene, high density	0.7
Polyethylene, low density	0.2
Rubbers	0.01–0.1
Cork	0.01–0.03
Foamed polymers	0.001–0.01

we have to examine the *structure* of materials, and *the nature of the forces* holding the atoms together. In the next two chapters we will examine these, and then return to the modulus, and to our bar chart, with new understanding.

Examples

3.1 (a) Define *Poisson's ratio*, ν, and the *dilatation*, Δ, in the straining of an elastic solid.

(b) Calculate the dilatation Δ in the uniaxial elastic extension of a bar of material, assuming strains are small, in terms of ν and the tensile strain, ϵ. Hence find the value of ν for which the volume change during elastic deformation is zero.

(c) Poisson's ratio for most metals is about 0.3. For cork it is close to zero; for rubber it is close to 0.5. What are the approximate volume changes in each of these materials during an elastic tensile strain of ϵ?

Answers

(b) 0.5; (c) "most metals": 0.4ϵ, cork: ϵ, rubber: 0.

3.2 The sole of a shoe is to be surfaced with soft synthetic rubber having a Poisson's ratio of 0.5. The cheapest solution is to use a solid rubber slab of uniform thickness. However, a colleague suggests that the sole would give better cushioning if it were moulded as shown in the diagram. Is your colleague correct? If so, why?

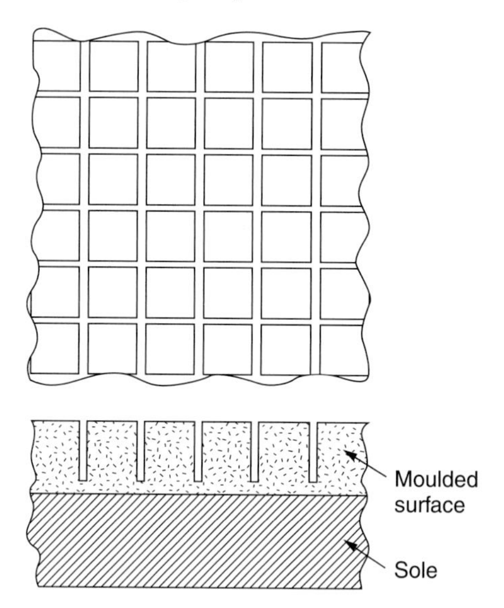

3.3 Explain why it is much easier to push a cork into a wine bottle than a rubber bung. Poisson's ratio is zero for cork and 0.5 for rubber.

Chapter 4

Bonding between atoms

Chapter contents

4.1 Introduction

In order to understand the origin of material properties like *Young's modulus*, we need to focus on materials at the *atomic* level. Two things are especially important in influencing the modulus:

(1) The forces which hold atoms together (the *interatomic bonds*) which act like little springs, linking one atom to the next in the solid state (Figure 4.1).
(2) The ways in which atoms pack together (the *atom packing*), since this determines how many little springs there are per unit area, and the angle at which they are pulled (Figure 4.2).

In this chapter we shall look at the forces which bind atoms together—the springs. In the next we shall examine the arrangements in which they can be packed.

The various ways in which atoms can be bound together involve

(1) *Primary bonds—ionic, covalent* or *metallic* bonds, which are all relatively *strong* (they generally melt between 1000 and 4000 K) and
(2) *Secondary bonds—Van der Waals* and *hydrogen* bonds, which are both relatively *weak* (they melt between 100 and 500 K).

We should remember, however, when drawing up a list of distinct bond types like this, that many atoms are really bound together by bonds which are a hybrid, so to speak, of the simpler types (mixed bonds).

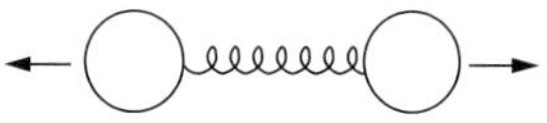

Figure 4.1 The spring-like bond between two atoms.

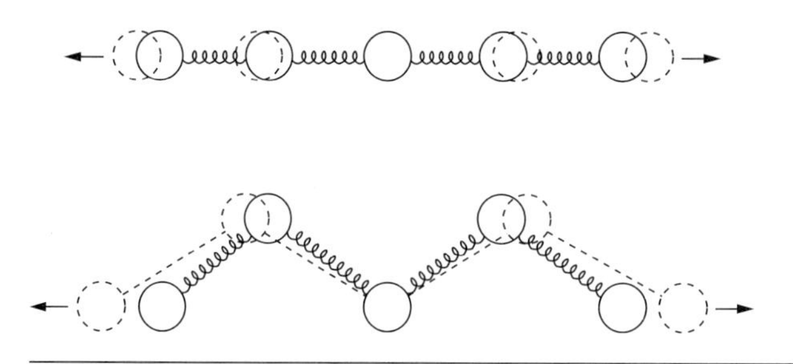

Figure 4.2 Atom packing and bond-angle.

4.2 Primary bonds

Ceramics and metals are entirely held together by primary bonds—the ionic and covalent bond in ceramics, and the metallic and covalent bond in metals. These strong, stiff bonds give high moduli.

The *ionic bond* is the most obvious sort of electrostatic attraction between positive and negative charges. It is typified by cohesion in sodium chloride. Other alkali halides (such as lithium fluoride), oxides (magnesia, alumina), and components of cement (hydrated carbonates and oxides) are wholly or partly held together by ionic bonds.

Let us start with the *sodium atom*. It has a nucleus of 11 *protons*, each with a + charge (and 12 neutrons with no charge at all) surrounded by 11 *electrons* each carrying a – charge (Figure 4.3).

The electrons are attracted to the nucleus by electrostatic forces and therefore have negative energies. But the energies of the electrons are not all the same. Those furthest from the nucleus naturally have the highest (least negative) energy. The electron that we can most easily remove from the sodium atom is therefore the outermost one: we can remove it by expending 5.14 eV* of work. This electron can most profitably be transferred to a vacant position on a distant chlorine atom, giving us back

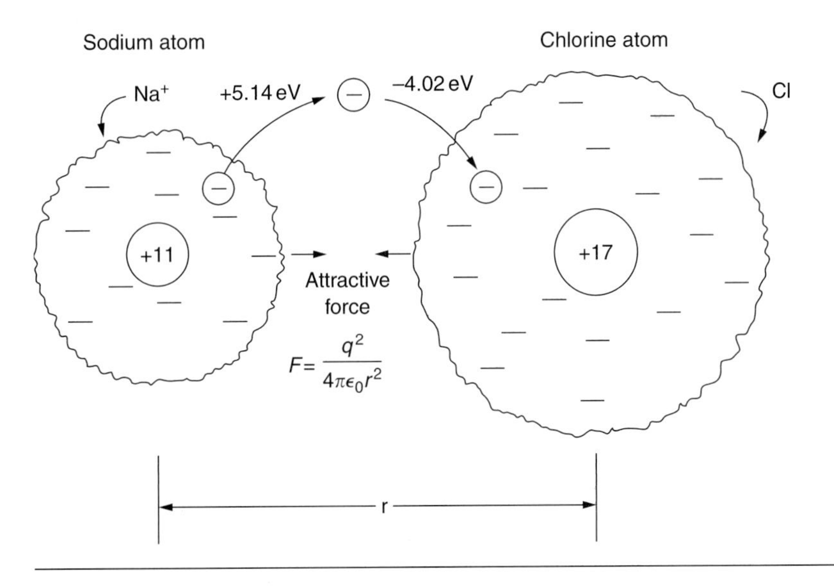

Figure 4.3 The formation of an ionic bond—in this case between a sodium atom and a chlorine atom, making sodium chloride.

* The eV is a convenient unit for energy when dealing with atoms because the values generally lie in the range 1–10. 1 eV is equal to 1.6×10^{-19} joules.

4.02 eV of energy. Thus, we can make isolated Na^+ and Cl^- by doing 5.14 eV – 4.02 eV = 1.12 eV of work, U_i.

So far, we have had to *do* work to create the ions which will make the ionic bond: it does not seem to be a very good start. However, the + and – charges attract each other and if we now bring them together, the force of attraction does work. This force is simply that between two opposite point charges:

$$F = q^2/4\pi\epsilon_0 r^2 \tag{4.1}$$

where q is the charge on each ion, ϵ_0 is the permittivity of vacuum, and r is the separation of the ions. The work done as the ions are brought to a separation r (from infinity) is:

$$U = \int_r^\infty F\,\mathrm{d}r = q^2/4\pi\epsilon_0 r \tag{4.2}$$

Figure 4.4 shows how the energy of the pair of ions falls as r decreases, until, at $r \approx 1$ nm for a typical ionic bond, we have paid off the 1.12 eV of work borrowed to form Na^+ and Cl^- in the first place. For $r < 1$ nm ($1\ \text{nm} = 10^{-9}$ m), it is all gain, and the ionic bond now becomes more and more stable.

Why does r not decrease indefinitely, releasing more and more energy, and ending in the *fusion* of the two ions? Well, when the ions get close enough together, the electronic charge distributions start to overlap one another, and this causes a very large repulsion. Figure 4.4 shows the potential energy increase that this causes. Clearly, the ionic bond is most stable at the minimum

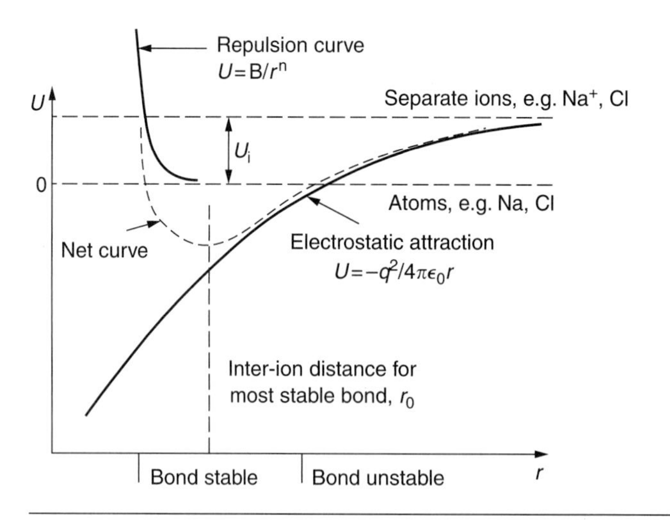

Figure 4.4 The formation of an ionic bond, viewed in terms of energy.

in the *U(r)* curve, which is well approximated by

$$U(r) = U_i - \underbrace{\frac{q^2}{4\pi\epsilon_0 r}}_{\text{attractive part}} + \underbrace{\frac{B}{r^n}}_{\text{repulsive part}} \tag{4.3}$$

where n is a large power—typically about 12.

How much can we bend this bond? Well, the electrons of each ion occupy complicated three-dimensional regions (or "orbitals") around the nuclei. But at an approximate level we can assume the ions to be spherical, and there is then considerable freedom in the way we pack the ions round each other. The ionic bond therefore *lacks directionality*, although in packing ions of opposite sign, it is obviously necessary to make sure that the total charge (+ and −) adds up to zero, and that positive ions (which repel each other) are always separated by negative ions.

Covalent bonding appears in its pure form in diamond, silicon, and germanium—all materials with large moduli (that of diamond is the highest known). It is the dominant bond type in silicate ceramics and glasses (stone, pottery, brick, all common glasses, components of cement) and contributes to the bonding of the high-melting-point metals (tungsten, molybdenum, tantalum, etc.). It appears, too, in polymers, linking carbon atoms to each other along the polymer chain; but because polymers also contain bonds of other, much weaker, types (see below) their moduli are usually small.

The simplest example of covalent bonding is the hydrogen molecule. The proximity of the two nuclei creates a new electron orbital, shared by the two atoms, into which the two electrons go (Figure 4.5). This sharing of electrons leads to a reduction in energy, and a stable bond, as Figure 4.6 shows. The energy of a covalent bond is well described by the empirical equation

$$U = -\underbrace{\frac{A}{r^m}}_{\text{attractive part}} + \underbrace{\frac{B}{r^n}}_{\text{repulsive part}} \quad (m < n) \tag{4.4}$$

Hydrogen is hardly an engineering material. A more relevant example of the covalent bond is that of diamond, one of several solid forms of carbon. It is

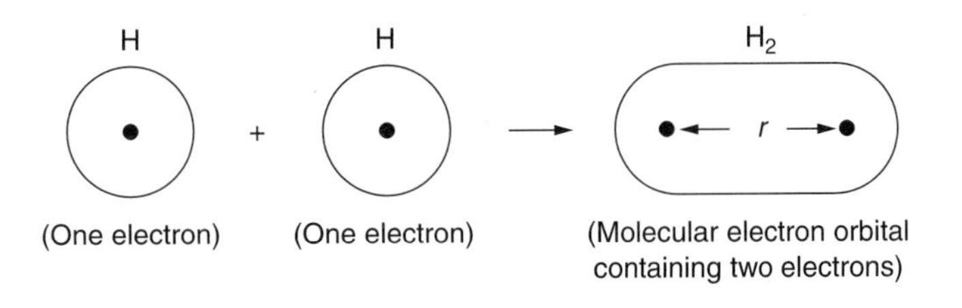

Figure 4.5 The formation of a covalent bond—in this case between two hydrogen atoms, making a hydrogen *molecule*.

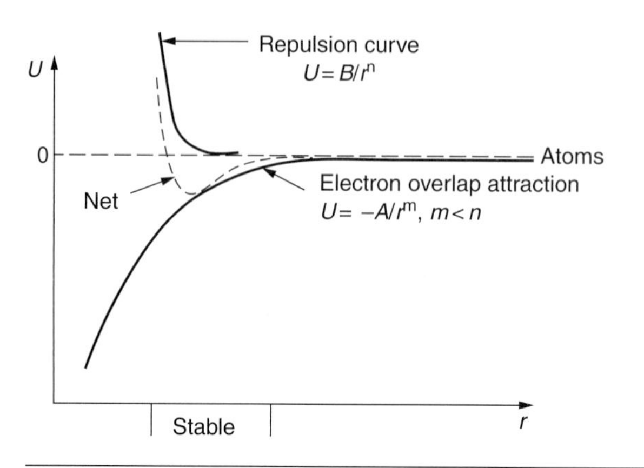

Figure 4.6 The formation of a covalent bond, viewed in terms of energy.

more of an engineering material than you might at first think, finding wide application for rock-drilling bits, cutting tools, grinding wheels, and precision bearings. Here, the shared electrons occupy regions that point to the corners of a tetrahedron, as shown in Figure 4.7(a). The unsymmetrical shape of these orbitals leads to a very *directional* form of bonding in diamond, as Figure 4.7(b) shows. All covalent bonds have directionality which, in turn, influences the ways in which atoms pack together to form crystals — more on that topic in the next chapter.

The *metallic bond*, as the name says, is the dominant (though not the only) bond in metals and their alloys. In a solid (or, for that matter, a liquid) metal, the highest energy electrons tend to leave the parent atoms (which become ions) and combine to form a "sea" of freely wandering electrons, not attached to any ion in particular (Figure 4.8). This gives an energy curve that is very similar to that for covalent bonding; it is well described by equation (4.4) and has a shape like that of Figure 4.6.

The easy movement of the electrons gives the high electrical conductivity of metals. The metallic bond has no directionality, so that metal ions tend to pack to give simple, high-density structures, like ball bearings shaken down in a box.

4.3 Secondary bonds

Although much weaker than primary bonds, secondary bonds are still very important. They provide the links between polymer molecules in polyethylene (and other polymers) which make them solids. Without them, water would boil at −80°C, and life as we know it on earth would not exist.

Van der Waals bonding describes a dipolar attraction between *uncharged* atoms. The electronic charge on an atom is in motion; one can think of the electrons as little charged blobs whizzing round the nucleus like the

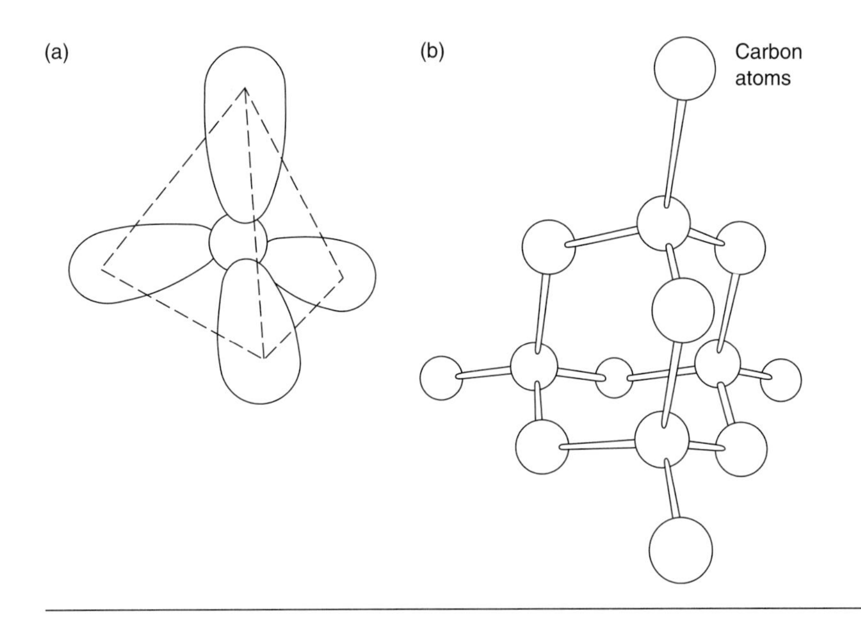

Figure 4.7 Directional covalent bonding in diamond.

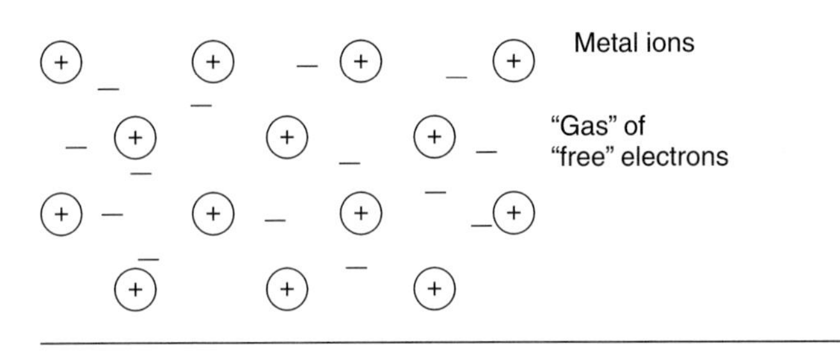

Figure 4.8 Bonding in a metal — metallic bonding.

moon around the earth. Averaged over time, the electron charge has spherical symmetry, but at any given instant it is unsymmetric relative to the nucleus. The effect is a bit like that which causes the tides. The instantaneous distribution has a dipole moment; this moment induces a like moment on a nearby atom and the two dipoles attract (Figure 4.9).

Dipoles attract such that their energy varies as $1/r^6$. Thus the energy of the Van der Waals bond has the form

$$U = -\underbrace{\frac{A}{r^6}}_{\text{attractive part}} + \underbrace{\frac{B}{r^n}}_{\text{repulsive part}} \quad (n \approx 12) \tag{4.5}$$

A good example is liquid nitrogen, which liquifies, at atmospheric pressure, at $-198°C$ glued by Van der Waals forces between the covalently bonded

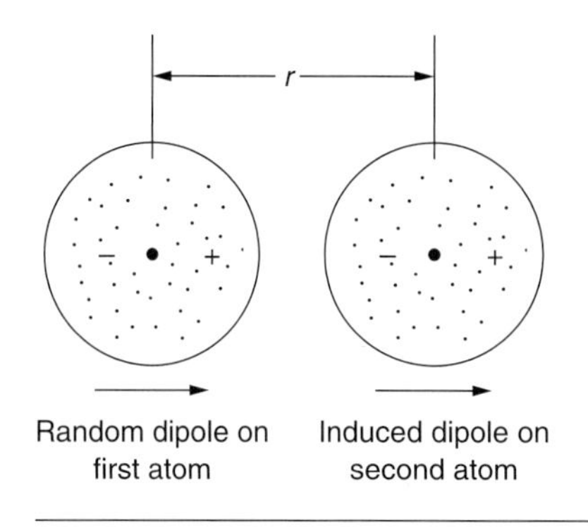

Figure 4.9 Van der Waals bonding; the atoms are held together by the dipole charge distribution.

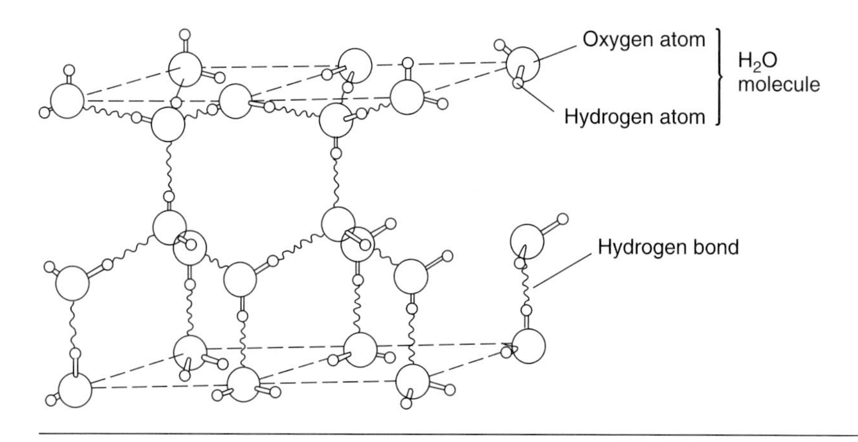

Figure 4.10 The arrangement of H_2O molecules in the common form of ice, showing the hydrogen bonds. The hydrogen bonds keep the molecules well apart, which is why ice has a lower density than water.

N_2 molecules. The thermal agitation produced when liquid nitrogen is poured on the floor at room temperature is more than enough to break the Van der Waals bonds, showing how weak they are. But without these bonds, most gases would not liquefy at attainable temperatures, and we should not be able to separate industrial gases from the atmosphere.

Hydrogen bonds keep water liquid at room temperature, and bind polymer chains together to give solid polymers. Ice (Figure 4.10) is hydrogen bonded. Each hydrogen atom shares its charge with the nearest oxygen atom. The hydrogen, having lost part of its share, acquires a + charge; the oxygen, having a share in more electrons than it should, is −ve. The positively charged H atom acts as a bridging bond between neighboring oxygen ions, because the charge redistribution gives each H_2O molecule a dipole moment which attracts other H_2O dipoles.

4.4 The condensed states of matter

It is because these primary and secondary bonds can form that matter condenses from the gaseous state to give liquids and solids. Five distinct *condensed states of matter*, differing in their structure and the state of their bonding, can be identified (Table 4.1). The bonds in ordinary liquids have melted, and for this reason the liquid resists compression, but not shear; the bulk modulus, K, is large compared to the gas because the atoms are in contact, so to speak; but the shear modulus, G, is zero because they can slide past each other. The other states of matter, listed in Table 4.1, are distinguished by the state of their bonding (molten versus solid) and their structure (crystalline versus noncrystalline). These differences are reflected in the relative magnitudes of their bulk modulus and shear modulus — the more liquid like the material becomes, the smaller is its ratio of G/K.

4.5 Interatomic forces

Having established the various types of bonds that can form between atoms, and the shapes of their potential energy curves, we are now in a position to explore the *forces* between atoms. Starting with the $U(r)$ curve, we can find this force F for any separation of the atoms, r, from the relationship

$$F = \frac{dU}{dr} \tag{4.6}$$

Figure 4.11 shows the shape of the *force*/distance curve that we get from a typical energy/distance curve in this way. Points to note are:

(1) F is zero at the equilibrium point $r = r_0$; however, if the atoms are pulled apart by distance $(r - r_0)$ a resisting force appears. For *small* $(r - r_0)$ the resisting force is proportional to $(r - r_0)$ for all materials, in both tension and compression.

Table 4.1 Condensed states of matter

State		*Bonds*		*Moduli*	
		Molten	*Solid*	*K*	*G and E*
1.	Liquids	*		Large	Zero
2.	Liquid crystals	*		Large	Some nonzero but very small
3.	Rubbers	* (secondary)	* (primary)	Large	Small ($E \ll K$)
4.	Glasses		*	Large	Large($E \approx K$)
5.	Crystals		*	Large	Large ($E \approx K$)

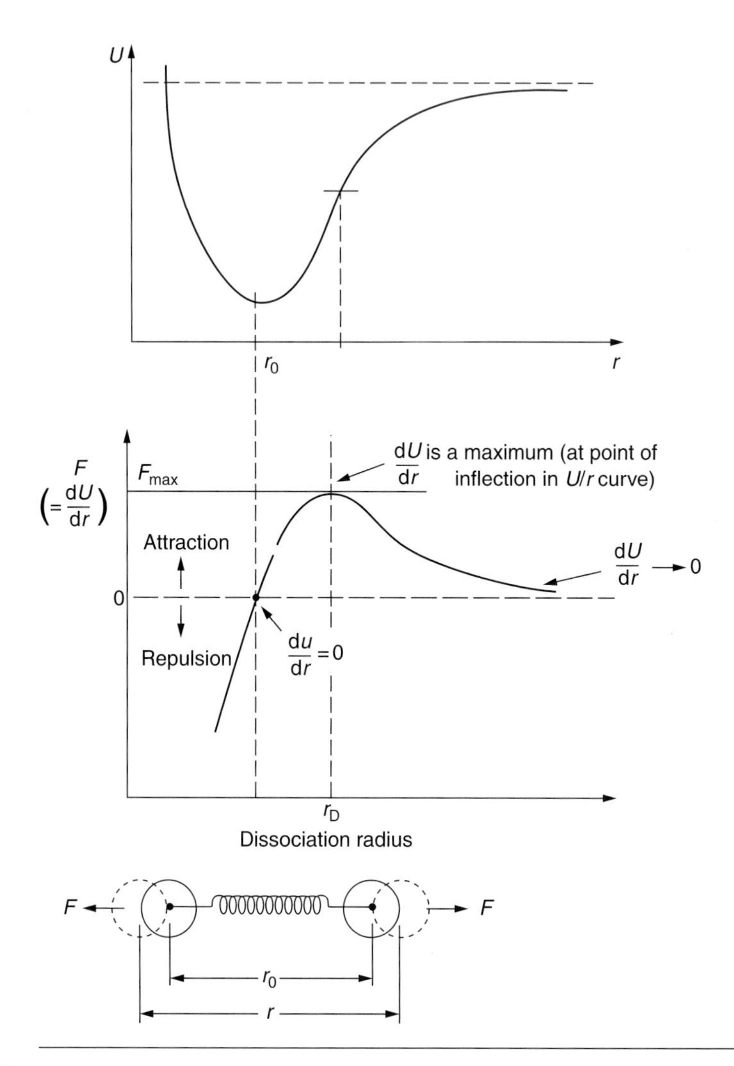

Figure 4.11 The energy–distance curve (top), when differentiated (equation (4.6)) gives the force–distance curve (centre).

(2) The *stiffness, S*, of the bond is given by

$$S = \frac{\mathrm{d}F}{\mathrm{d}r} = \frac{\mathrm{d}^2U}{\mathrm{d}r^2} \tag{4.7}$$

When the stretching is small, S is constant and equal to

$$S_0 = \left(\frac{\mathrm{d}^2U}{\mathrm{d}r^2}\right)_{r=r_0} \tag{4.8}$$

that is, the bond behaves in linear elastic manner—this is the physical origin of Hooke's law.

To conclude, the concept of bond stiffness, based on the energy/distance curves for the various bond types, goes a long way towards explaining the origin of the elastic modulus. But we need to find out how individual atom bonds build up to form whole pieces of material before we can fully explain experimental data for the modulus. The nature of the bonds we have mentioned influences the *packing* of atoms in engineering materials. This is the subject of the next chapter.

Examples

4.1 The potential energy U of two atoms, a distance r apart, is

$$U = -\frac{A}{r^m} + \frac{B}{r^n}, \qquad m = 2, \quad n = 10$$

Given that the atoms form a stable molecule at a separation of 0.3 nm with an energy of -4 eV, calculate A and B. Also find the force required to break the molecule, and the critical separation at which the molecule breaks. You should sketch an energy/distance curve for the atom, and sketch beneath this curve the appropriate *force*/distance curve.

Answers

A: 7.2×10^{-20} J nm^2; B: 9.4×10^{-25} J nm^{10}; Force: 2.39×10^{-9} N at 0.352 nm.

4.2 The potential energy U of a pair of atoms in a solid can be written as

$$U = -\frac{A}{r^m} + \frac{B}{r^n}$$

where r is the separation of the atoms, and A, B, m and n are positive constants. Indicate the physical significance of the two terms in this equation.

4.3 The table below gives the Young's modulus, E, the atomic volume, Ω, and the melting temperature, T_{M}, for a number of metals. If

$$E \simeq \frac{\tilde{A}kT_{\mathrm{M}}}{\Omega}$$

(where k is Boltzmann's constant and $\tilde{A}$ is a constant), calculate and tabulate the value of the constant $\tilde{A}$ for each metal. Hence find an arithmetic mean of $\tilde{A}$ for these metals.

Use the equation, with the average $\tilde{A}$, to calculate the approximate Young's modulus of (a) diamond and (b) ice. Compare these with the experimental values of 1.0×10^{12} N m^{-2} and 7.7×10^{9} N m^{-2}, respectively. Watch the units!

Material	$\Omega \times 10^{29}$(m^3)	T_m (K)	E (GN m^{-2})
Nickel	1.09	1726	214
Copper	1.18	1356	124
Silver	1.71	1234	76
Aluminum	1.66	933	69
Lead	3.03	600	14
Iron	1.18	1753	196
Vanadium	1.40	2173	130
Chromium	1.20	2163	289
Niobium	1.80	2741	100
Molybdenum	1.53	2883	360
Tantalum	1.80	3271	180
Tungsten	1.59	3683	406

Data for ice and for diamond.

Ice	Diamond
$\Omega = 3.27 \times 10^{-29}\,\mathrm{m}^3$	$\Omega = 5.68 \times 10^{-30}\,\mathrm{m}^3$
$T_M = 273\,\mathrm{K}$	$T_M = 4200\,\mathrm{K}$
$E = 7.7 \times 10^{9}\,\mathrm{N\,m}^{-2}$	$E = 1.0 \times 10^{12}\,\mathrm{N\,m}^{-2}$

Answers

Mean $\tilde{A} = 88$. Calculated moduli: diamond, $9.0 \times 10^{11}\,\mathrm{N\,m}^{-2}$; ice, $1.0 \times 10^{10}\,\mathrm{N\,m}^{-2}$.

Chapter 5

Packing of atoms in solids

Chapter contents

5.1 Introduction

In the previous chapter, as a first step in understanding the stiffness of solids, we examined the stiffnesses of the bonds holding atoms together. But bond stiffness alone does not fully explain the stiffness of solids; the way in which the atoms are packed together is equally important. In this chapter we examine how atoms are arranged in some typical engineering solids.

5.2 Atom packing in crystals

Many engineering materials (almost all metals and ceramics, for instance) are made up entirely of small crystals or *grains* in which atoms are packed in regular, repeating, three-dimensional patterns; the grains are stuck together, meeting at *grain boundaries*, which we will describe later. We focus now on the individual crystals, which can best be understood by thinking of the atoms as *hard spheres* (although, from what we said in the previous chapter, it should be obvious that this is a considerable, although convenient, simplification). To make things even simpler, let us for the moment consider a material which is *pure*—with only one size of hard sphere to consider—and which also has *non-directional bonding*, so that we can arrange the spheres subject only to geometrical constraints. Pure copper is a good example of a material satisfying these conditions.

In order to build up a three-dimensional packing pattern, it is easier, conceptually, to begin by

(i) packing atoms two-dimensionally in *atomic planes*,
(ii) stacking these planes on top of one another to give *crystals*.

5.3 Close-packed structures and crystal energies

An example of how we might pack atoms in a *plane* is shown in Figure 5.1; it is the arrangement in which the reds are set up on a billiard table before starting a game of snooker. The balls are packed in a triangular fashion so as to take up the least possible space on the table. This type of plane is thus called a *close-packed plane*, and contains three *close-packed directions;* they are the directions along which the balls touch. The figure shows only a small region of close-packed plane—if we had more reds we could extend the plane sideways and could, if we wished, fill the whole billiard table. The important thing to notice is the way in which the balls are packed in a *regularly repeating two-dimensional pattern*.

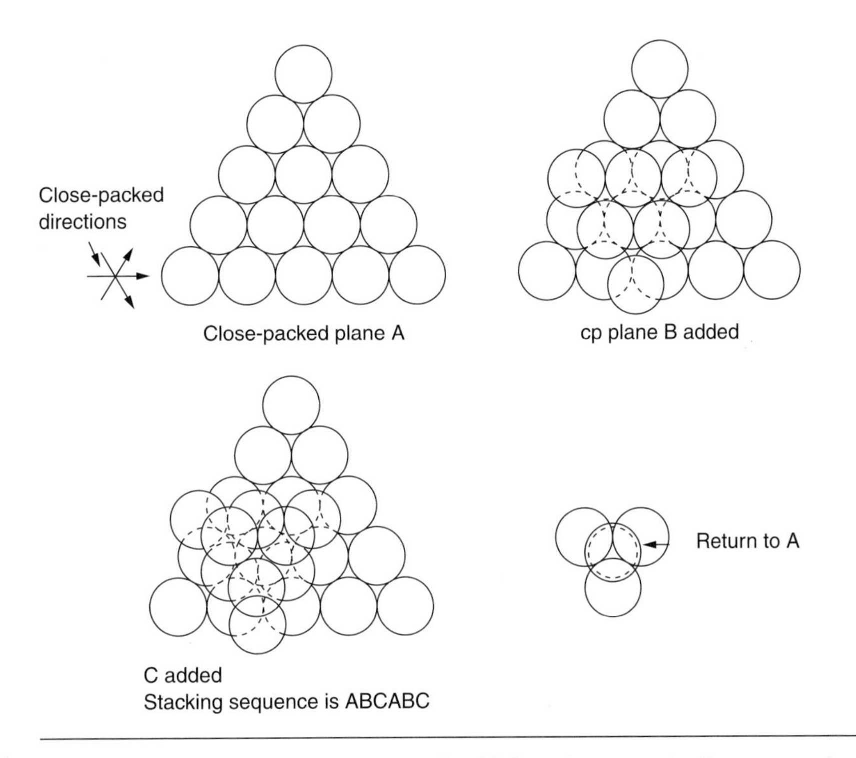

Figure 5.1 The close packing of hard-sphere atoms. The ABC stacking gives the "face-centered cubic" (f.c.c.) structure.

How could we add a second layer of atoms to our close-packed plane? As Figure 5.1 shows, the depressions where the atoms meet are ideal "seats" for the next layer of atoms. By dropping atoms into alternate seats, we can generate a second close-packed plane lying on top of the original one and having an identical packing pattern. Then a third layer can be added, and a fourth, and so on until we have made a sizeable piece of crystal—with, this time, a *regularly repeating pattern of atoms in three dimensions*. The particular structure we have produced is one in which the atoms take up the *least volume* and is therefore called a *close-packed structure*. The atoms in many solid metals are packed in this way.

There is a complication to this apparently simple story. There are two alternative and different *sequences* in which we can stack the close-packed planes on top of one another. If we follow the stacking sequence in Figure 5.1 rather more closely, we see that, by the time we have reached the *fourth* atomic plane, we are placing the atoms directly above the original atoms (although, naturally, separated from them by the two interleaving planes of atoms). We then carry on adding atoms as before, generating an ABCABC... sequence. In Figure 5.2 we show the alternative way of stacking, in which the atoms in the *third* plane are now directly above those in the first layer. This gives an

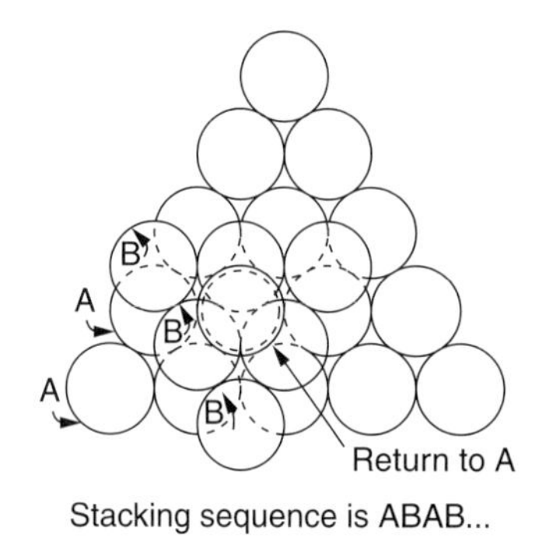

Figure 5.2 Close packing of hard-sphere atoms — an alternative arrangement, giving the "close-packed hexagonal" (c.p.h.) structure.

ABAB... sequence. These two different stacking sequences give two different three-dimensional packing structures — *face-centered cubic (f.c.c.)* and *close-packed hexagonal (c.p.h.)* respectively. Many common metals (e.g. Al, Cu, and Ni) have the f.c.c. structure and many others (e.g. Mg, Zn, and Ti) have the c.p.h. structure.

Why should Al choose to be f.c.c. while Mg chooses to be c.p.h.? The answer is that the f.c.c. structure is the one that gives an Al crystal the least *energy*, and the c.p.h. structure the one that gives a Mg crystal the least energy. In general, materials choose the crystal structure that gives minimum energy. This structure may not necessarily be close packed or, indeed, very simple geometrically, although, to be a crystal, it must still have some sort of three-dimensional repeating pattern.

The difference in energy between alternative structures is often slight. Because of this, the crystal structure which gives the minimum energy at one temperature may not do so at another. Thus tin changes its crystal structure if it is cooled enough; and, incidentally, becomes much more brittle in the process (causing the tin-alloy coat-buttons of Napoleon's army to fall apart during the harsh Russian winter; and the soldered cans of paraffin on Scott's South Pole expedition to leak, with disastrous consequences). Cobalt changes its structure at 450°C, transforming from a c.p.h. structure at lower temperatures to an f.c.c. structure at higher temperatures. More important, pure iron transforms from a b.c.c. structure (defined below) to one which is f.c.c. at 911°C, a process which is important in the heat-treatment of steels.

5.4 Crystallography

We have not yet explained why an ABCABC sequence is called "f.c.c." or why an ABAB sequence is referred to as "c.p.h.". And we have not even begun to describe the features of the more complicated crystal structures like those of ceramics such as alumina. In order to explain things such as the geometric differences between f.c.c. and c.p.h. or to ease the conceptual labor of constructing complicated crystal structures, we need an appropriate descriptive language. The methods of *crystallography* provide this language, and give us also an essential shorthand way of describing crystal structures.

Let us illustrate the crystallographic approach in the case of f.c.c. Figure 5.3 shows that the *atom centers* in f.c.c. can be placed at the corners of a cube and in the centers of the cube faces. The cube, of course, has no physical significance but is merely a constructional device. It is called a *unit cell*. If we look along the cube diagonal, we see the view shown in Figure 5.3 (top center): a triangular pattern which, with a little effort, can be seen to be that of bits of close-packed planes stacked in an ABCABC sequence. This unit-cell visualization of the atomic positions is thus exactly equivalent to our earlier approach based on stacking of close-packed planes, but is much more powerful as a descriptive aid. For example, we can see how our complete f.c.c. crystal is built up by attaching further unit cells to the first one (like assembling a set of children's building cubes) so as to fill space without leaving awkward gaps—something you cannot so easily do with 5-sided shapes (in a plane) or 7-sided shapes (in three dimensions). Beyond this, inspection of the unit cell reveals planes in which the atoms are packed in other than a close-packed way. On the "cube" faces the atoms are packed in a square array, and on the cube-diagonal planes in separated rows, as shown in Figure 5.3. Obviously,

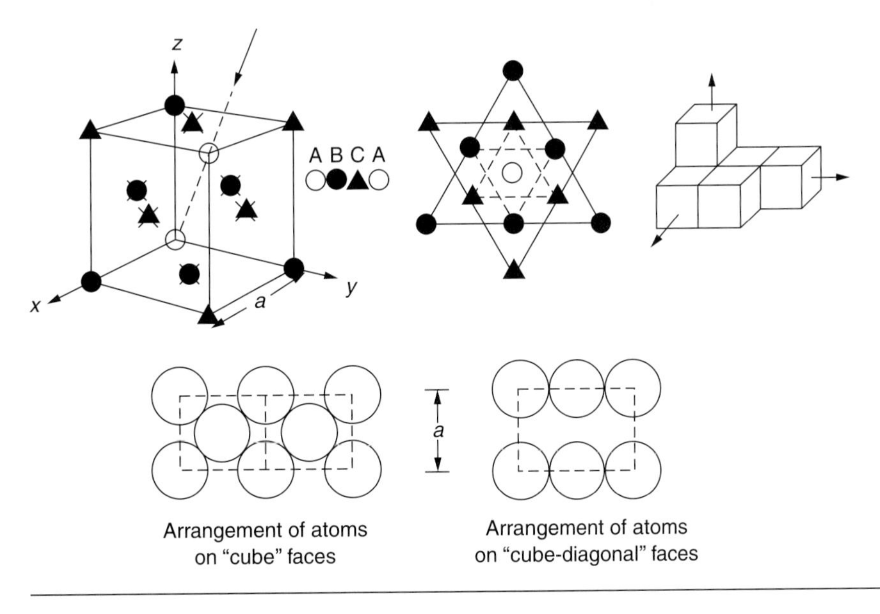

Figure 5.3 The face-centered cubic (f.c.c.) structure.

properties like the shear modulus might well be different for close-packed planes and cube planes, because the number of bonds attaching them per unit area is different. This is one of the reasons that it is important to have a method of describing various planar packing arrangements.

Let us now look at the c.p.h. unit cell as shown in Figure 5.4. A view looking down the vertical axis reveals the ABA stacking of close-packed planes. We build up our c.p.h. crystal by adding hexagonal building blocks to one another: hexagonal blocks also stack so that they fill space. Here, again, we can use the unit cell concept to "open up" views of the various types of planes.

5.5 Plane indices

We could make scale drawings of the many types of planes that we see in all unit cells; but the concept of a unit cell also allows us to describe any plane by a set of numbers called *Miller Indices*. The two examples given in Figure 5.5 should enable you to find the Miller index of any plane in a cubic unit cell, although they take a little getting used to. The indices (for a plane) are the *reciprocals* of the intercepts the plane makes with the three axes, reduced to the smallest integers (reciprocals are used simply to avoid infinities when planes are parallel to axes). As an example, the six individual *"cube" planes* are called (100), (010), (001). Collectively this type of plane is called {100}, with curly brackets. Similarly the six cube *diagonal* planes are (110), $(1\bar{1}0)$, (101), $(\bar{1}01)$, (011) and $(0\bar{1}1)$, or, collectively, {110}. (Here the sign $\bar{1}$ means an intercept of -1.) As a final example, our original close-packed planes — the ones of the ABC stacking — are of {111} type. Obviously the unique structural description of "{111} f.c.c." is a good deal more succinct than a scale drawing of close-packed billiard balls.

Different indices are used in hexagonal cells (we build a c.p.h. crystal up by adding bricks in four directions, not three as in cubic). We do not need them here — the crystallography books listed under "References" at the end of the book do them more than justice.

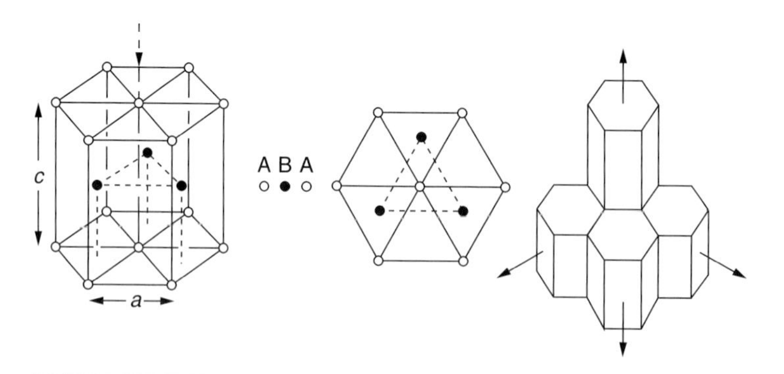

Figure 5.4 The close-packed hexagonal (c.p.h.) structure.

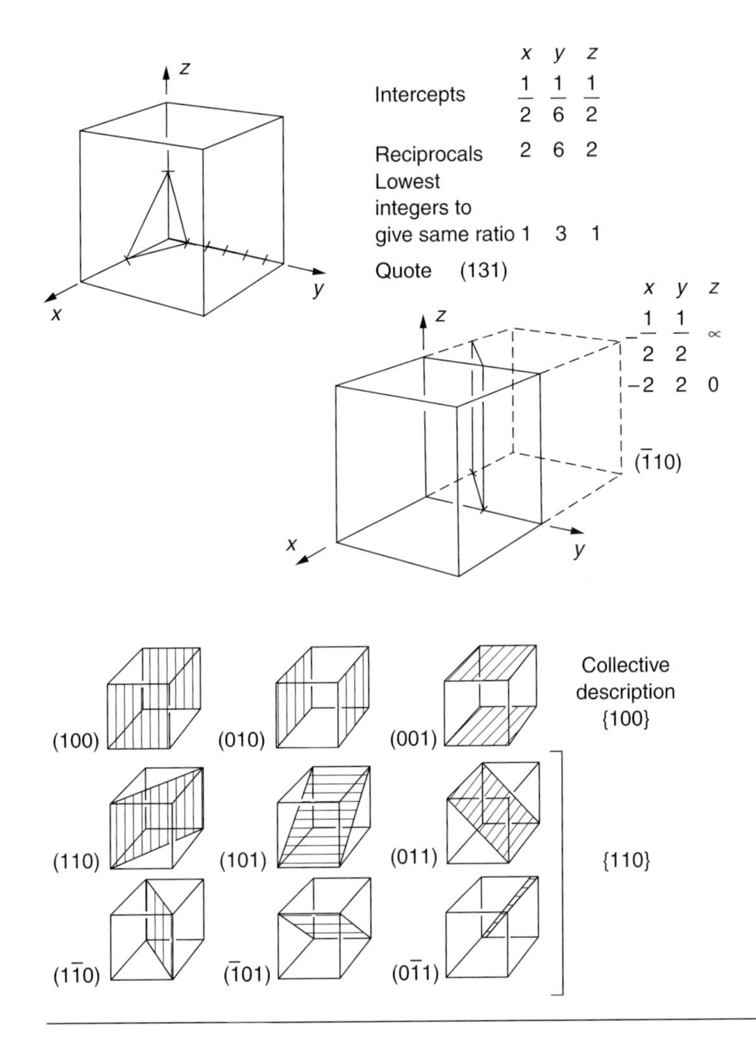

Figure 5.5 Miller indices for identifying crystal planes, showing how the (1 3 1) plane and the ($\bar{1}$ 1 0) planes are defined. The lower part of the figure shows the family of {1 0 0} and of {1 1 0} planes.

5.6 Direction indices

Properties like Young's modulus may well vary with *direction* in the unit cell; for this (and other) reasons we need a succinct description of crystal directions. Figure 5.6 shows the method and illustrates some typical directions. The indices of direction are the components of a vector (*not* reciprocals, because infinities do not crop up here), starting from the origin, along the desired direction, again reduced to the smallest integer set. A single direction (like the "111" direction which links the origin to the corner of the cube furthest from

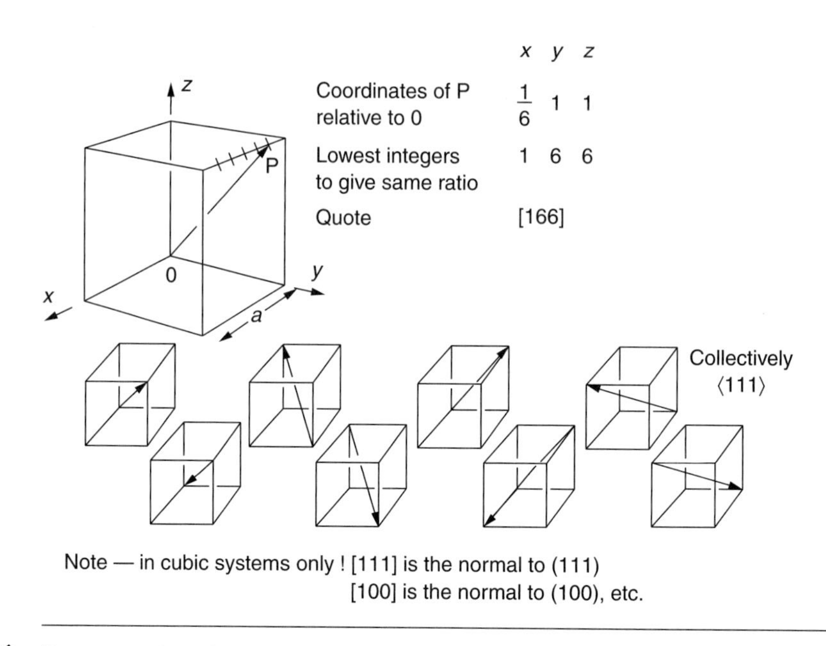

Figure 5.6 Direction indices for identifying crystal directions, showing how the [1 6 6] direction is defined. The lower part of the figure shows the family of ⟨1 1 1⟩ directions.

the origin) is given square brackets (i.e. [111]), to distinguish it from the Miller index of a plane. The family of directions of this type (illustrated in Figure 5.6) is identified by angled brackets: ⟨111⟩.

5.7 Other simple important crystal structures

Figure 5.7 shows a new crystal structure, and an important one: it is the body-centered cubic (b.c.c.) structure of tungsten, of chromium, of iron, and many steels. The ⟨111⟩ directions are close packed (that is to say: the atoms touch along this direction) but there are no close packed planes. The result is that b.c.c. packing is less dense than either f.c.c. or c.p.h. It is found in materials which have *directional bonding:* the directionality distorts the structure, preventing the atoms from dropping into one of the two close-packed structures we have just described. There are other structures involving only one sort of atom which are not close packed, for the same reason, but we don't need them here.

In compound materials — in the ceramic sodium chloride, for instance — there are two (sometimes more) species of atoms, packed together. The crystal structures of such compounds can still be simple. Figure 5.8(a) shows that the ceramics NaCl, KCl, and MgO, for example, also form a cubic structure. Naturally, when two species of atoms are not in the ratio 1 : 1, as in compounds

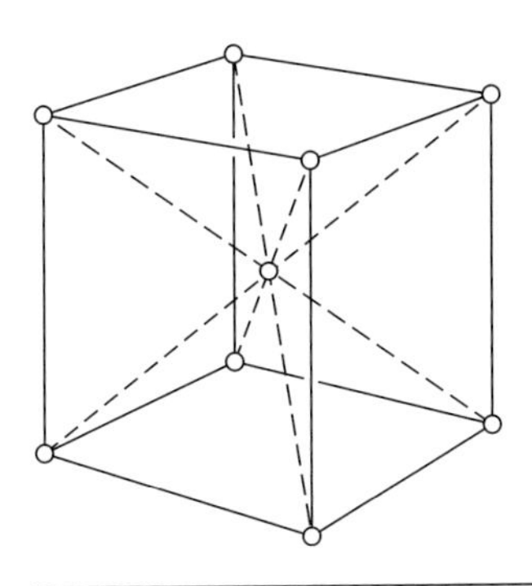

Figure 5.7 The body-centered cubic (b.c.c.) structure.

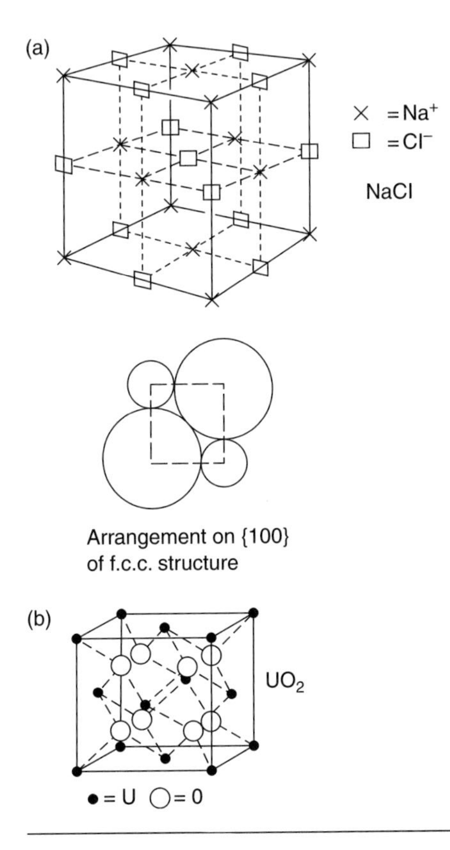

Figure 5.8 (a) Packing of the unequally sized ions of sodium chloride to give a f.c.c. structure; KCl and MgO pack in the same way. (b) Packing of ions in uranium dioxide; this is more complicated than in NaCl because the U and O ions are not in a 1 : 1 ratio.

like the nuclear fuel UO_2 (a ceramic too) the structure is more complicated (it is shown in Figure 5.8(b)), although this, too, has a cubic unit cell.

5.8 Atom packing in polymers

As we saw in the first chapter, polymers have become important engineering materials. They are much more complex structurally than metals, and because of this they have very special mechanical properties. The extreme elasticity of a rubber band is one; the formability of polyethylene is another.

Polymers are huge chain-like molecules (huge, that is, by the standards of an atom) in which the atoms forming the backbone of the chain are linked by *covalent* bonds. The chain backbone is usually made from carbon atoms (although a limited range of silicon-based polymers can be synthesized—they are called "silicones"). A typical high polymer ("high" means "of large molecular weight") is polyethylene. It is made by the catalytic polymerization of ethylene, shown on the left, to give a chain of ethylenes, minus the double bond:

```
H  H        H   H   H   H   H   H
|  |        |   |   |   |   |   |
C = C →   — C — C — C — C — C — C — etc.
|  |        |   |   |   |   |   |
H  H        H   H   H   H   H   H
```

Polystyrene, similarly, is made by the polymerization of styrene (left), again by sacrificing the double bond to provide the hooks which give the chain:

```
H   C6H5      H   C6H5  H    H      H    C6H5
|   |         |   |     |    |      |    |
C = C  →   — C — C   — C — C     — C  — C — etc.
|   |         |   |     |    |      |    |
H   H         H   H     H    C6H5   H    H
```

A copolymer is made by polymerization of two monomers, adding them randomly (a random copolymer) or in an ordered way (a block copolymer). An example is styrene-butadiene rubber, SBR. Styrene, extreme left, loses its double bond in the marriage; butadiene, richer in double bonds to start with, keeps one.

```
H   C6H5   H   H   H   H        H   C6H5  H   H   H   H
|   |      |   |   |   |        |   |     |   |   |   |
C = C    + C = C — C = C  →   — C — C —   C — C = C — C — etc.
|   |      |           |        |   |     |           |
H   H      H           H        H   H     H           H
```

Molecules such as these form long, flexible, spaghetti-like chains like that of Figure 5.9. Figure 5.10 shows how they pack to form bulk material. In some polymers the chains can be folded carefully backwards and forwards over one another so as to look like the firework called the "jumping jack". The regularly repeating symmetry of this chain folding leads to crystallinity, so polymers can be crystalline. More usually the chains are arranged *randomly* and *not*

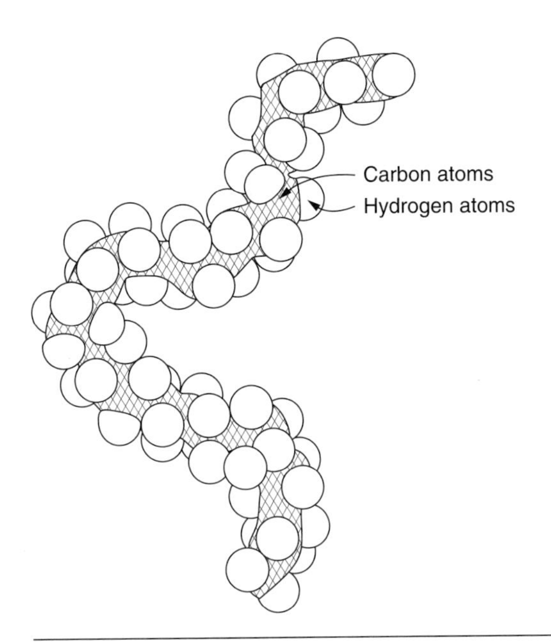

Figure 5.9 The three-dimensional appearance of a short bit of a polyethylene molecule.

in regularly repeating three-dimensional patterns. These polymers are thus *noncrystalline*, or *amorphous*. Many contain both amorphous and crystalline regions, as shown in Figure 5.10, that is, they are *partly crystalline*.

There is a whole science called *molecular architecture* devoted to making all sorts of chains and trying to arrange them in all sorts of ways to make the final material. There are currently thousands of different polymeric materials, all having different properties—and new ones are under development. This sounds like bad news, but we need only a few: six basic polymers account for almost 95 percent of all current production. We will meet them later.

5.9 Atom packing in inorganic glasses

Inorganic glasses are mixtures of oxides, almost always with silica, SiO_2, as the major ingredient. As the name proclaims, the atoms in glasses are packed in a noncrystalline (or amorphous) way. Figure 5.11(a) shows schematically the structure of silica glass, which is solid to well over 1000°C because of the strong covalent bonds linking the Si to the O atoms. Adding soda (Na_2O) breaks up the structure and lowers the *softening temperature* (at which the glass can be worked) to about 600°C. This soda glass (Figure 5.11(b)) is the material of which milk bottles and window panes are made. Adding boron oxide (B_2O_3) instead gives boro-silicate glasses (pyrex is one) which withstand higher temperatures than ordinary window-glass.

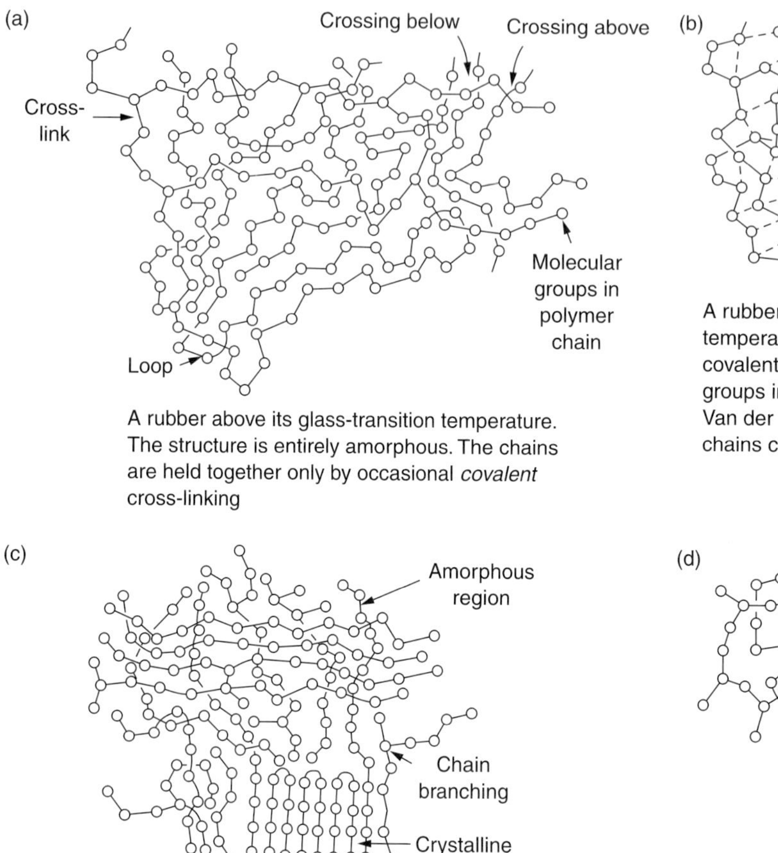
(a)
Crossing below
Crossing above
Cross-link
Molecular groups in polymer chain
Loop
A rubber above its glass-transition temperature. The structure is entirely amorphous. The chains are held together only by occasional *covalent* cross-linking
(b)
Van der Waals bonds
A rubber below its glass-transition temperature. In addition to occasional covalent cross-linking the molecular groups in the polymer chains attract by Van der Waals bonding, tying the chains closely to one another.
(c)
Amorphous region
Chain branching
Crystalline region
Side groups
Low-density polyethylene, showing both amorphous and crystalline regions

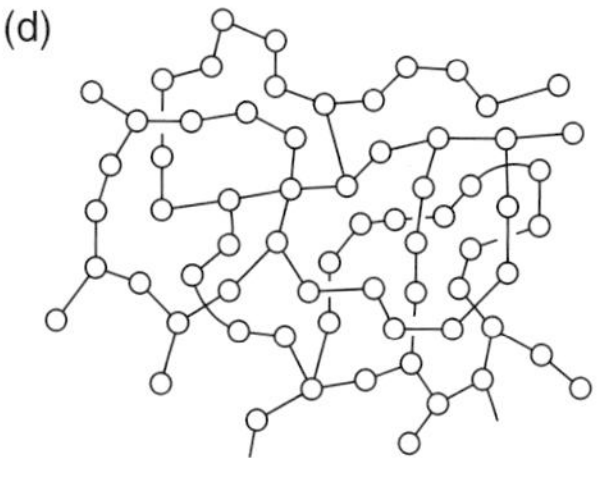
(d)
A polymer (e.g. epoxy resin) where the chains are tied tightly together by frequent *covalent* cross-links

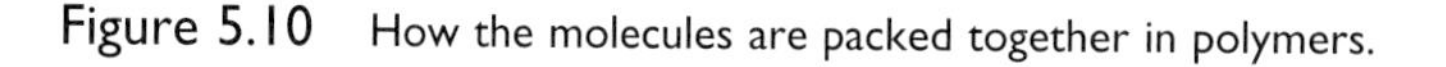

Figure 5.10 How the molecules are packed together in polymers.

5.10 The density of solids

The densities of common engineering materials are listed in Table 5.1 and shown in Figure 5.12. These reflect the mass and diameter of the atoms that make them up and the efficiency with which they are packed to fill space. Metals, most of them, have high densities because the atoms are heavy and closely packed. Polymers are much less dense because the atoms of which they are made (C, H, O) are light, and because they generally

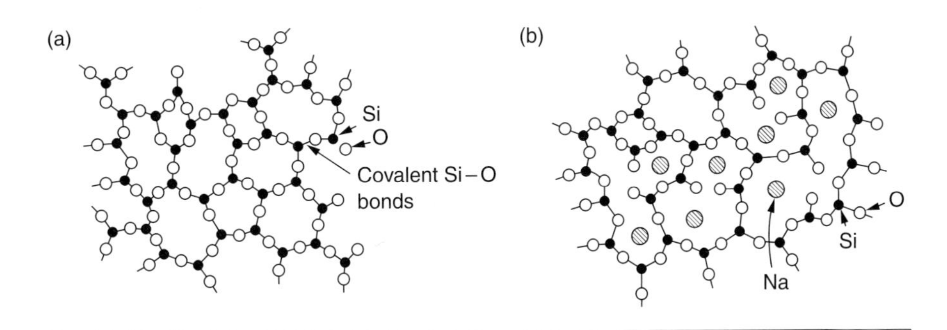

Figure 5.11 (a) Atom packing in amorphous (glassy) silica. (b) How the addition of soda breaks up the bonding in amorphous silica, giving soda glass.

Table 5.1 Data for density, ρ

Material	ρ ($Mg\,m^{-3}$)
Osmium	22.7
Platinum	21.4
Tungsten and alloys	13.4–19.6
Gold	19.3
Uranium	18.9
Tungsten carbide, WC	14.0–17.0
Tantalum and alloys	16.6–16.9
Molybdenum and alloys	10.0–13.7
Cobalt/tungsten-carbide cermets	11.0–12.5
Lead and alloys	10.7–11.3
Silver	10.5
Niobium and alloys	7.9–10.5
Nickel	8.9
Nickel alloys	7.8–9.2
Cobalt and alloys	8.1–9.1
Copper	8.9
Copper alloys	7.5–9.0
Brasses and bronzes	7.2–8.9
Iron	7.9
Iron-based super-alloys	7.9–8.3
Stainless steels, austenitic	7.5–8.1
Tin and alloys	7.3–8.0
Low-alloy steels	7.8–7.85
Mild steel	7.8–7.85
Stainless steel, ferritic	7.5–7.7
Cast iron	6.9–7.8
Titanium carbide, TiC	7.2
Zinc and alloys	5.2–7.2

Table 5.1 (*Continued*)

Material	ρ (Mg m^{-3})
Chromium	7.2
Zirconium carbide, ZrC	6.6
Zirconium and alloys	6.6
Titanium	4.5
Titanium alloys	4.3–5.1
Alumina, Al_2O_3	3.9
Alkali halides	3.1–3.6
Magnesia, MgO	3.5
Silicon carbide, SiC	2.5–3.2
Silicon nitride, Si_3N_4	3.2
Mullite	3.2
Beryllia, BeO	3.0
Common rocks	2.2–3.0
Calcite (marble, limestone)	2.7
Aluminum	2.7
Aluminum alloys	2.6–2.9
Silica glass, SiO_2 (quartz)	2.6
Soda glass	2.5
Concrete/cement	2.4–2.5
GFRPs	1.4–2.2
Carbon fibers	2.2
PTFE	2.3
Boron fiber/epoxy	2.0
Beryllium and alloys	1.85–1.9
Magnesium and alloys	1.74–1.88
Fiberglass (GFRP/polyester)	1.55–1.95
Graphite, high strength	1.8
PVC	1.3–1.6
CFRPs	1.5–1.6
Polyesters	1.1–1.5
Polyimides	1.4
Epoxies	1.1–1.4
Polyurethane	1.1–1.3
Polycarbonate	1.2–1.3
PMMA	1.2
Nylon	1.1–1.2
Polystyrene	1.0–1.1
Polyethylene, high-density	0.94–0.97
Ice, H_2O	0.92
Natural rubber	0.83–0.91
Polyethylene, low-density	0.91
Polypropylene	0.88–0.91
Common woods	0.4–0.8
Cork	0.1–0.2
Foamed plastics	0.01–0.6

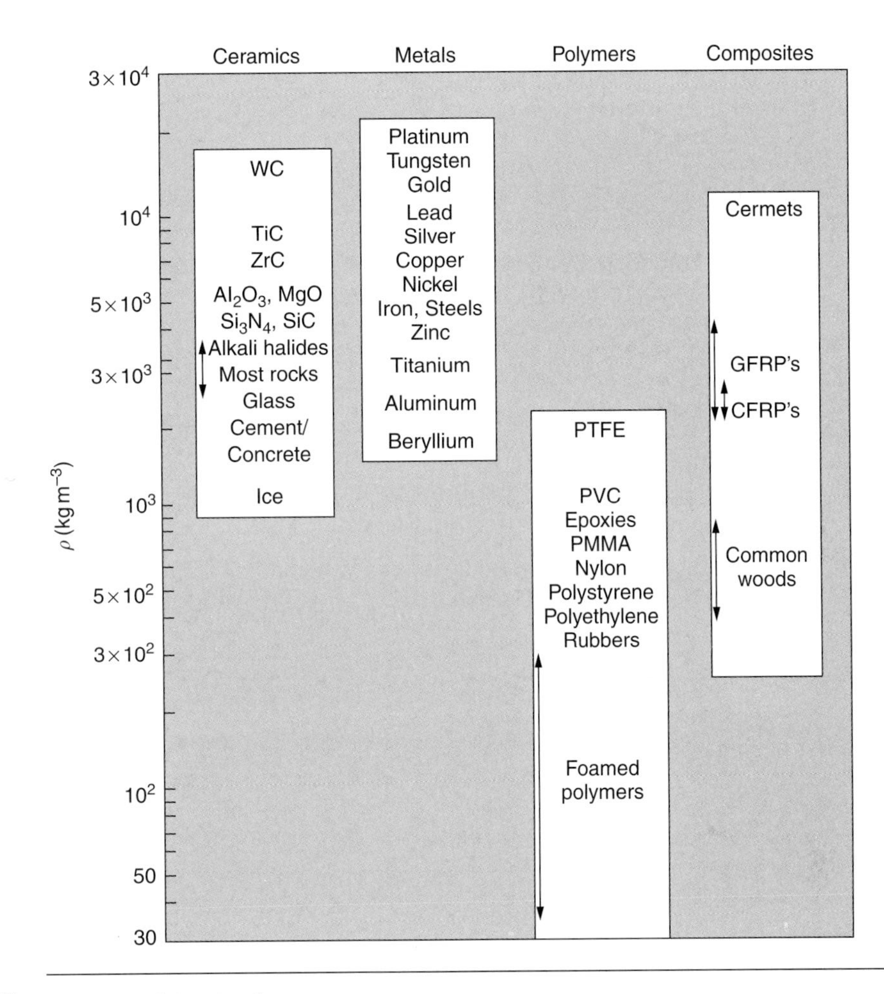

Figure 5.12 Bar chart of data for density, ρ.

adopt structures which are not close packed. Ceramics—even the ones in which atoms are packed closely—are, on average, a little less dense than metals because most of them contain light atoms like O, N and C. Composites have densities which are simply an average of the materials of which they are made.

Examples

5.1 (a) Calculate the density of an f.c.c. packing of spheres of unit density.
(b) If these same spheres are packed to form a *glassy* structure, the arrangement is called "dense random packing" and has a density of

0.636. If crystalline f.c.c. nickel has a density of 8.90 Mg m^{-3}, calculate the density of glassy nickel.

Answers

(a) 0.740; (b) 7.65 Mg m^{-3}.

5.2 (a) Sketch three-dimensional views of the unit cell of a b.c.c. crystal, showing a (100) plane, a (110) plane, a (111) plane and a (210) plane.
(b) The slip planes of b.c.c. iron are the {110} planes: sketch the atom arrangement in these planes, and mark the ⟨111⟩ slip directions.
(c) Sketch three-dimensional views of the unit cell of an f.c.c. crystal, showing a [100], a [110], a [111] and a [211] direction.
(d) The slip planes of f.c.c. copper are the {111} planes: sketch the atom arrangement in these planes and mark the ⟨110⟩ slip directions.

5.3 (a) The atomic diameter of an atom of nickel is 0.2492 nm. Calculate the lattice constant a of f.c.c. nickel.
(b) The atomic weight of nickel is 58.71 kg kmol^{-1}. Calculate the density of nickel. (Calculate first the mass per atom, and the number of atoms in a unit cell.)
(c) The atomic diameter of an atom of iron is 0.2482 nm. Calculate the lattice constant a of b.c.c. iron.
(d) The atomic weight of iron is 55.85 kg kmol^{-1}. Calculate the density of iron.

Answers

(a) 0.352 nm; (b) 8.91 Mg m^{-3}; (c) 0.287 nm; (d) 7.88 Mg m^{-3}.

5.4 Crystalline copper and magnesium have f.c.c and c.p.h structures respectively.

(a) Assuming that the atoms can be represented as hard spheres, calculate the percentage of the volume occupied by atoms in each material.
(b) Calculate, from first principles, the dimensions of the unit cell in copper and in magnesium.

(The densities of copper and magnesium are 8.96 Mg m^{-3} and 1.74 Mg m^{-3}, respectively.)

Answers

(a) 74% for both; (b) copper: $a = 0.361$ nm, magnesium: $a = 0.320$ nm, $c = 0.523$ nm.

5.5 What are the factors which determine the densities of solids? Why are metals more dense than polymers?

5.6 The density of a crystal of polyethylene (PE) is 1.014 (all densities are in Mg m^{-3} at 20°C). The density of amorphous PE is 0.84. Estimate the percentage crystallinity in:

(a) a low-density PE with a density of 0.92,
(b) a high-density PE with a density of 0.97.

Answers

(a) 46%; (b) 75%.

Chapter 6

The physical basis of Young's modulus

Chapter contents

6.1 Introduction

We are now in a position to bring together the factors underlying the moduli of materials. First, let us look back to Figure 3.5, the bar chart showing the moduli of materials. Recall that most ceramics and metals have moduli in a comparatively narrow range: 30–300 GN m^{-2}. Cement and concrete (45 GN m^{-2}) are near the bottom of that range. Aluminum (69 GN m^{-2}) is higher up; and steels (200 GN m^{-2}) are near the top. Special materials, it is true, lie outside it — diamond and tungsten lie above; ice and lead lie a little below — but most crystalline materials lie in that fairly narrow range. Polymers are quite different: all of them have moduli which are smaller, some by several orders of magnitude. Why is this? What determines the general level of the moduli of solids? And is there the possibility of producing stiff polymers?

We shall now examine the modulus of ceramics, metals, polymers, and composites, relating it to their structure.

6.2 Moduli of crystals

As we showed in Chapter 4, atoms in crystals are held together by bonds which behave like little springs. We defined the stiffness of one of these bonds as

$$S_0 = \left(\frac{d^2U}{dr^2}\right)_{r=r_0} \qquad (6.1)$$

For *small* strains, S_0 stays constant (it is the *spring constant* of the bond). This means that the force between a pair of atoms, stretched apart to a distance r ($r \approx r_0$), is

$$F = S_0(r - r_0) \qquad (6.2)$$

Imagine, now, a solid held together by such little springs, linking atoms between two planes within the material as shown in Figure 6.1. For simplicity we shall put atoms at the corners of cubes of side r_0. To be correct, of course, we should draw out the atoms in the positions dictated by the *crystal structure* of a particular material, but we shall not be *too* far out in our calculations by making our simplifying assumption — and it makes drawing the physical situation considerably easier!

Now, the total force exerted across *unit area*, if the two planes are pulled apart a distance $(r - r_0)$ is defined as the stress σ, with

$$\sigma = NS_0(r - r_0) \qquad (6.3)$$

N is the number of bonds/unit area, equal to $1/r_0^2$ (since r_0^2 is the average area per atom). We convert displacement $(r - r_0)$ into strain ϵ_n by dividing by the *initial* spacing, r_0, so that

$$\sigma = \left(\frac{S_0}{r_0}\right)\epsilon_n \qquad (6.4)$$

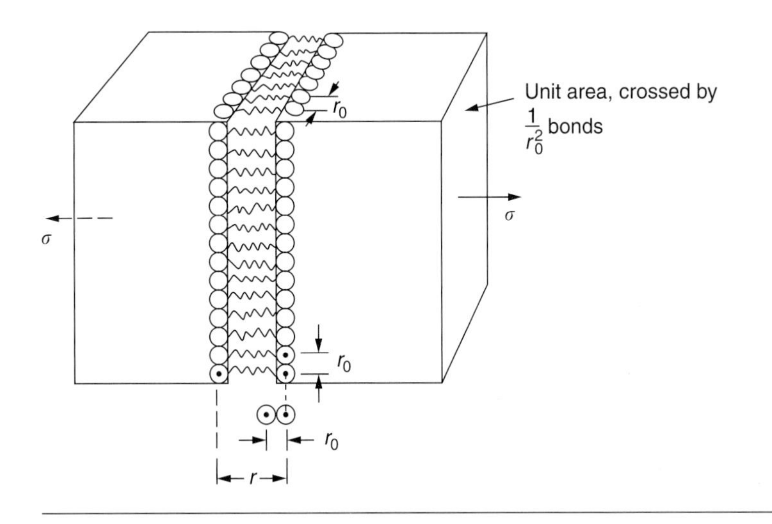

Figure 6.1 The method of calculating Young's modulus from the stiffnesses of individual bonds.

Young's modulus, then, is just

$$E = \frac{\sigma}{\epsilon_n} = \left(\frac{S_0}{r_0}\right) \tag{6.5}$$

S_0 can be calculated from the theoretically derived $U(r)$ curves of the sort described in Chapter 4. This is the realm of the solid-state physicist and quantum chemist, but we shall consider one example: the ionic bond, for which $U(r)$ is given in equation (4.3). Differentiating once with respect to r gives the *force* between the atoms, which must, of course, be zero at $r = r_0$ (because the material would not otherwise be in equilibrium, but would move). This gives the value of the constant B in equation (4.3):

$$B = \frac{q^2 r_0^{n-1}}{4\pi n \epsilon_0} \tag{6.6}$$

where q is the electron charge and ϵ_0 the permittivity of vacuum.
Then equation (6.1) for S_0 gives

$$S_0 = \frac{\alpha q^2}{4\pi\epsilon_0 r_0^3} \tag{6.7}$$

where $\alpha = (n-1)$. But the coulombic attraction is a *long-range* interaction (it varies as $1/r$; an example of a short-range interaction is one which varies as $1/r^{10}$). Because of this, a given Na^+ ion not only interacts (attractively) with its shell of six neighboring Cl^- ions, it also interacts (repulsively) with the 12 slightly more distant Na^+ ions, with the eight Cl^- ions beyond that, and with the six Na^+ ions which form the shell beyond *that*. To calculate S_0 properly, we must sum over all these bonds, taking attractions and repulsions properly into account. The result is identical with equation (6.7), with $\alpha = 0.58$.

Table 6.1 Hierarchy of bond stiffnesses

Bond type	S_0 ($N\,m^{-1}$)	E(GPa); *from equation (6.5)* (*with* $r_0 = 2.5 \times 10^{-10}$ m)
Covalent, e.g. C–C	50–180	200–1000
Metallic, e.g. Cu–Cu	15–75	60–300
Ionic, e.g. Na–Cl	8–24	32–96
H-bond, e.g. H_2O–H_2O	2–3	8–12
Van der Waals, e.g. polymers	0.5–1	2–4

The Table of Physical Constants on the inside front cover gives values for q and ϵ_0; and r_0, the atom spacing, is close to 2.5×10^{-10} m. Inserting these values gives:

$$S_0 = \frac{0.58(1.6 \times 10^{-19})^2}{4\pi \times 8.85 \times 10^{-12}(2.5 \times 10^{-10})^3} = 8.54\,\text{N}\,\text{m}^{-1}$$

The stiffnesses of other bond types are calculated in a similar way (in general, the cumbersome sum described above is not needed because the interactions are of *short range*). The resulting hierarchy of bond stiffnesses is as shown in Table 6.1.

A comparison of these predicted values of E with the measured values plotted in the bar chart of Figure 3.5 shows that, for metals and ceramics, the values of E we calculate are about right: the bond-stretching idea explains the stiffness of these solids. We can be happy that we can explain the moduli of these classes of solid. But a paradox remains: *there exists a whole range of polymers and rubbers which have moduli which are lower—by up to a factor of 100—than the lowest we have calculated.* Why is this? What determines the moduli of these floppy polymers if it is not the springs between the atoms? We shall explain this under our next heading.

6.3 Rubbers and the glass transition temperature

All polymers, if really solid, should have moduli above the lowest level we have calculated—about $2\,\text{GN}\,\text{m}^{-2}$—since they are held together partly by Van der Waals and partly by covalent bonds. If you take ordinary rubber tubing (a polymer) and cool it down in liquid nitrogen, it becomes stiff—its modulus rises rather suddenly from around $10^{-2}\,\text{GN}\,\text{m}^{-2}$ to a "proper" value of $2\,\text{GN}\,\text{m}^{-2}$. But if you warm it up again, its modulus drops back to $10^{-2}\,\text{GN}\,\text{m}^{-2}$.

This is because rubber, like many polymers, is composed of long spaghetti-like chains of carbon atoms, all tangled together as we showed in Chapter 5. In the case of rubber, the chains are also lightly cross-linked, as shown in Figure 5.10. There are covalent bonds along the carbon chain, and where there are occasional cross-links. These are very stiff, but they contribute very little to the overall modulus because when you load the structure it is the flabby Van der Waals bonds *between* the chains which stretch, and it is these which determine the modulus.

Well, that is the case at the low temperature, when the rubber has a "proper" modulus of a few GPa. As the rubber warms up to room temperature, the Van der Waals bonds *melt*. (In fact, the stiffness of the bond is proportional to its melting point: that is why diamond, which has the highest melting point of any material, also has the highest modulus.) The rubber remains solid because of the cross-links which form a sort of skeleton: but when you load it, the chains now slide over each other in places where there are no cross-linking bonds. This, of course, gives extra strain, and the modulus goes down (remember, $E = \sigma/\epsilon_n$).

Many of the most floppy polymers have half-melted in this way at room temperature. The temperature at which this happens is called the *glass temperature*, T_G, for the polymer. Some polymers, which have no cross-links, melt completely at temperatures above T_G, becoming viscous liquids. Others, containing cross-links, become *leathery* (like PVC) or rubbery (as polystyrene butadiene does). Some typical values for T_G are: polymethylmethacrylate (PMMA, or perspex), 100°C; polystyrene (PS), 90°C; polyethylene (low-density form), −20°C; *natural* rubber, −40°C. To summarize, above T_G, the polymer is leathery, rubbery or molten; below, it is a true solid with a modulus of at least 2 GN m^{-2}. This behavior is shown in Figure 6.2 which also shows how the

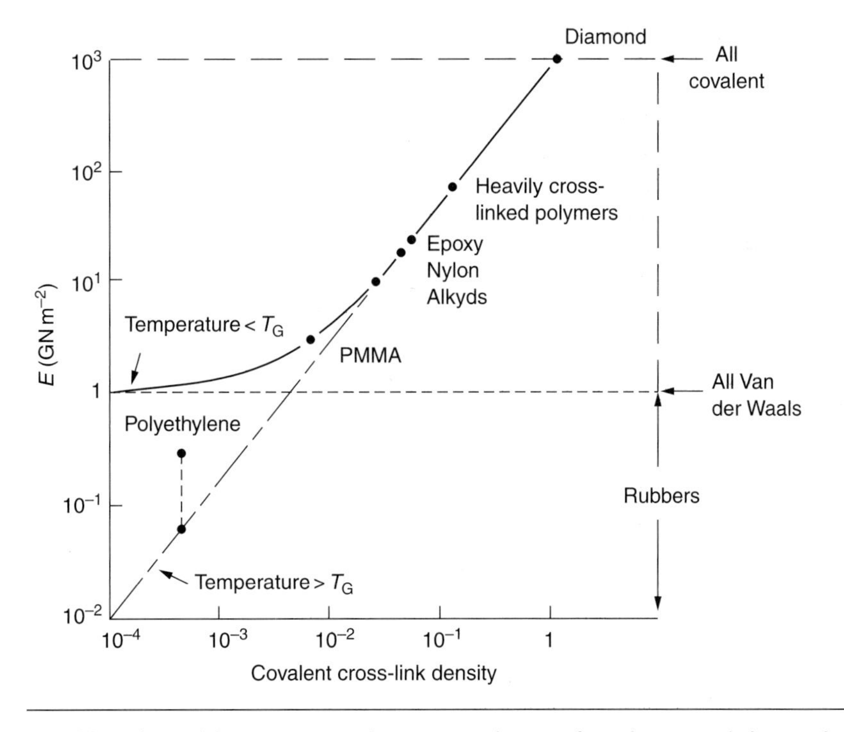

Figure 6.2 How Young's modulus increases with increasing density of covalent cross-links in polymers, including rubbers above the glass temperature. Below T_G, the modulus of rubbers increases markedly because the Van der Waals bonds take hold. Above T_G they melt, and the modulus drops.

stiffness of polymers increases as the covalent cross-link density increases, towards the value for diamond (which is simply a polymer with 100 percent of its bonds cross-linked, Figure 4.7). Stiff polymers, then, *are* possible; the stiffest now available have moduli comparable with that of aluminum.

6.4 Composites

Is it possible to make polymers stiffer than the Van der Waals bonds which usually hold them together? The answer is yes—if we mix into the polymer a second, stiffer, material. Good examples of materials stiffened in this way are:

(a) GFRP—glass-fiber reinforced polymers, where the polymer is stiffened or reinforced by long fibers of soda glass;
(b) CFRP—carbon-fiber reinforced polymers, where the reinforcement is achieved with fibers of graphite;
(c) KFRP—Kevlar-fiber reinforced polymers, using Kevlar fibers (a unique polymer with a high density of covalent bonds oriented along the fiber axis) as stiffening;
(d) FILLED POLYMERS—polymers into which glass powder or silica flour has been mixed to stiffen them;
(e) WOOD—a natural composite of lignin (an amorphous polymer) stiffened with fibers of cellulose.

The bar chart of moduli (Figure 3.5) shows that composites can have moduli much higher than those of their matrices. And it also shows that they can be *very* anisotrophic, meaning that the modulus is higher in some directions than others. Wood is an example: its modulus, measured parallel to the fibers, is about $10\,\mathrm{GN\,m^{-2}}$; at right angles to this, it is less than $1\,\mathrm{GN\,m^{-2}}$.

There is a simple way to estimate the modulus of a fiber-reinforced composite. Suppose we stress a composite, containing a volume fraction V_f of fibers, parallel to the fibers (see Figure 6.3(a)). Loaded in this direction, the strain, ϵ_n, in the fibers and the matrix is the same. The stress carried by the composite is

$$\sigma = V_f \sigma_f + (1 - V_f)\sigma_m,$$

where the subscripts f and m refer to the fiber and matrix, respectively.

Since $\sigma = E\epsilon_n$, we can rewrite this as:

$$\sigma = E_f V_f \epsilon_n + E_m (1 - V_f)\epsilon_n$$

But since

$$E_{composite} = \sigma / \epsilon_n$$

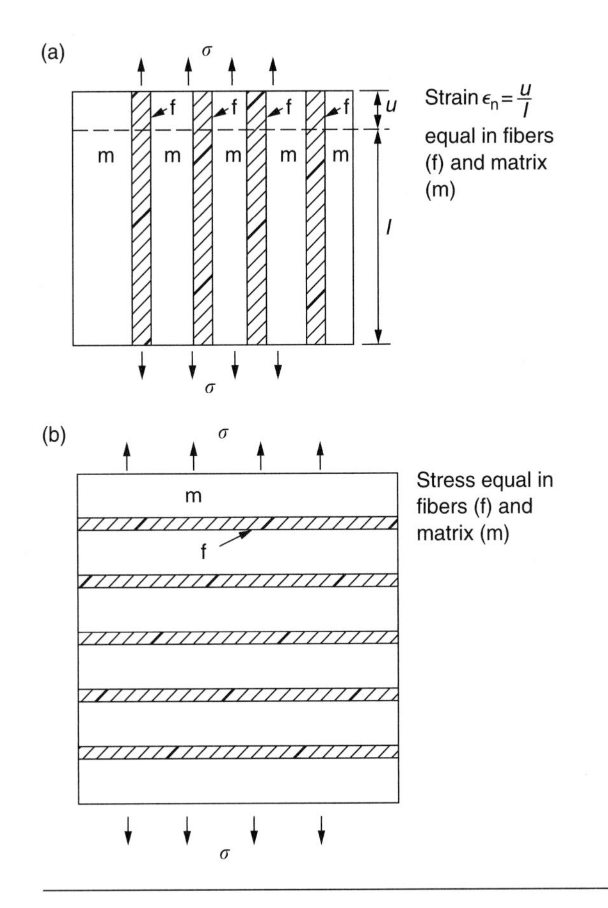

Figure 6.3 A fiber-reinforced composite loaded in the direction in which the modulus is (a) a maximum, (b) a minimum.

we find

$$E_{composite} = V_f E_f + (1 - V_f) E_m \tag{6.8}$$

This gives us an upper estimate for the modulus of our fiber-reinforced composite. The modulus cannot be greater than this, since the strain in the stiff fibers can never be greater than that in the matrix.

How is it that the modulus can be less? Suppose we had loaded the composite in the opposite way, at right angles to the fibers (as in Figure 6.3(b)). It now becomes much more reasonable to assume that the *stresses*, not the strains, in the two components are equal. If this is so, then the total nominal strain ϵ_n is the weighted sum of the individual strains:

$$\epsilon_n = V_f \epsilon_{nf} + (1 - V_f) \epsilon_{nm}$$

Using $\epsilon_n = \sigma/E$ gives

$$\epsilon_n = \frac{V_f \sigma}{E_f} + \left(\frac{1 - V_f}{E_m}\right)\sigma$$

The modulus is still σ/ε_n, so that

$$E_{composite} = 1 \Bigg/ \left\{\frac{V_f}{E_f} + \frac{(1 - V_f)}{E_m}\right\} \tag{6.9}$$

Although it is not obvious, this is a lower limit for the modulus—it cannot be less than this.

The two estimates, if plotted, look as shown in Figure 6.4. This explains why fiber-reinforced composites like wood and GFRP are so stiff along the reinforced direction (the upper line of the figure) and yet so floppy at right angles to the direction of reinforcement (the lower line), that is, it explains their *anisotropy*. Anisotropy is sometimes what you want—as in the shaft of a squash racquet or a vaulting pole. Sometimes it is not, and then the layers of fibers can be *laminated* in a criss-cross way, as they are in the body shell of a Formula 1 racing car.

Not all composites contain fibers. Materials can also be stiffened by (roughly spherical) *particles*. The theory is, as one might imagine, more difficult than for fiber-reinforced composites; and is too advanced to talk about here. But it is useful to know that the moduli of these so-called *particulate* composites lie between the upper and lower limits of equations (6.8) and (6.9), nearer the lower one than the upper one, as shown in Figure 6.4. Now, it is much cheaper to mix sand into a polymer than to carefully align specially produced glass fibers in the

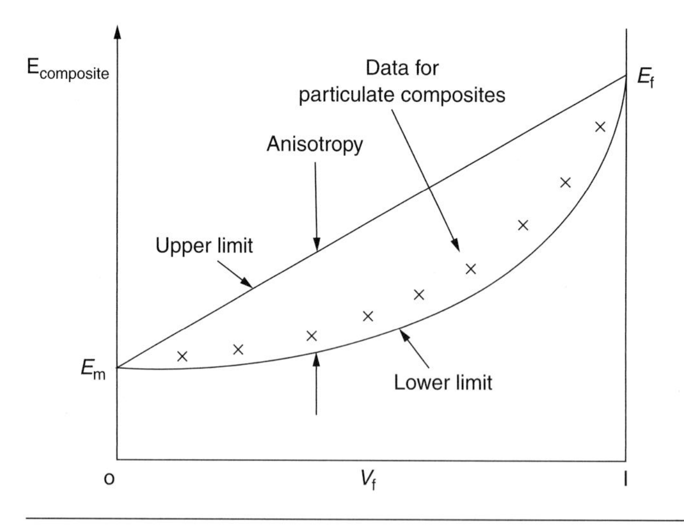

Figure 6.4 Composite modulus for various volume fractions of stiffener, showing the upper and lower limits of equations (6.8) and (6.9).

same polymer. Thus the modest increase in stiffness given by particles is economically worthwhile. Naturally the resulting particulate composite is *isotropic*, rather than *anisotropic* as would be the case for the fiber-reinforced composites; and this, too, can be an advantage. These filled polymers can be formed and moulded by normal methods (most fiber-composites cannot) and so are cheap to fabricate. Many of the polymers of everyday life — bits of cars and bikes, household appliances and so on — are, in fact, filled.

6.5 Summary

The moduli of metals, ceramics, and glassy polymers below T_G reflect the stiffness of the bonds which link the atoms. Glasses and glassy polymers above T_G are leathers, rubbers or viscous liquids, and have much lower moduli. Composites have moduli which are a weighted average of those of their components.

Examples

6.1 The table lists Young's modulus, $E_{composite}$, for a glass-filled epoxy composite. The material consists of a volume fraction V_f of glass particles (Young's modulus, E_f, 80 GN m^{-2}) dispersed in a matrix of epoxy (Young's modulus, E_m, 5 GN m^{-2}).

Calculate the upper and lower values for the modulus of the composite material, and plot them, together with the data, as a function of V_f. Which set of values most nearly describes the results? Why? How does the modulus of a random chopped-fiber composite differ from those of an aligned continuous-fiber composite?

6.2 A composite material consists of parallel fibers of Young's modulus E_f in a matrix of Young's modulus E_m. The volume fraction of fibers is V_f. Derive an

Volume fraction of glass, V_f	$E_{composite}$ (GN m^{-2})
0	5.0
0.05	5.5
0.10	6.4
0.15	7.8
0.20	9.5
0.25	11.5
0.30	14.0

expression for E_c, Young's modulus of the composite along the direction of the fibers, in terms of E_f, E_m and V_f. Obtain an analogous expression for the density of the composite, ρ_c Using material parameters given below, find ρ_c and E_c for the following composites: (a) carbon fiber–epoxy resin ($V_f = 0.5$), (b) glass fiber–polyester resin ($V_f = 0.5$), (c) steel–concrete ($V_f = 0.02$).

Material	*Density* ($Mg\,m^{-3}$)	*Young's modulus* ($GN\,m^{-2}$)
Carbon fiber	1.90	390
Glass fiber	2.55	72
Epoxy resin Polyester resin }	1.15	3
Steel	7.90	200
Concrete	2.40	45

Answers

$E_c = E_f V_f + (1 - V_f)E_m$; $\rho_c = \rho_f V_f + (1 - V_f)\rho_m$.

(a) $\rho_c = 1.53\,Mg\,m^{-3}$, $E_c = 197\,GN\,m^{-2}$;

(b) $\rho_c = 1.85\,Mg\,m^{-3}$, $E_c = 37.5\,GN\,m^{-2}$;

(c) $\rho_c = 2.51\,Mg\,m^{-3}$, $E_c = 48.1\,GN\,m^{-2}$.

6.3 A composite material consists of flat, thin metal plates of uniform thickness glued one to another with a thin, epoxy-resin layer (also of uniform thickness) to form a "multi-decker-sandwich" structure. Young's modulus of the metal is E_1, that of the epoxy resin is E_2 (where $E_2 < E_1$) and the volume fraction of metal is V_1. Find the ratio of the maximum composite modulus to the minimum composite modulus in terms of E_1, E_2 and V_1. Which value of V_1 gives the largest ratio?

Answer

Largest ratio when $V_1 = 0.5$.

6.4 (a) Define a high polymer; list three engineering polymers.
(b) Define a thermoplastic and a thermoset.
(c) Distinguish between a glassy polymer, a crystalline polymer, and a rubber.
(d) Distinguish between a cross-linked and a non-cross-linked polymer.
(e) What is a copolymer?
(f) List the monomers of polyethylene (PE), polyvinyl chloride (PVC), and polystyrene (PS).
(g) What is the glass transition temperature, T_G?

(h) Explain the change of moduli of polymers at the glass transition temperature.
(i) What is the order of magnitude of the number of carbon atoms in a single molecule of a high polymer?
(j) What is the range of temperature in which T_G lies for most engineering polymers?
(k) How would you increase the modulus of a polymer?

Chapter 7

Case studies in modulus-limited design

Chapter contents

7.1 Case study 1: a telescope mirror — involving the selection of a material to minimize the deflection of a disc under its own weight

Introduction

The world's largest single mirror reflecting telescope is sited on Mount Semivodrike, near Zelenchukskaya in the Caucasus Mountains. The mirror is 6 m (236 inches) in diameter, but it has never worked very well. The largest satisfactory single mirror reflector is that at Mount Palomar in California; it is 5.08 m (200 inches) in diameter. To be sufficiently rigid, the mirror (which is made of glass) is about 1 m thick and weighs 70 tonnes.*

The cost of a 5 m telescope is, like the telescope itself, astronomical — about UK£120m or US$220m. This cost varies roughly with the square of the weight of the mirror so it rises very steeply as the diameter of the mirror increases. The mirror itself accounts for about 5 percent of the total cost of the telescope. The rest goes on the mechanism which holds, positions, and moves the mirror as it tracks across the sky (Figure 7.1). This must be so stiff that it can position the mirror relative to the collecting system with a precision about equal to that of the wavelength of light. At first sight, if you double the mass M of the mirror, you need only double the sections of the structure which holds it in order to keep the stresses (and hence the strains and deflections) the same, but this is incorrect because the heavier structure deflects under its *own* weight. In practice, you have to add more section to allow for this so that the volume (and thus the cost) of the structure goes as M^2. The main obstacle to building such large telescopes is the cost.

Before the turn of the century, mirrors were made of speculum metal, a copper–tin alloy (the Earl of Rosse (1800–1867), who lived in Ireland, used one to discover spiral galaxies) but they never got bigger than 1 m because of the weight. Since then, mirrors have been made of glass, silvered on the *front* surface, so none of the optical properties of the glass are used. Glass is chosen for its mechanical properties only; the 70 tonnes of glass is just a very elaborate support for 100 nm (about 30 g) of silver. Could one, by taking a radically new look at mirror design, suggest possible routes to the construction of larger mirrors which are much lighter (and therefore cheaper) than the present ones?

* The world's *largest* telescope is the 10 m Keck reflector. It is made of 36 separate segments each of which is independently controlled.

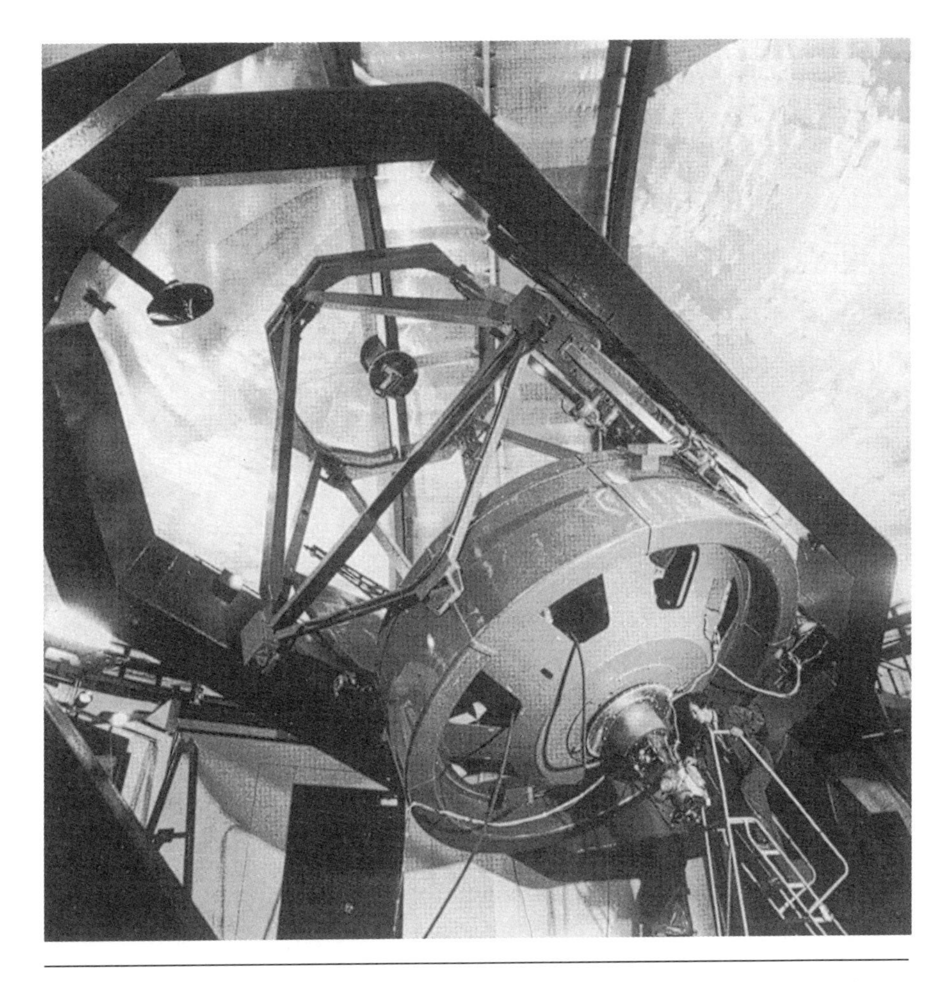

Figure 7.1 The British infrared telescope at Mauna Kea, Hawaii. The picture shows the housing for the 3.8 m diameter mirror, the supporting frame, and the interior of the aluminum dome with its sliding "window".

Optimum combination of elastic properties for the mirror support

Consider the selection of the material for the mirror backing of a 200-in. (5 m) diameter telescope. We want to identify the material that gives a mirror which will distort by less than the wavelength of light when it is moved, and has minimum weight. We will limit ourselves to these criteria alone for the moment—we will leave the problem of grinding the parabolic shape and getting an optically perfect surface to the development research team.

At its simplest, the mirror is a circular disc, of diameter $2a$ and mean thickness t, simply supported at its periphery (Figure 7.2). When horizontal, it

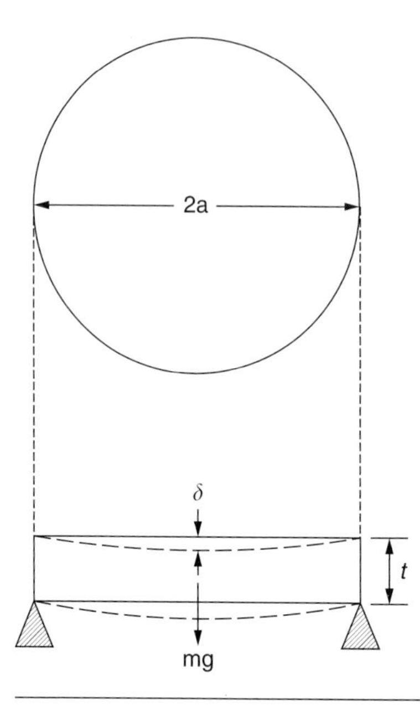

Figure 7.2 The elastic deflection of a telescope mirror, shown for simplicity as a flat-faced disc, under its own weight.

will deflect under its own weight M; when vertical it will not deflect significantly. We want this distortion (which changes the focal length and introduces aberrations into the mirror) to be small enough that it does not significantly reduce the performance of the mirror. In practice, this means that the deflection δ of the midpoint of the mirror must be less than the wavelength of light. We shall require, therefore, that the mirror deflect less than $\approx 1\,\mu m$ at its center. This is an exceedingly stringent limitation. Fortunately, it can be *partially* overcome by engineering design without reference to the material used. By using counterbalanced weights or hydraulic jacks, the mirror can be supported by distributed forces over its back surface which are made to vary automatically according to the attitude of the mirror (Figure 7.3). Nevertheless, the limitations of this compensating system still require that the mirror have a stiffness such that δ be less than $10\,\mu m$.

You will find the formulae for the elastic deflections of plates and beams under their own weight in standard texts on mechanics or structures (one is listed in the References at the end of this book). We need only one formula here: it is that for the deflection, δ, of the center of a horizontal disc, simply supported at its periphery (meaning that it rests there but is not clamped) due to its own weight. It is:

$$\delta = \frac{0.67}{\pi} \frac{Mga^2}{Et^3} \qquad (7.1)$$

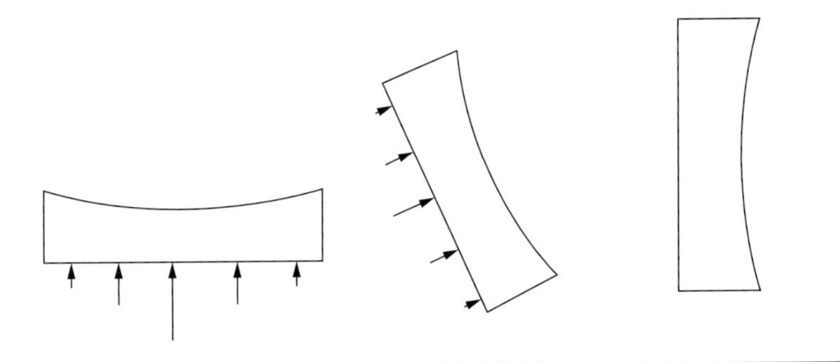

Figure 7.3 The distortion of the mirror under its own weight can be corrected by applying forces (shown as arrows) to the back surface.

(for a material having a Poisson's ratio fairly close to 0.33). The quantity g in this equation is the acceleration due to gravity. We need to minimize the mass for fixed values of $2a$ (5 m) and δ (10 μm). The mass can be expressed in terms of the dimensions of the mirror:

$$M = \pi a^2 t \rho \tag{7.2}$$

where ρ is the density of the material. We can make it smaller by reducing the thickness t—but there is a constraint: if we reduce it too much the deflection δ of equation (7.1) will be too great. So we solve equation (7.1) for t (giving the t which is just big enough to keep the deflection down to δ) and we substitute this into equation (7.2) giving

$$M = \left(\frac{0.67g}{\delta}\right)^{1/2} \pi a^4 \left(\frac{\rho^3}{E}\right)^{1/2} \tag{7.3}$$

Clearly, the only variables left on the right-hand side of equation (7.3) are the *material* properties ρ and E. To minimize M, we must choose a material having the smallest possible value of

$$M_1 = (\rho^3/E)^{1/2} \tag{7.4}$$

where M_1 is called the "material index".

Let us now examine its values for some materials. Data for E we can take from Table 3.1 in Chapter 3; those for density, from Table 5.1 in Chapter 5. The resulting values of the index M_1 are as shown in Table 7.1

Conclusions

The optimum material is CFRP. The next best is polyurethane foam. Wood is obviously impractical, but beryllium is good. Glass is better than steel,

Table 7.1 Mirror backing for 200-in. telescope

Material	$E(GN\,m^{-2})$	ρ $(Mg\,m^{-3})$	M_1	*M (tonne)*	*t* (m)
Steel	200	7.8	1.54	199	0.95
Concrete	47	2.5	0.58	53	1.1
Aluminum	69	2.7	0.53	50	0.95
Glass	69	2.5	0.48	45	0.91
GFRP	40	2.0	0.45	42	1.0
Beryllium	290	1.85	0.15	15	0.4
Wood	12	0.6	0.13	13	1.1
Foamed polyurethane	0.06	0.1	0.13	12	6.8
CFRP	270	1.5	0.11	10	0.36

aluminum or concrete (that is why most mirrors are made of glass), but a lot less good than beryllium, which is used for mirrors when cost is not a concern.

We should, of course, examine other aspects of this choice. The mass of the mirror can be calculated from equation (7.3) for the various materials listed in Table 7.1. Note that the polyurethane foam and the CFRP mirrors are roughly one-fifth the weight of the glass one, and that the structure needed to support a CFRP mirror could thus be as much as 25 times less expensive than the structure needed to support an orthodox glass mirror.

Now that we have the mass *M*, we can calculate the thickness *t* from equation (7.2). Values of *t* for various materials are given in Table 7.1. The glass mirror has to be about 1 m thick (and real mirrors are about this thick); the CFRP-backed mirror need only be 0.38 m thick. The polyurethane foam mirror has to be very thick—although there is no reason why one could not make a 6 m cube of such a foam.

Some of the above solutions—such as the use of polyurethane foam for mirrors—may at first seem ridiculously impractical. But the potential cost-saving (UK£5m or US$9m per telescope in place of UK£120m or US$220m) is so attractive that they are worth examining closely. There are ways of casting a thin film of silicone rubber, or of epoxy, onto the surface of the mirror backing (the polyurethane or the CFRP) to give an optically smooth surface which could be silvered. The most obvious obstacle is the lack of stability of polymers—they change dimensions with age, humidity, temperature, and so on. But glass itself can be foamed to give a material with a density not much larger than polyurethane foam, and the same stability as solid glass, so a study of this sort can suggest radically new solutions to design problems by showing how new classes of materials might be used. In fact, the mirror of the 2m Canadian–French–Hawaain telescope is made from a special polymer.

7.2 Case study 2: materials selection to give a beam of a given stiffness with minimum weight

Introduction

Many structures require that a beam sustain a certain force F without deflecting more than a given amount, δ. If, in addition, the beam forms part of a transport system—a plane or rocket, or a train—or something which has to be carried or moved—a rucksack for instance—then it is desirable, also, to minimize the weight.

In the following, we shall consider a single cantilever beam, of square section, and will analyze the material requirements to minimize the weight for a given stiffness. The results are quite general in that they apply equally to any sort of beam of square section, and can easily be modified to deal with beams of other sections: tubes, I-beams, box-sections, and so on.

Analysis

The square-section beam of length l (determined by the design of the structure, and thus fixed) and thickness t (a variable) is held rigidly at one end while a force F (the maximum service force) is applied to the other, as shown in Figure 7.4 The same texts that list the deflection of discs give equations for the elastic deflection of beams. The formula we want is

$$\delta = \frac{4l^3F}{Et^4} \tag{7.5}$$

(ignoring self weight).

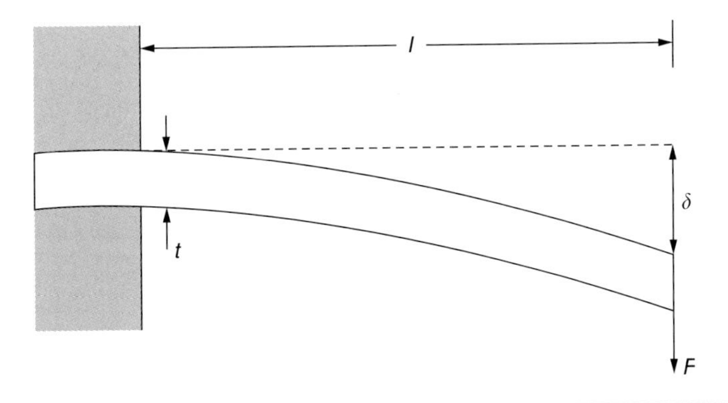

Figure 7.4 The elastic deflection δ of a cantilever beam of length l under an externally imposed force F.

The mass of the beam is given by

$$M = lt^2\rho \tag{7.6}$$

As before, the mass of the beam can be reduced by reducing t, but only so far that it does not deflect too much. The thickness is therefore *constrained* by equation (7.5). Solving this for t and inserting it into the last equation gives:

$$M = \left(\frac{4l^5 F}{\delta}\right)^{1/2}\left(\frac{\rho^2}{E}\right)^{1/2} \tag{7.7}$$

The mass of the beam, for given stiffness F/δ, is minimized by selecting a material with the minimum value of the material index

$$M_2 = \left(\frac{\rho^2}{E}\right)^{1/2} \tag{7.8}$$

The second column of numbers in Table 7.2 gives values for M_2.

Conclusions

The table shows that wood is one of the best materials for stiff beams — that is why it is so widely used in small-scale building, for the handles of rackets, and shafts of golf clubs, for vaulting poles, even for building aircraft. Polyurethane foam is no good at all — the criteria here are quite different from those of the first case study. The only material which is clearly superior to wood is CFRP — and it would reduce the mass of the beam very substantially: by the factor 0.17/0.09, or very nearly a factor of 2. That is why CFRP is used when weight saving is the overriding design criterion. But as we shall see in a moment, it is very expensive.

Why, then, are bicycles not made of wood? (There was a time when they were.) That is because metals, and polymers too, can readily be made into tubes; with wood it is more difficult. The formula for the bending of a tube depends

Table 7.2 Data for beam of given stiffness

Material	M_2	$\tilde{p}$	M_3
Steel	0.55	100	55
Polyurethane foam	0.41	1000	410
Concrete	0.36	50	18
Aluminum	0.32	400	128
GFRP	0.31	1000	310
Wood	0.17	70	12
CFRP	0.09	20,000	1800

on the mass of the tube in a different way than does that of a solid beam, and the optimization we have just performed—which is easy enough to redo—favors the tube.

7.3 Case study 3: materials selection to minimize the cost of a beam of given stiffness

Introduction

Often it is not the weight, but the *cost* of a structure which is the overriding criterion. Suppose that had been the case with the cantilever beam that we have just considered—would our conclusion have been the same? Would we still select wood? And how much more expensive would a replacement by CFRP be?

Analysis

The relative price per tonne, $\tilde{p}$, of materials is the first of the properties that we talked about in this book. The relative price of the beam, crudely, is the weight of the beam times $\tilde{p}$ (although this may neglect certain aspects of manufacture). Thus

$$\text{Price} = \left(\frac{4l^5F}{\delta}\right)^{1/2} \tilde{p}\left(\frac{\rho^2}{E}\right)^{1/2} \tag{7.9}$$

The beam of minimum price is therefore the one with the lowest value of the index

$$M_3 = \tilde{p}\left(\frac{\rho^2}{E}\right)^{1/2} \tag{7.10}$$

Values for M_3 are given in Table 7.2, with relative prices taken from Table 2.1.

Conclusions

Concrete and wood are the cheapest materials to use for a beam having a given stiffness. Steel costs more; but it can be rolled to give I-section beams which have a much better stiffness-to-weight ratio than the solid square-section beam we have been analyzing here. This compensates for steel's rather high cost, and accounts for the interchangeable use of steel, wood, and concrete that we talked about in bridge construction in Chapter 1. Finally, the lightest beam (CFRP) costs more than 100 times that of a wooden one—and this cost at present rules out CFRP for all but the most specialized applications like aircraft components or sophisticated sporting equipment. But the cost of CFRP falls as the market for it expands. If (as now seems possible) its market continues to grow, its price could fall to a level at which it would compete with metals in many applications.

Examples

7.1 A uniform, rectangular-section beam of fixed width w, unspecified depth d, and fixed length l rests horizontally on two simple supports at either end of the beam. A concentrated force F acts vertically downwards through the center of the beam. The deflection, δ, of the loaded point is

$$\delta = \frac{Fl^3}{4E_c w d^3}$$

ignoring the deflection due to self weight. Which of the three composites in example 6.2 will give the lightest beam for a given force and deflection?

Answer

Carbon fiber–epoxy resin.

7.2 (a) Select the material for the frame of the ultimate bicycle—meaning the lightest possible for a given stiffness. You may assume that the tubes of which the frame is made are cantilever beams (of length l) and that the elastic bending displacement δ of one end of a tubular beam under a force F (the other end being rigidly clamped) is

$$\delta = \frac{Fl^3}{3E\pi r^3 t}$$

$2r$ is the diameter of the tube (fixed by the designer) and t is the wall thickness of the tube, which you may vary. t is much less than r. Find the combination of material properties which determine the mass of the tube for a given stiffness, and hence make your material selection using data given in Chapters 3 and 5. Try steel, aluminum alloy, wood, GFRP, and CFRP.

(b) Which of the following materials leads to the cheapest bicycle frame, for a given stiffness: mild steel, aluminum alloy, titanium alloy, GFRP, CFRP or hardwood?

Answers

(a) CFRP; (b) steel.

7.3 You have been asked to prepare an outline design for the pressure hull of a deep-sea submersible vehicle capable of descending to the bottom of the Mariana Trench in the Pacific Ocean. The external pressure at this depth is approximately 100 MN m^{-2}, and the design pressure is to be taken as 200 MN m^{-2}. The pressure hull is to have the form of a thin-walled sphere with a specified radius r of 1 m and a uniform thickness t. The sphere can fail by external-pressure buckling at a pressure p_b given by

$$p_b = 0.3E\left(\frac{t}{r}\right)^2$$

where E is Young's modulus.

The basic design requirement is that the pressure hull shall have the minimum possible mass compatible with surviving the design pressure.

By eliminating t from the equations, show that the minimum mass of the hull is given by the expression

$$M_b = 22.9r^3 p_b^{0.5}\left(\frac{\rho}{E^{0.5}}\right)$$

Hence obtain a merit index to meet the design requirement for the failure mechanism. [You may assume that the surface area of the sphere is $4\pi r^2$.]

Answer

$E^{0.5}/\rho$

Part C

Yield strength, tensile strength and ductility

Chapter 8

The yield strength, tensile strength and ductility

Chapter contents

8.1 Introduction

All solids have an *elastic limit* beyond which something happens. A totally brittle solid will fracture, either suddenly (like glass) or progressively (like cement or concrete). Most engineering materials do something different; they deform *plastically* or change their shapes in a *permanent* way. It is important to know when, and how, they do this — both so that we can design structures which will withstand normal service loads without any permanent deformation, and so that we can design rolling mills, sheet presses, and forging machinery which will be strong enough to impose the desired deformation onto materials we wish to form. To study this, we pull carefully prepared samples in a tensile testing machine, or compress them in a compression machine (which we will describe in a moment), and record the *stress* required to produce a given *strain*.

8.2 Linear and nonlinear elasticity; anelastic behavior

Figure 8.1 shows the *stress–strain* curve of a material exhibiting *perfectly linear elastic* behavior. This is the behavior characterized by Hooke's law (Chapter 3). All solids are linear elastic at small strains — by which we usually mean less than 0.001, or 0.1 percent. The slope of the stress–strain line, which is the same in compression as in tension, is of course Young's Modulus, E. The area (shaded) is the elastic energy stored, per unit volume: since it is an elastic solid, we can get it all back if we unload the solid, which behaves like a linear spring.

Figure 8.2 shows a *nonlinear* elastic solid. *Rubbers* have a stress–strain curve like this, extending to very large strains (of order 5). The material is still

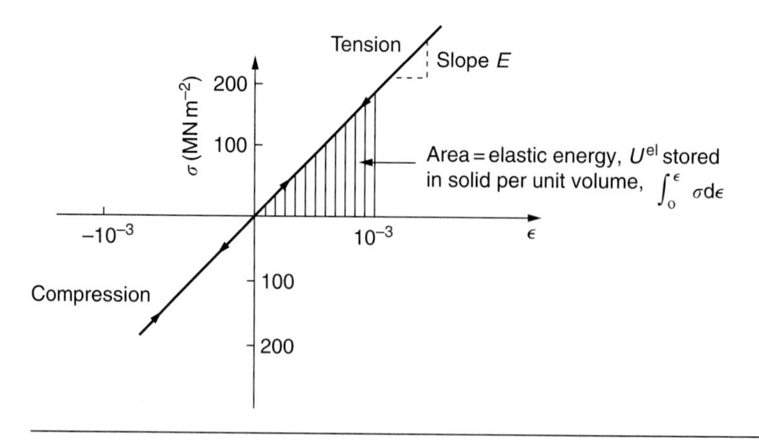

Figure 8.1 Stress–strain behavior for a *linear elastic solid.* The axes are calibrated for a material such as steel.

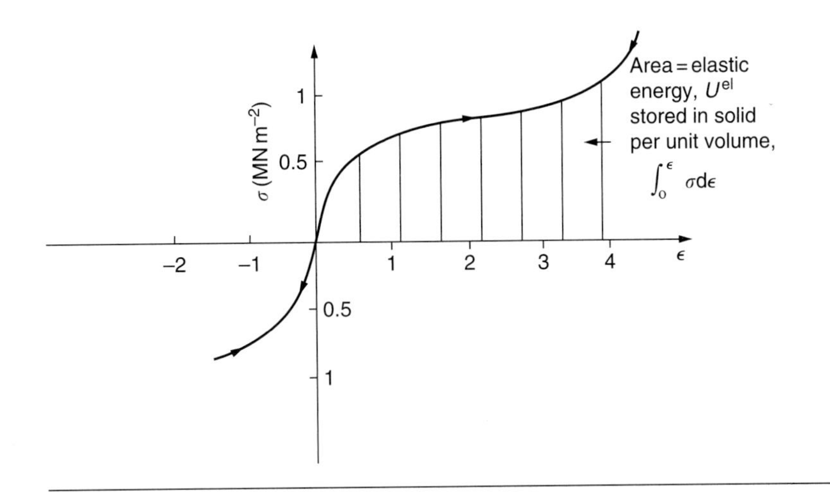

Figure 8.2 Stress–strain behavior for a *nonlinear elastic solid.* The axes are calibrated for a material such as rubber.

elastic: if unloaded, it follows the same path down as it did up, and all the energy stored, per unit volume, during loading is recovered on unloading—that is why catapults can be as lethal as they are.

Finally, Figure 8.3 shows a third form of *elastic* behavior found in certain materials. This is called *anelastic* behavior. All solids are anelastic to a small extent: even in the régime where they are nominally elastic, the loading curve does not *exactly* follow the unloading curve, and energy is dissipated (equal to the shaded area) when the solid is cycled. Sometimes this is useful—if you wish to damp out vibrations or noise, for example; you can do so with polymers or with soft metals (like lead) which have a high *damping capacity* (high anelastic loss). But often such damping is undesirable—springs and bells, for instance, are made of materials with the lowest possible damping capacity (spring steel, bronze, glass).

8.3 Load–extension curves for non-elastic (plastic) behavior

Rubbers are exceptional in behaving reversibly, or *almost* reversibly, to high strains; as we said, *almost all materials, when strained by more than about* 0.001 (0.1 percent), *do something irreversible*: and most engineering materials deform *plastically* to change their shape *permanently.* If we load a piece of ductile metal (like copper), for example in tension, we get the following relationship between the load and the extension (Figure 8.4). This can be demonstrated nicely by pulling a piece of plasticine (a ductile nonmetallic material). Initially, the plasticine deforms elastically, but at a small strain begins to deform plastically, so that if the load is removed, the piece of

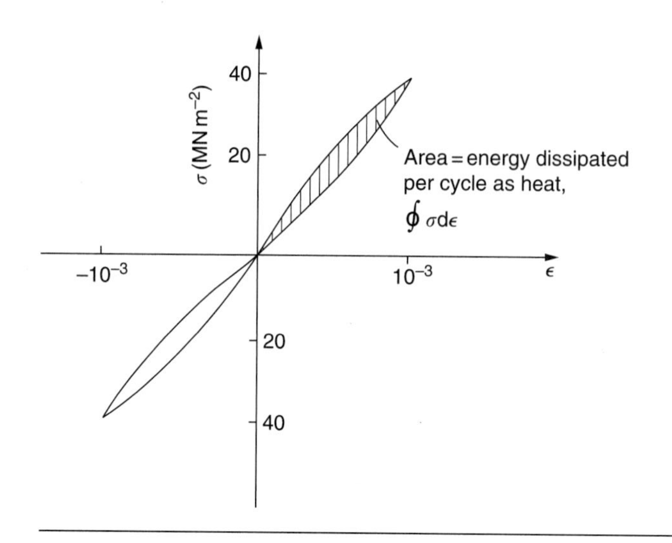

Figure 8.3 Stress–strain behavior for an *anelastic solid.* The axes are calibrated for fiberglass.

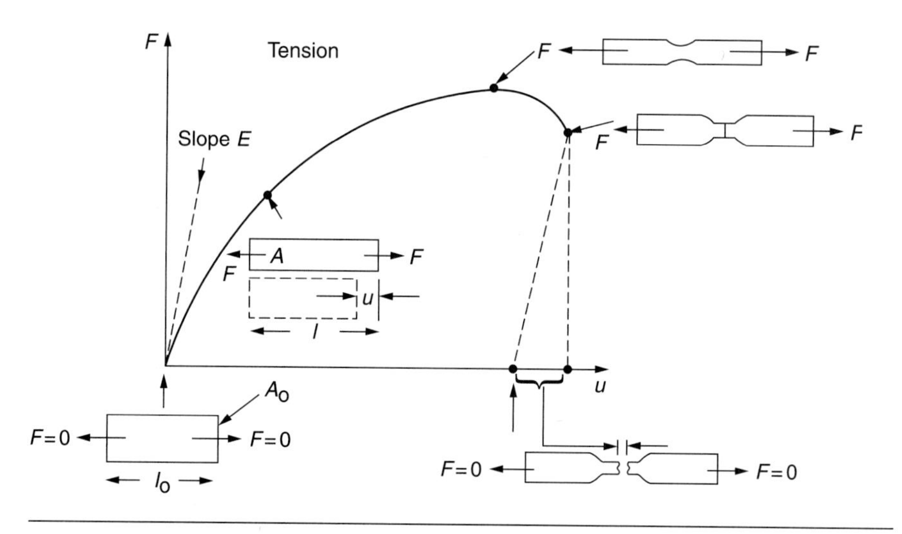

Figure 8.4 Load–extension curve for a bar of ductile metal (e.g. annealed copper) pulled in tension.

plasticine is permanently longer than it was at the beginning of the test: it has undergone *plastic* deformation (Figure 8.5). If you continue to pull, it continues to get longer, at the same time getting thinner because in plastic deformation *volume is conserved* (matter is just flowing from place to place). Eventually, the plasticine becomes unstable and begins to *neck* at the maximum load point in the force–extension curve (Figure 8.4). Necking is an *instability* which we shall look at in more detail in Chapter 11. The neck then

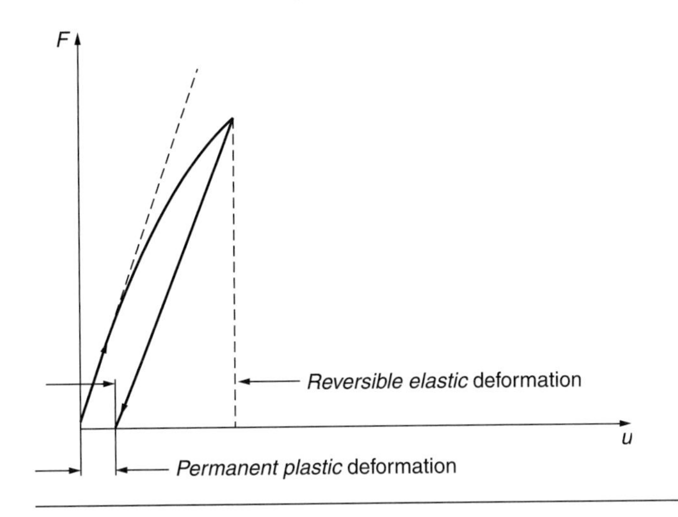

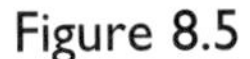
Figure 8.5

grows quite rapidly, and the load that the specimen can bear through the neck decreases until breakage takes place. The two pieces produced *after* breakage have a total length that is slightly *less* than the length *just* before breakage by the amount of the *elastic* extension produced by the terminal load.

If we load a material in *compression*, the force–displacement curve is simply the reverse of that for tension *at small strains*, but it becomes different at larger strains. As the specimen squashes down, becoming shorter and fatter to conserve volume, the load needed to keep it flowing rises (Figure 8.6). No instability such as necking appears, and the specimen can be squashed almost indefinitely, this process only being limited eventually by severe cracking in the specimen or the plastic flow of the compression plates.

Why this great difference in behavior? After all, we are dealing with the same material in either case.

8.4 True stress–strain curves for plastic flow

The apparent difference between the curves for tension and compression is due solely to the geometry of testing. If, instead of plotting *load*, we plot *load divided by the actual area of the specimen, A, at any particular elongation or compression*, the two curves become much more like one another. In other words, we simply plot *true stress* (see Chapter 3) as our vertical coordinate (Figure 8.7). This method of plotting allows for the *thinning* of the material when pulled in tension, or the *fattening* of the material when compressed.

But the two curves still do not exactly match, as Figure 8.7 shows. The reason is a displacement of (for example) $u = l_0/2$ in tension and compression

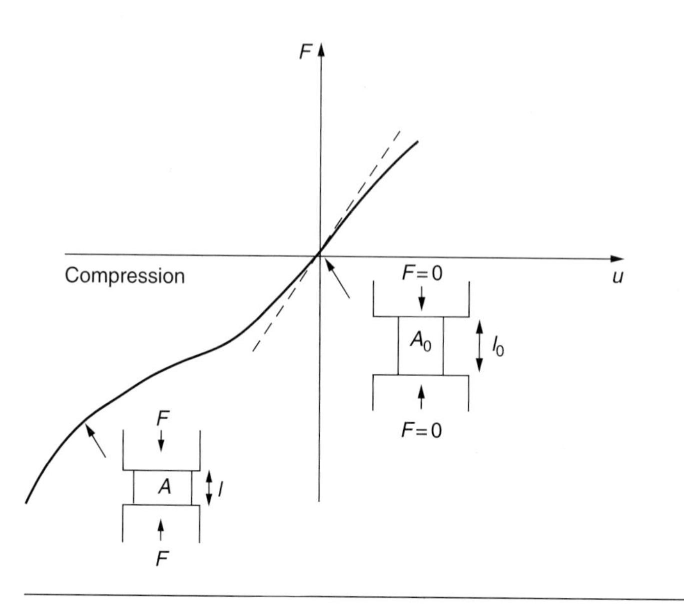

Figure 8.6

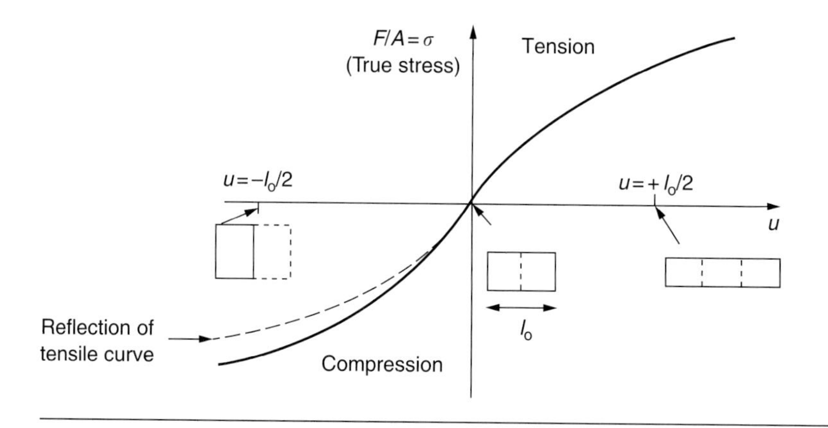

Figure 8.7

gives different *strains;* it represents a drawing out of the tensile specimen from l_0 to $1.5l_0$, but a squashing down of the compressive specimen from l_0 to $0.5l_0$. The material of the compressive specimen has thus undergone *much* more plastic deformation than the material in the tensile specimen, and can hardly be expected to be in the same state, or to show the same resistance to plastic deformation. The two conditions can be compared properly by taking small *strain increments*

$$\delta\epsilon = \frac{\delta u}{l} = \frac{\delta l}{l} \tag{8.1}$$

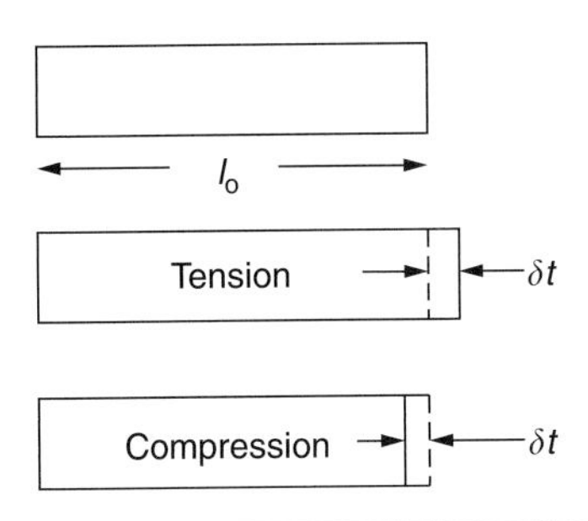

Figure 8.8

about which the state of the material is the same for either tension or compression (Figure 8.8). This is the same as saying that a decrease in length from 100 mm (l_0) to 99 mm (l), or an increase in length from 100 mm (l_0) to 101 mm (l) both represent a 1 percent change in the state of the material. Actually, they do not *quite* give exactly 1 percent in both cases, of course, but they *do* in the limit

$$d\epsilon = \frac{dl}{l} \tag{8.2}$$

Then, if the stresses in compression and tension are plotted against

$$\epsilon = \int_{l_0}^{l} \frac{dl}{l} = \ln\left(\frac{l}{l_0}\right) \tag{8.3}$$

the two curves *exactly* mirror one another (Figure 8.9). The quantity ϵ is called the *true* strain (to be contrasted with the *nominal* strain u/l_0 defined in Chapter 3) and the matching curves are *true stress/true strain* (σ/ϵ) curves. Now, a final catch. We can, from our original load–*extension* or load–*compression* curves easily calculate ϵ, simply by knowing l_0 and taking natural logs. But how do we calculate σ? Because volume is conserved during plastic deformation we can write, at any strain,

$$A_0 l_0 = Al$$

provided the extent of plastic deformation is much greater than the extent of elastic deformation (this is usually the case, but the qualification must be mentioned because volume is only conserved during *elastic* deformation if Poisson's ratio $v = 0.5$; and, as we showed in Chapter 3, it is near 0.33 for most materials). Thus

$$A = \frac{A_0 l_0}{l} \tag{8.4}$$

and

$$\sigma = \frac{F}{A} = \frac{Fl}{A_0 l_0} \tag{8.5}$$

all of which we know or can measure easily.

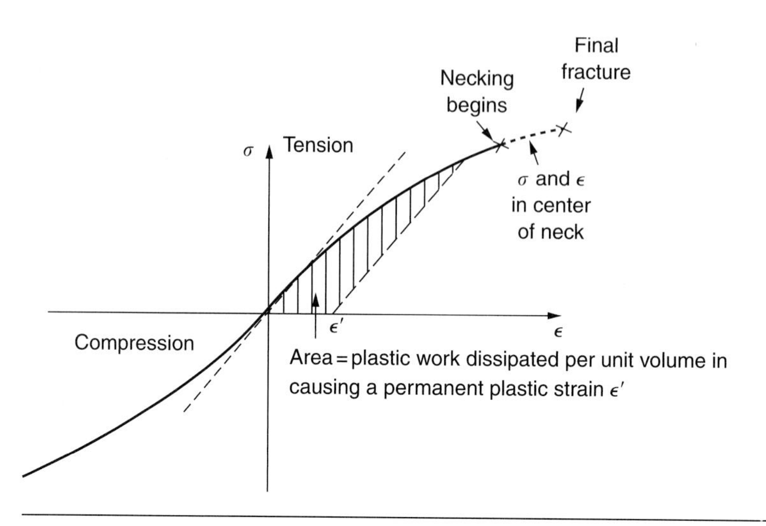

Figure 8.9

8.5 Plastic work

When metals are rolled or forged, or drawn to wire, or when polymers are injection molded or pressed or drawn, energy is absorbed. The work done on a material to change its shape permanently is called the *plastic work;* its value, per unit volume, is the area of the crosshatched region shown in Figure 8.9; it may easily be found (if the stress–strain curve is known) for any amount of permanent plastic deformation, ϵ'. Plastic work is important in metal- and polymer-forming operations because it determines the forces that the rolls, or press, or molding machine must exert on the material.

8.6 Tensile testing

The plastic behavior of a material is usually measured by conducting a tensile test. Tensile testing equipment is standard in all engineering laboratories. Such equipment produces a load/displacement (*F*/*u*) curve for the material, which is then converted to a nominal stress/nominal strain, or σ_n/ϵ_n, curve (Figure 8.10), where

$$\sigma_n = \frac{F}{A_0} \tag{8.6}$$

and

$$\epsilon_n = \frac{u}{l_0} \tag{8.7}$$

(see Chapter 3, and above). Naturally, because A_0 and l_0 are constant, the *shape* of the σ_n/ϵ_n curve is identical to that of the load-extension curve. But the

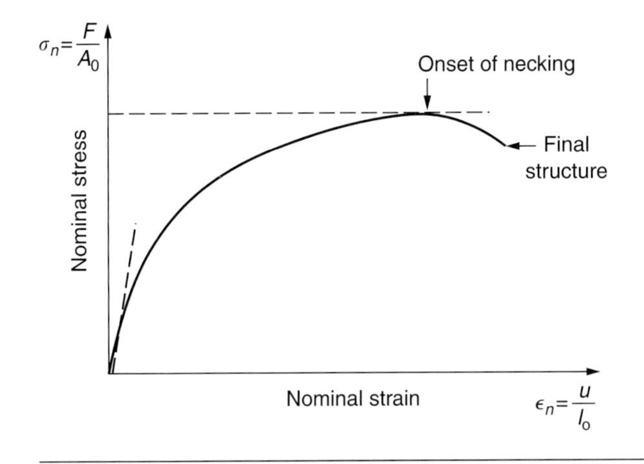

Figure 8.10

σ_n/ϵ_n plotting method allows one to compare data for specimens having different (though now standardized) A_0 and l_0, and thus to examine the properties of *material*, unaffected by specimen size. The advantage of keeping the stress in *nominal* units and not converting to *true* stress (as shown above) is that the onset of necking can clearly be seen on the σ_n/ϵ_n curve.

Now, let us define the quantities usually listed as the results of a *tensile test*. The easiest way to do this is to show them on the σ_n/ϵ_n curve itself (Figure 8.11). They are:

σ_y *Yield strength* (F/A_0 at onset of plastic flow).

$\sigma_{0.1\%}$ *0.1 percent proof stress* (F/A_0 at a permanent strain of 0.1 percent) (0.2 percent proof stress is often quoted instead. Proof stress is useful for characterizing yield of a material that yields gradually, and does not show a distinct yield point.)

σ_{TS} *Tensile strength* (F/A_0 at onset of necking).

ε_f (Plastic) *strain after fracture*, or tensile ductility. The broken pieces are put together and measured, and ϵ_f calculated from $(l - l_0)/l_0$, where l is the length of the assembled pieces.

8.7 Data

Data for the *yield strength*, *tensile strength*, and the *tensile ductility* are given in Table 8.1 and shown on the bar chart (Figure 8.12). Like moduli, they span a range of about 10^6: from about 0.1 MN m^{-2} (for polystyrene foams) to nearly 10^5 MN m^{-2} (for diamond).

Most ceramics have enormous yield stresses. In a tensile test, at room temperature, ceramics almost all fracture long before they yield: this is because

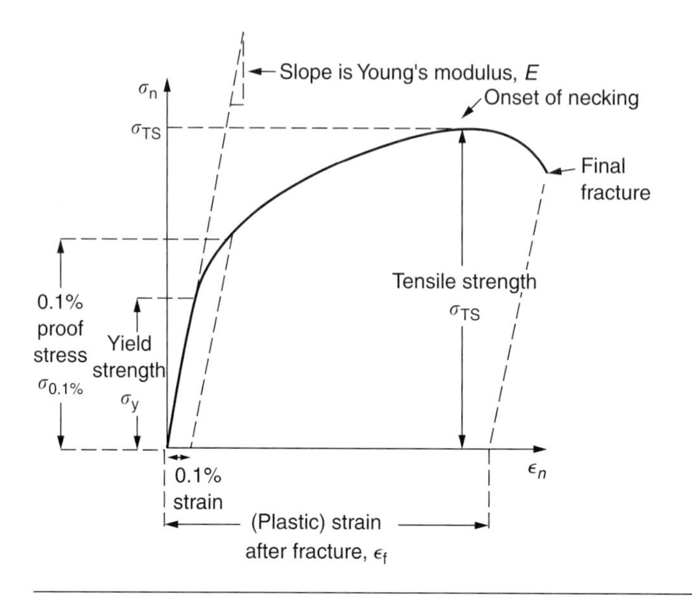

Figure 8.11

their fracture toughness, which we will discuss later, is very low. Because of this, you cannot measure the yield strength of a ceramic by using a tensile test. Instead, you have to use a test which somehow suppresses fracture: a compression test, for instance. The best and easiest is the hardness test: the data shown here are obtained from hardness tests, which we shall discuss in a moment.

Pure metals are very soft indeed, and have a high ductility. This is what, for centuries, has made them so attractive at first for jewellery and weapons, and then for other implements and structures: they can be *worked* to the shape that you want them in; furthermore, their ability to work-harden means that, after you have finished, the metal is much stronger than when you started. By alloying, the strength of metals can be further increased, though—in yield strength—the strongest metals still fall short of most ceramics.

Polymers, in general, have lower yield strengths than metals. The very strongest (and, at present, these are produced only in small quantities, and are expensive) barely reach the strength of aluminum alloys. They can be strengthened, however, by making composites out of them: GFRP has a strength only slightly inferior to aluminum, and CFRP is substantially stronger.

8.8 The hardness test

This consists of loading a pointed diamond or a hardened steel ball and pressing it into the surface of the material to be examined. The further into the

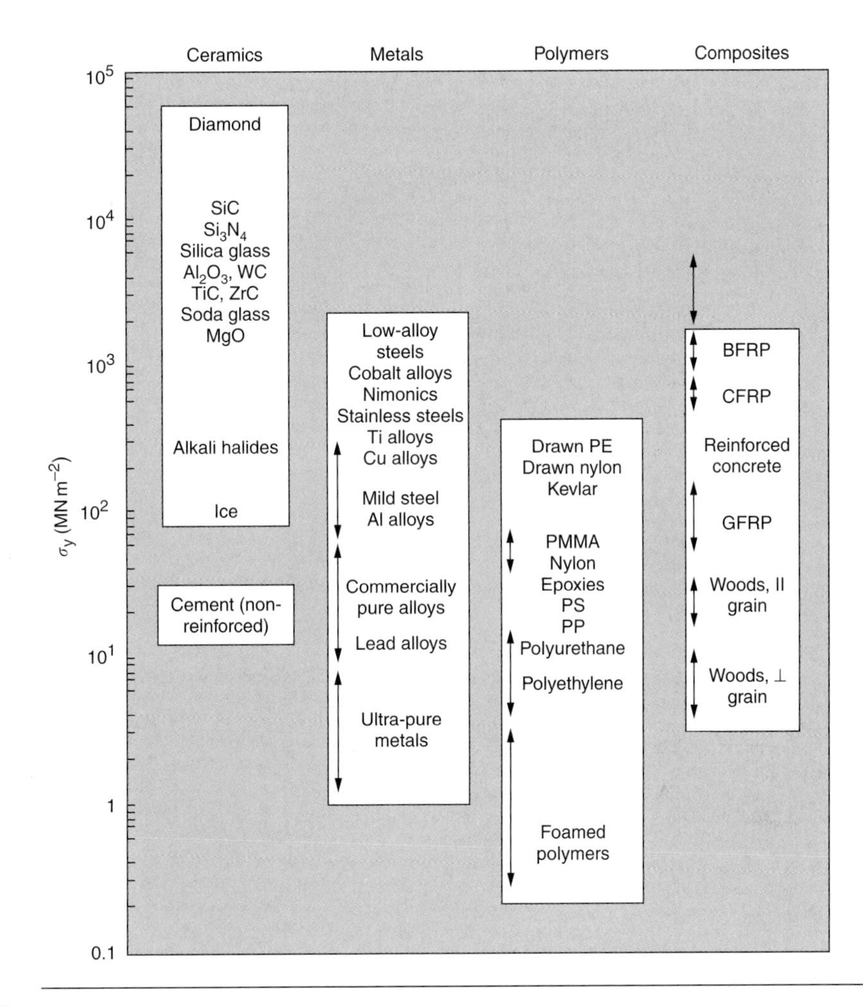

Figure 8.12 Bar chart of data for yield strength, σ_y.

material the 'indenter' (as it is called) sinks, the *softer* is the material and the lower its yield strength. The *true hardness* is defined as the load (F) divided by the projected area of the 'indent', A. (The Vickers hardness, H_v, unfortunately was, and still is, defined as F divided by the total surface area of the 'indent'. Tables are available to relate H to H_v.)

The yield strength can be found from the relation (derived in Chapter 11)

$$H = 3\sigma_y \tag{8.8}$$

but a correction factor is needed for materials which work-harden appreciably.

As well as being a good way of measuring the yield strengths of materials like ceramics, as we mentioned above, the hardness test is also a very simple and cheap *nondestructive test for* σ_y. There is no need to go to the expense of

Table 8.1 Yield strength, σ_y, tensile strength, σ_{TS}, and tensile ductility, ϵ_f

Material	σ_y(MN m^{-2})	σ_{TS}(MN m^{-2})	ϵ_f
Diamond	50,000	–	0
Boron carbide, B_4C	14,000	(330)	0
Silicon carbide, SiC	10,000	(200–800)	0
Silicon nitride, Si_3N_4	9600	(200–900)	0
Silica glass, SiO_2	7200	(110)	0
Tungsten carbide, WC	5600–8000	(80–710)	0
Niobium carbide, NbC	6000	–	0
Alumina, Al_2O_3	5000	(250–550)	0
Beryllia, BeO	4000	(130–280)	0
Syalon(Si-Al-O-N ceramic)	4000	(945)	0
Mullite	4000	(128–140)	0
Titanium carbide, TiC	4000	–	0
Zirconium carbide, ZrC	4000	–	0
Tantalum carbide, TaC	4000	–	0
Zirconia, ZrO_2	4000	(100–700)	0
Soda glass (standard)	3600	(50–70)	0
Magnesia, MgO	3000	(100)	0
Cobalt and alloys	180–2000	500–2500	0.01–6
Low-alloy steels (water-quenched and tempered)	500–1900	680–2400	0.02–0.3
Pressure-vessel steels	1500–1900	1500–2000	0.3–0.6
Stainless steels, austenitic	286–500	760–1280	0.45–0.65
Boron/epoxy composites	–	725–1730	–
Nickel alloys	200–1600	400–2000	0.01–0.6
Nickel	70	400	0.65
Tungsten	1000	1510	0.01–0.6
Molybdenum and alloys	560–1450	665–1650	0.01–0.36
Titanium and alloys	180–1320	300–1400	0.06–0.3
Carbon steels (water-quenched and tempered)	260–1300	500–1880	0.2–0.3
Tantalum and alloys	330–1090	400–1100	0.01–0.4
Cast irons	220–1030	400–1200	0–0.18
Copper alloys	60–960	250–1000	0.01–0.55
Copper	60	400	0.55
Cobalt/tungsten carbide cermets	400–900	900	0.02
CFRPs	–	670–640	–
Brasses and bronzes	70–640	230–890	0.01–0.7
Aluminum alloys	100–627	300–700	0.05–0.3
Aluminum	40	200	0.5
Stainless steels, ferritic	240–400	500–800	0.15–0.25
Zinc alloys	160–421	200–500	0.1–1.0
Concrete, steel reinforced	–	410	0.02
Alkali halides	200–350	–	0
Zirconium and alloys	100–365	240–440	0.24–0.37

Table 8.1 *(Continued)*

Material	σ_y(MN m^{-2})	σ_{TS}(MN m^{-2})	ϵ_f
Mild steel	220	430	0.18–0.25
Iron	50	200	0.3
Magnesium alloys	80–300	125–380	0.06–0.20
GFRPs	–	100–300	–
Beryllium and alloys	34–276	380–620	0.02–0.10
Gold	40	220	0.5
PMMA	60–110	110	0.03–0.05
Epoxies	30–100	30–120	–
Polyimides	52–90	–	–
Nylons	49–87	100	–
Ice	85	(6)	0
Pure ductile metals	20–80	200–400	0.5–1.5
Polystyrene	34–70	40–70	–
Silver	55	300	0.6
ABS/polycarbonate	55	60	–
Common woods (‖ to grain)	–	35–55	–
Lead and alloys	11–55	14–70	0.2–0.8
Acrylic/PVC	45–48	–	–
Tin and alloys	7–45	14–60	0.3–0.7
Polypropylene	19–36	33–36	–
Polyurethane	26–31	58	–
Polyethylene, high density	20–30	37	–
Concrete, non-reinforced	20–30	(1–5)	0
Natural rubber	–	30	5.0
Polyethylene, low density	6–20	20	–
Common woods (⊥ to grain)	–	4–10	–
Ultrapure f.c.c. metals	1–10	200–400	1–2
Foamed polymers, rigid	0.2–10	0.2–10	0.1–1
Polyurethane foam	1	1	0.1–1

Note: bracketed σ_{TS} data for brittle materials refer to the modulus of rupture σ_r (see Chapter 16).

making tensile specimens, and the hardness indenter is so small that it scarcely damages the material. So it can be used for routine batch tests on materials to see if they are up to specification on σ_y without damaging them.

8.9 Revision of the terms mentioned in this chapter, and some useful relations

σ_n, nominal stress (see Figure 8.14)

$$\sigma_n = F/A_0 \tag{8.9}$$

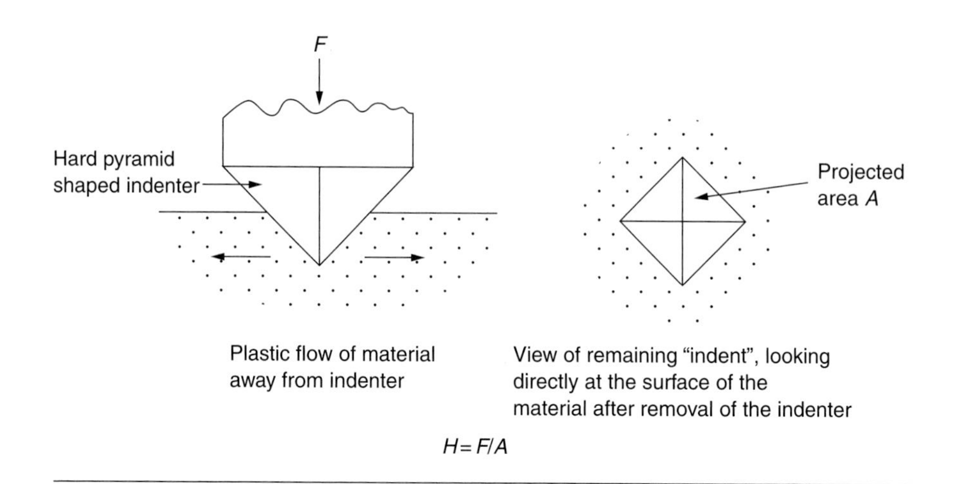

Figure 8.13 The hardness test for yield strength.

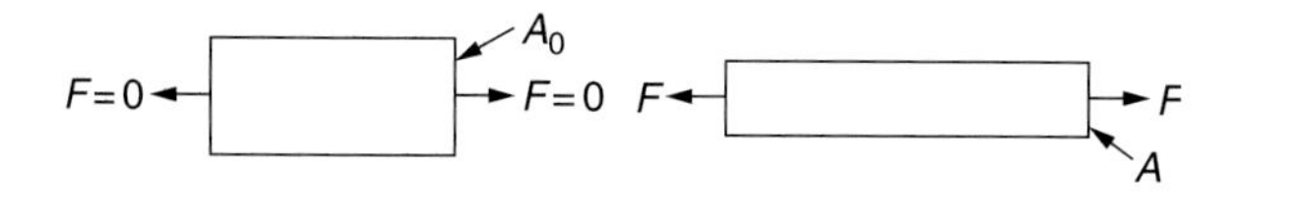

Figure 8.14

σ, *true stress* (see Figure 8.14)

$$\sigma = F/A \tag{8.10}$$

ϵ_n, nominal strain (see Figure 8.15)

$$\epsilon_n = \frac{u}{l_0}, \quad \text{or} \quad \frac{l - l_0}{l_0}, \quad \text{or} \quad \frac{l}{l_0} - 1 \tag{8.11}$$

Relations between σ_n, σ, and ϵ_n

Assuming constant volume (valid if $\upsilon = 0.5$ *or*, if not, plastic deformation $\gg$ elastic deformation):

$$A_0 l_0 = Al; \quad A_0 = \frac{Al}{l_0} = A(1 + \epsilon_n) \tag{8.12}$$

Thus

$$\sigma = \frac{F}{A} = \frac{F}{A_o}(1 + \epsilon_n) = \sigma_n(1 + \epsilon_n) \tag{8.13}$$

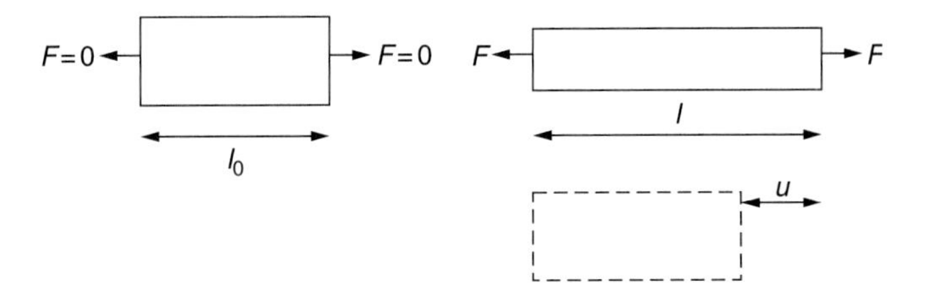

Figure 8.15

ϵ, true strain and the relation between ϵ and ϵ_n

$$\epsilon = \int_{l_o}^{l} \frac{dl}{l} = \ln\left(\frac{l}{l_o}\right) \tag{8.14}$$

Thus

$$\epsilon = \ln(1 + \epsilon_n) \tag{8.15}$$

Small strain condition

For small ϵ_n

$$\epsilon \approx \epsilon_n, \quad \text{from} \quad \epsilon = \ln(1 + \epsilon_n), \tag{8.16}$$

$$\sigma \approx \sigma_n, \quad \text{from} \quad \sigma = \sigma_n(1 + \epsilon_n) \tag{8.17}$$

Thus, when dealing with most *elastic* strains (but not in rubbers), it is immaterial whether ϵ or ϵ_n, or σ or σ_n, are chosen.

Energy

The energy expended in deforming a material *per unit volume* is given by the area under the stress–strain curve. For example, see Figure 8.16.

For *linear elastic strains,* and *only* linear elastic strains (see Figure 8.17),

$$\frac{\sigma_n}{\epsilon_n} = E, \quad \text{and} \quad U^{\text{el}} = \int \sigma_n \, d\epsilon_n = \int \sigma_n \frac{d\sigma_n}{E} = \left\{\frac{\sigma_n^2}{2E}\right\} \tag{8.20}$$

Elastic limit

In a tensile test, as the load increases, the specimen at first is strained *elastically,* that is reversibly. Above a limiting stress—the elastic limit—some of the strain is permanent; this is *plastic* deformation.

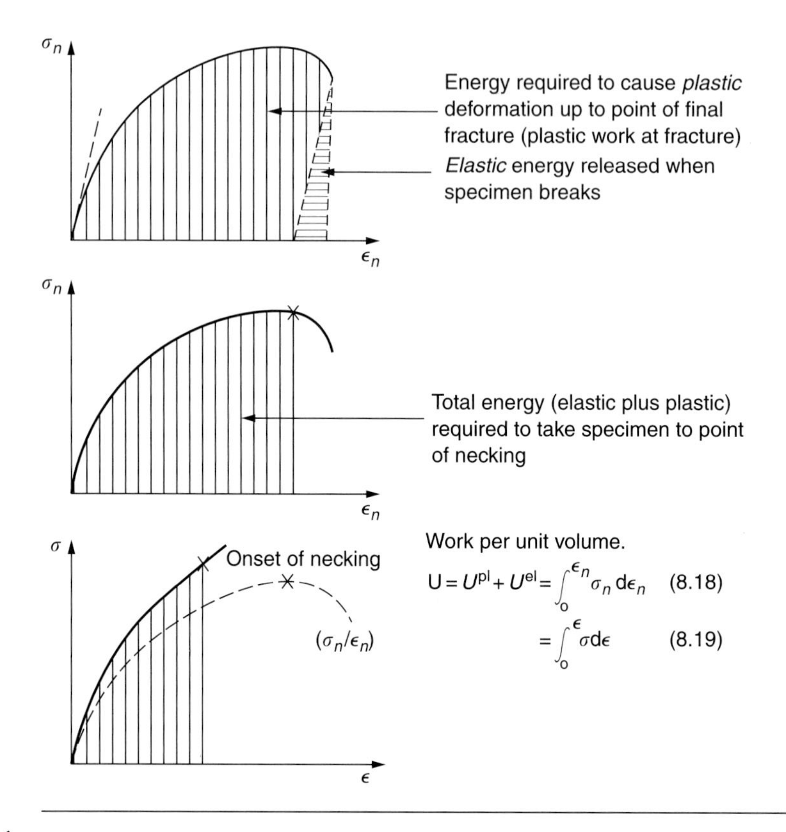

Figure 8.16

Yielding

The change from elastic to measurable plastic deformation.

Yield strength

The nominal stress at yielding. In many materials this is difficult to spot on the stress–strain curve and in such cases it is better to use a proof stress.

Proof stress

The stress which produces a permanent strain equal to a specified percentage of the specimen length. A common proof stress is one corresponding to 0.1 percent permanent strain.

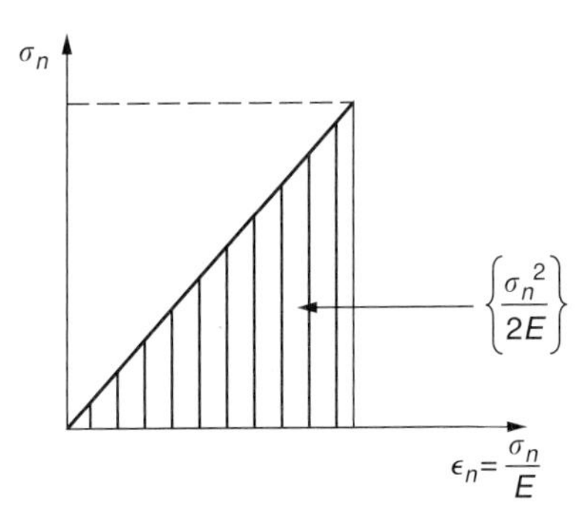

Figure 8.17

Strain hardening (work-hardening)

The increase in stress needed to produce further strain in the plastic region. Each strain increment strengthens or hardens the material so that a larger stress is needed for further strain.

σ_{TS}, tensile strength (in old books, ultimate tensile strength, or UTS) (see Figure 8.18).

ϵ_f, strain after fracture, or tensile ductility

The permanent extension in length (measured by fitting the broken pieces together) expressed as a percentage of the original gauge length.

$$\left\{\frac{l_{\text{break}} - l_o}{l_o}\right\} \times 100 \qquad (8.22)$$

Reduction in area at break

The maximum decrease in cross-sectional area at the fracture expressed as a percentage of the original cross-sectional area.

Strain after fracture and percentage reduction in area are used as measures of ductility, that is the ability of a material to undergo large plastic strain under stress before it fractures (see Figure 8.19).

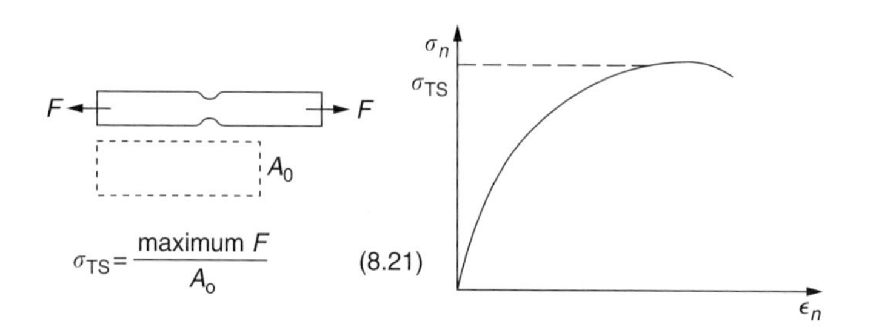

Figure 8.18

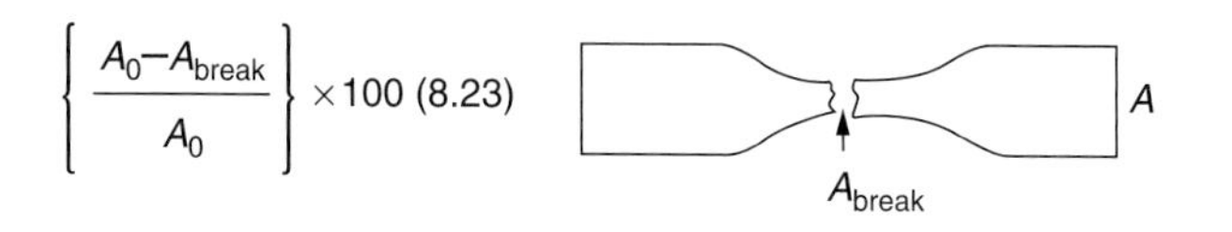

Figure 8.19

Examples

8.1 Nine strips of pure, fully annealed copper were deformed plastically by being passed between a pair of rotating rollers so that the strips were made thinner and longer. The increases in length produced were 1, 10, 20, 30, 40, 50, 60, 70 and 100 percent respectively. The diamond–pyramid hardness of each piece was measured after rolling. The results are given in the following table,

Nominal strain	0.01	0.1	0.2	0.3	0.4	0.5	0.6	0.7	1.0
Hardness (MN m^{-2})	423	606	756	870	957	1029	1080	1116	1170

Assuming that a diamond–pyramid hardness test creates a further nominal strain, on average, of 0.08, and that the hardness value is 3.0 times the true stress, construct the curve of *nominal* stress against nominal strain.
[Hint: add 0.08 to each value of nominal strain in the table.]

8.2 Using the plot of nominal stress against nominal strain from example 8.1, find:

(a) the tensile strength of copper;
(b) the strain at which tensile failure commences;
(c) the percentage reduction in cross-sectional area at this strain;
(d) the work required to initiate tensile failure in a cubic metre of annealed copper.

Answers

(a) 217 MN m^{-2}; (b) 0.6 approximately; (c) 38%, (d) 109 MJ.

8.3 Why can copper survive a much higher extension during rolling than during a tensile test?

8.4 The following data were obtained in a tensile test on a specimen with 50 mm gauge length and a cross-sectional area of 160 mm^2.

Extension (mm)	0.050	0.100	0.150	0.200	0.250	0.300	1.25	2.50	3.75	5.00	6.25	7.50
Load (kN)	12	25	32	36	40	42	63	80	93	100	101	90

The total elongation of the specimen just before final fracture was 16 percent, and the reduction in area at the fracture was 64 percent.

Find the maximum allowable working stress if this is to equal

(a) $0.25 \times$ tensile strength,
(b) 0.6×0.1 percent proof stress.

Answers

(a) 160 MN m^{-2}; (b) 131 MN m^{-2}

8.5 One type of hardness test involves pressing a hard sphere (radius r) into the test material under a fixed load F, and measuring the *depth*, h, to which the sphere sinks into the material, plastically deforming it. Derive an expression for the indentation hardness, H, of the material in terms of h, F, and r. Assume $h \ll r$.

Answer

$$H = \frac{F}{2\pi rh}$$

8.6 The diagram shows the force-extension characteristics of a bungee rope. The length of the rope under zero load is 15 m. One end of the rope is attached to a bridge deck, and the other end is attached to a person standing on the deck. The person then jumps off the deck, and descends vertically until arrested by the rope.

(a) Find the maximum mass of the person for a successful arrest, m_{max}.
(b) The diagram shows that the unloading line is 0.5 kN below the loading line. Comment on the practical significance of this to bungee jumping.

Answer

89.2 kg.

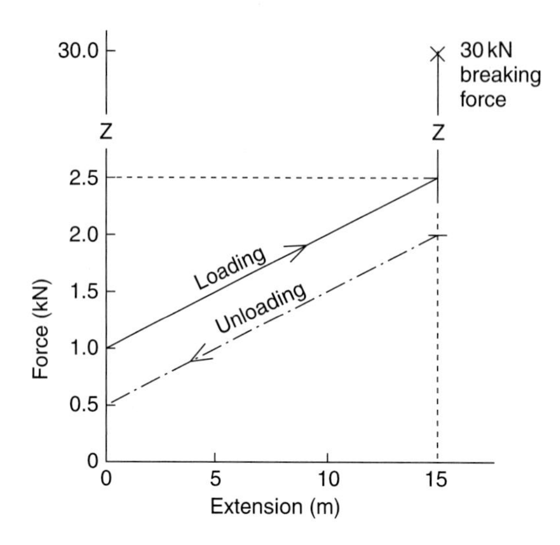
30.0
2.5
2.0
1.5
1.0
0.5
0
Force (kN)
30 kN
breaking
force
Loading
Unloading
0
5
10
15
Extension (m)

Chapter 9
Dislocations and yielding in crystals

Chapter contents

9.1 Introduction

In the last chapter we examined data for the yield strengths exhibited by materials. But what would we expect? From our understanding of the structure of solids and the stiffness of the bonds between the atoms, can we estimate what the yield strength should be? A simple calculation (given in the next section) overestimates it grossly. This is because real crystals contain defects, *dislocations*, which move easily. When they move, the crystal deforms; the stress needed to move them is the yield strength. Dislocations are the *carriers* of deformation, much as electrons are the carriers of charge.

9.2 The strength of a perfect crystal

As we showed in Chapter 6 (on the modulus), the slope of the interatomic force–distance curve at the equilibrium separation is proportional to Young's modulus E. Interatomic forces typically drop off to negligible values at a distance of separation of the atom centers of $2r_0$. The maximum in the force–distance curve is typically reached at $1.25r_0$ separation, and if the stress applied to the material is sufficient to exceed this maximum force per *bond*, fracture is bound to occur. We will denote the stress at which this bond rupture takes place by $\tilde{\sigma}$, the *ideal strength;* a material cannot be stronger than this. From Figure 9.1

$$
\begin{aligned}
\sigma &= E\epsilon \\
2\tilde{\sigma} &\approx E\,\frac{0.25r_0}{r_0} \approx \frac{E}{4} \\
\tilde{\sigma} &\approx \frac{E}{8}
\end{aligned}
\tag{9.1}
$$

More refined estimates of $\tilde{\sigma}$ are possible, using real interatomic potentials (Chapter 4): they give about $E/15$ instead of $E/8$.

Let us now see whether materials really show this strength. The bar chart (Figure 9.2) shows values of σ_y/E for materials. The heavy broken line at the top is drawn at the level $\sigma/E = 1/15$. Glasses, and some ceramics, lie close to this line — they exhibit their ideal strength, and we could not expect them to be stronger than this. Most polymers, too, lie near the line — although they have low yield strengths, these are low because the *moduli* are low.

All metals, on the other hand, have yield strengths far below the levels predicted by our calculation — as much as a factor of 10^5 smaller. Even ceramics, many of them, yield at stresses which are as much as a factor of 10 below their ideal strength. Why is this?

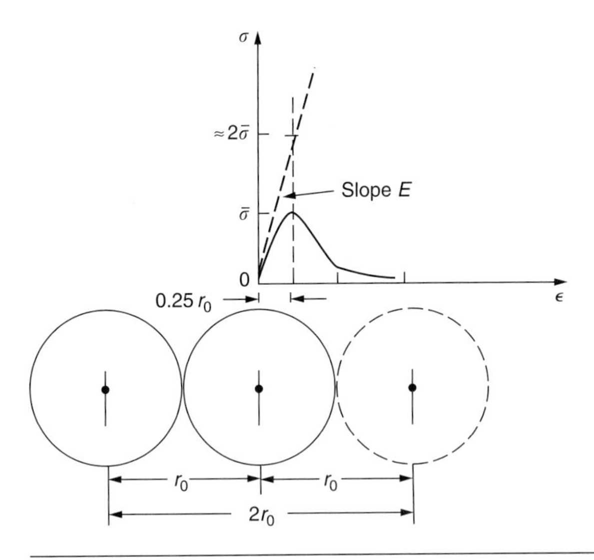

Figure 9.1 The ideal strength, $\tilde{\sigma}$.

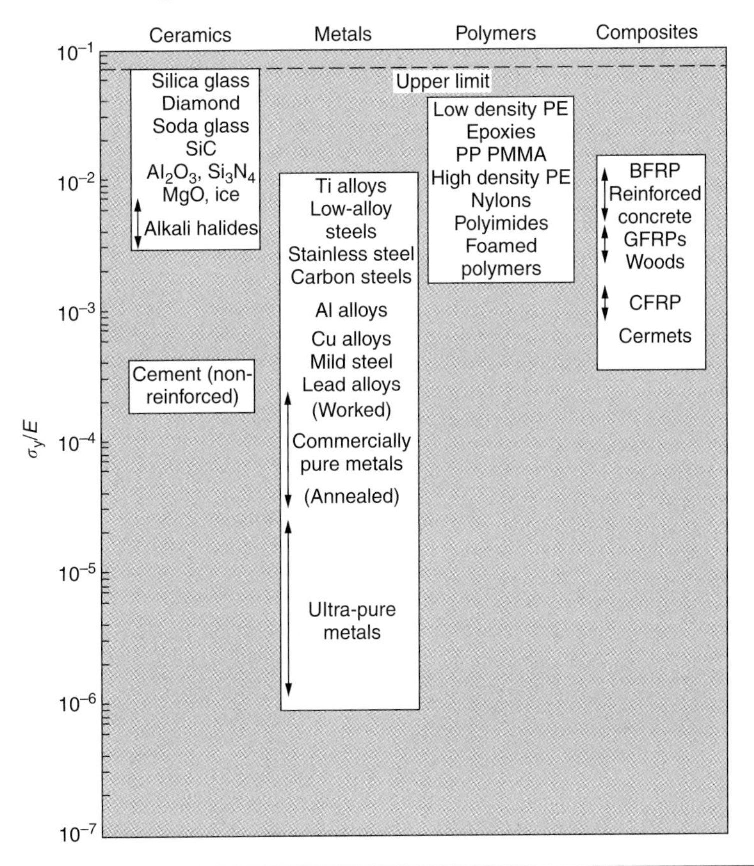

Figure 9.2 Bar chart of data for normalized yield strength, σ_y/E.

9.3 Dislocations in crystals

In Chapter 5 we said that many important engineering materials (e.g. metals) were normally made up of crystals, and explained that a perfect crystal was an assembly of *atoms packed together in a regularly repeating pattern*.

But crystals (like everything in this world) are not perfect; they have *defects* in them. Just as the strength of a chain is determined by the strength of the weakest link, so the strength of a crystal—and thus of our material—is usually limited by the defects that are present in it. The *dislocation* is a particular type of defect that has the effect of allowing materials to deform plastically (that is, they yield) at stress levels that are much less than $\tilde{\sigma}$.

Figure 9.3(a) shows an *edge dislocation* from a continuum viewpoint (i.e. ignoring the atoms). Such a dislocation is made in a block of material by cutting the block up to the line marked $\perp - \perp$, then displacing the material below the cut relative to that above by a distance b (the atom size) normal to the line $\perp - \perp$, and finally gluing the cut and displaced surfaces back

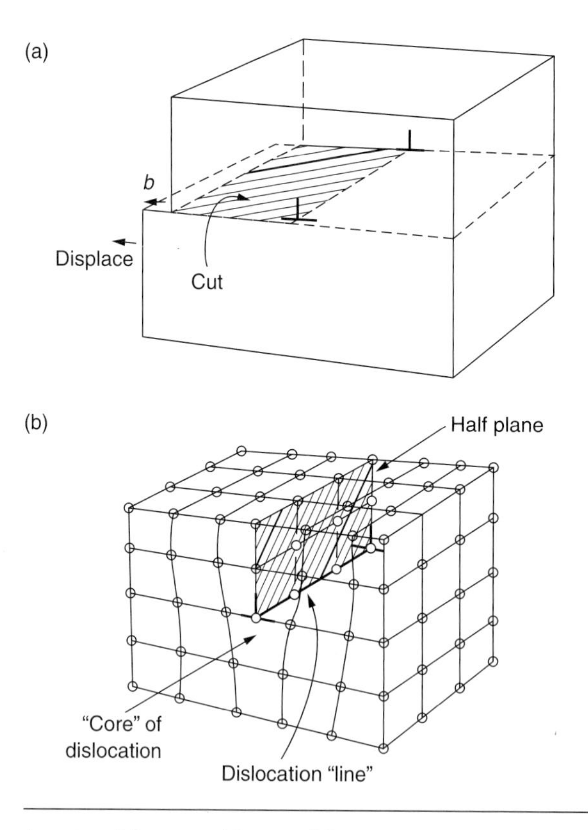

Figure 9.3 An edge dislocation, (a) viewed from a continuum standpoint (i.e. ignoring the atoms) and (b) showing the positions of the atoms near the dislocation.

together. The result, on an atomic scale, is shown in the adjacent diagram (Figure 9.3(b)); the material in the middle of the block now contains a *half plane* of atoms, with its lower edge lying along the line ⊥ – ⊥: the *dislocation line*. This defect is called an edge dislocation because it is formed by the edge of the half plane of atoms; and it is written briefly by using the symbol ⊥.

Dislocation motion produces plastic strain. Figure 9.4 shows how the atoms rearrange as the dislocation moves through the crystal, and that, when one dislocation moves entirely through a crystal, the lower part is displaced under the upper by the distance b (called the Burgers vector). The same process is drawn, without the atoms, and using the symbol ⊥ for the position of the dislocation line, in Figure 9.5. The way in which this dislocation works can be likened to the way in which a ballroom carpet can be moved across a large dance floor simply by moving rucks along the carpet—a very much easier process than pulling the whole carpet across the floor at one go (Figure 9.6).

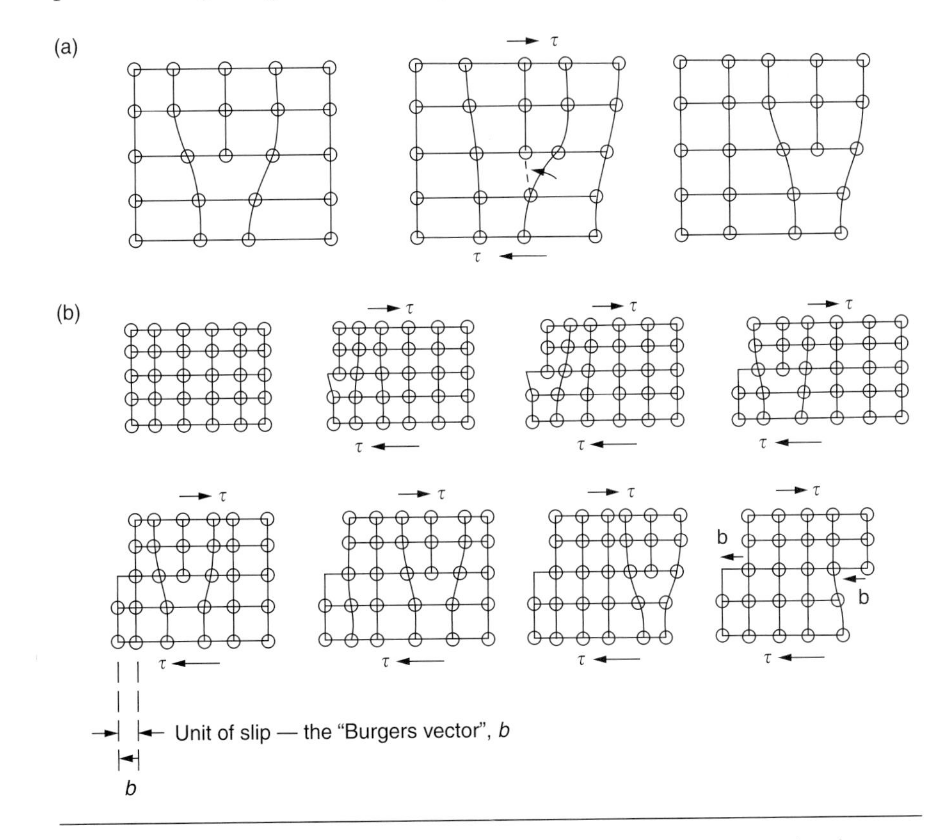

Figure 9.4 How an edge dislocation moves through a crystal. (a) Shows how the atomic bonds at the center of the dislocation break and reform to allow the dislocation to move. (b) Shows a complete sequence for the introduction of a dislocation into a crystal from the left-hand side, its migration through the crystal, and its expulsion on the right-hand side; this process causes the lower half of the crystal to slip by a distance b under the upper half.

In making the edge dislocation of Figure 9.3 we could, after making the cut, have displaced the lower part of the crystal under the upper part in a direction *parallel* to the bottom of the cut, instead of normal to it. Figure 9.7 shows the result; it, too, is a dislocation, called a *screw dislocation* (because it converts

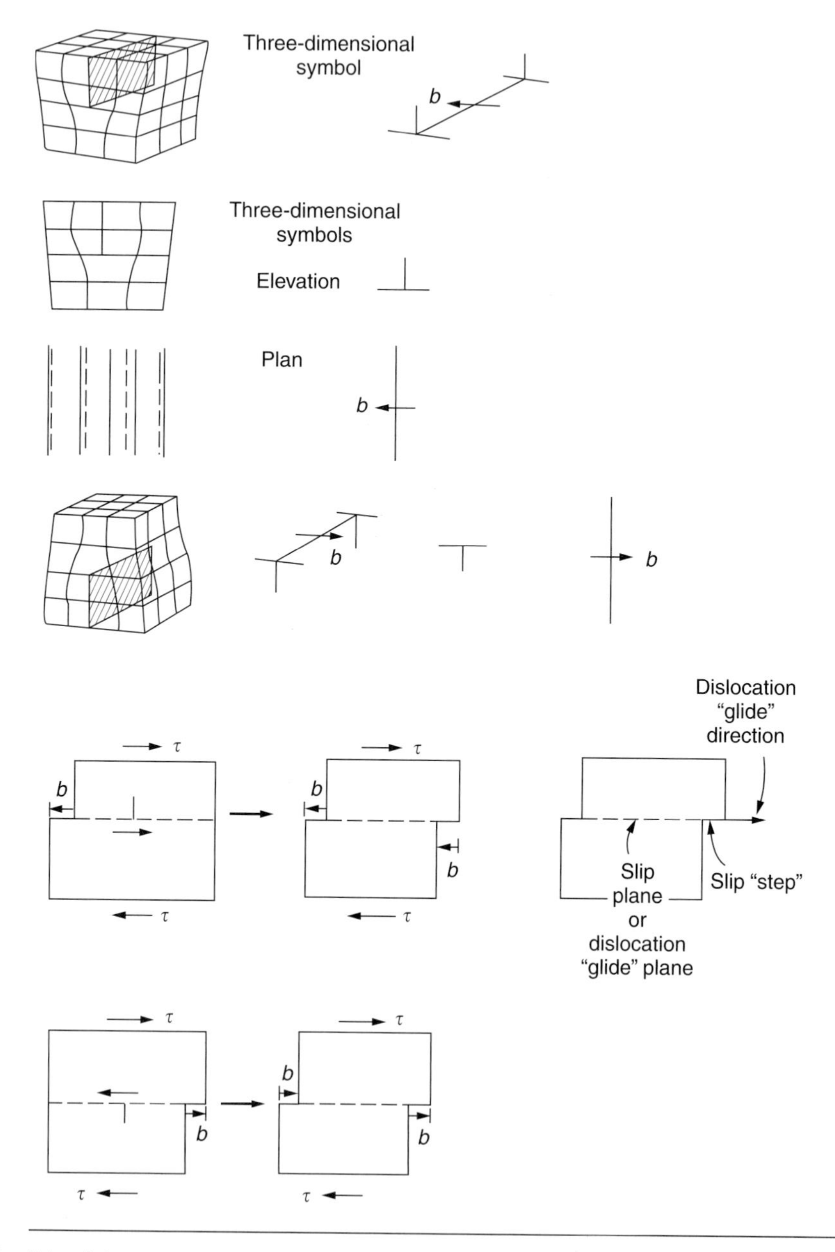

Figure 9.5 Edge dislocation conventions.

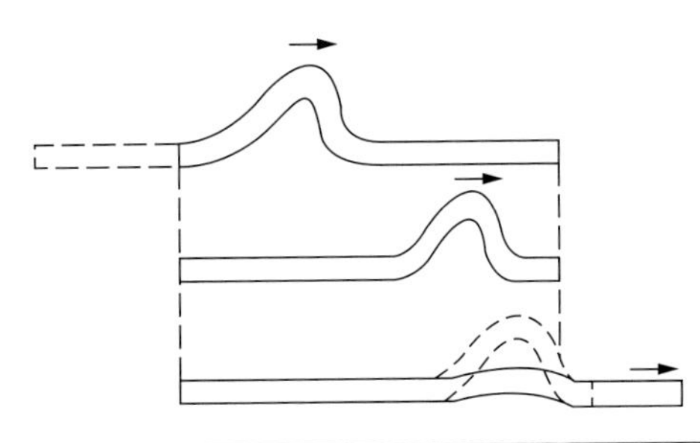

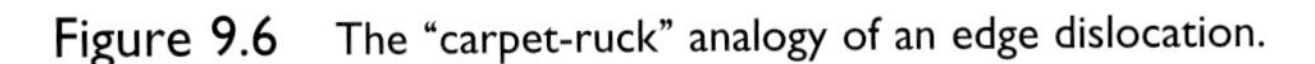

Figure 9.6 The "carpet-ruck" analogy of an edge dislocation.

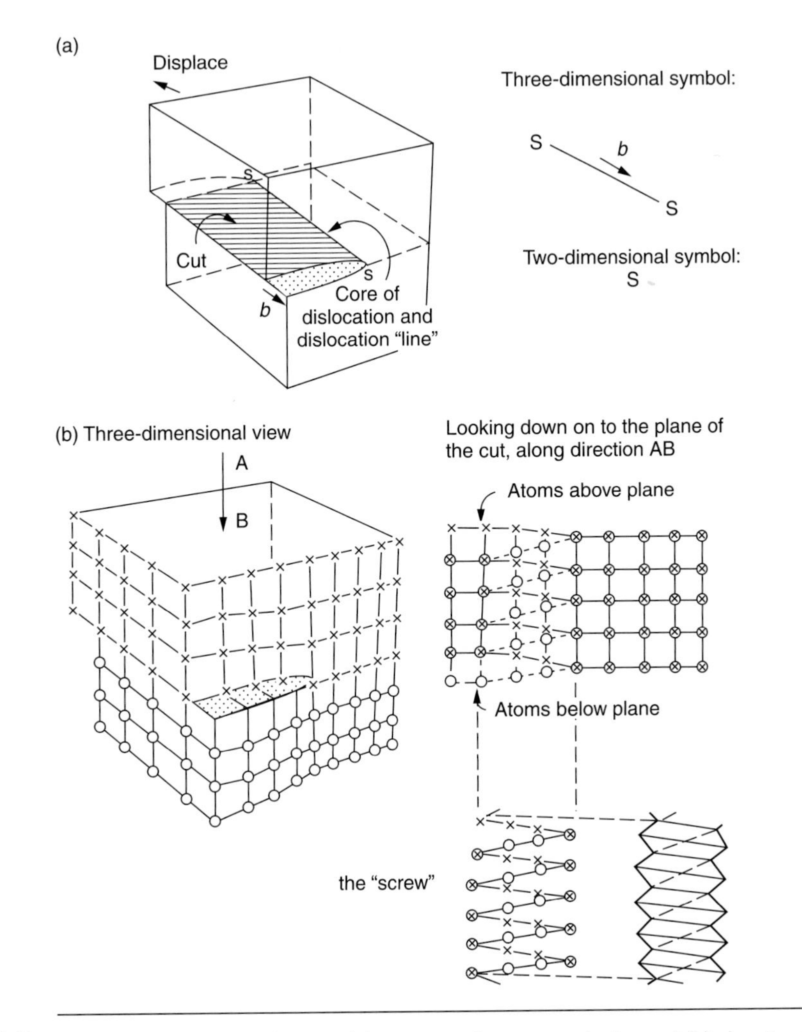

Figure 9.7 A screw dislocation, (a) viewed from a continuum standpoint and (b) showing the atom positions.

the planes of atoms into a helical surface, or screw). Like an edge dislocation, it produces plastic strain when it moves (Figures 9.8–9.10). Its geometry is a little more complicated but its properties are otherwise just like those of the edge. Any dislocation, in a real crystal, is either a screw or an edge; or can be thought

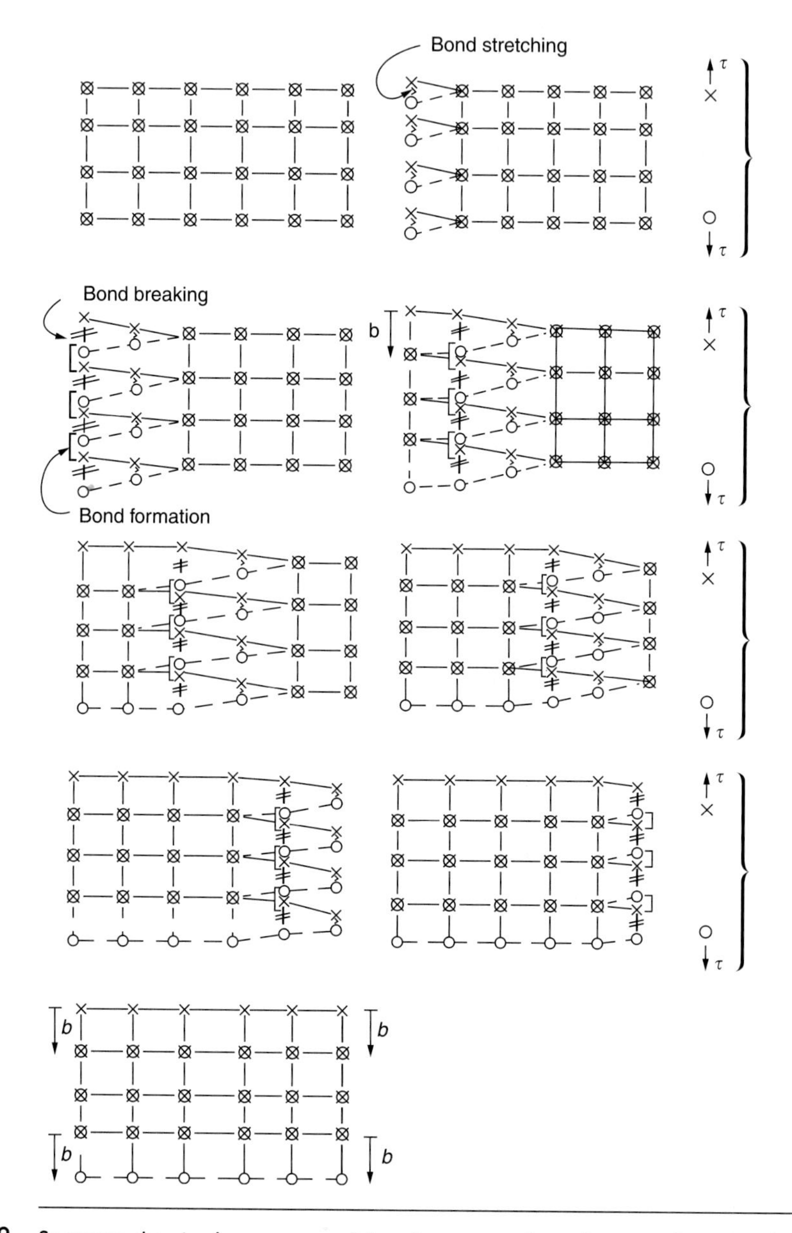

Figure 9.8 Sequence showing how a screw dislocation moves through a crystal causing the lower half of the crystal (○) to slip by a distance b under the upper half (×).

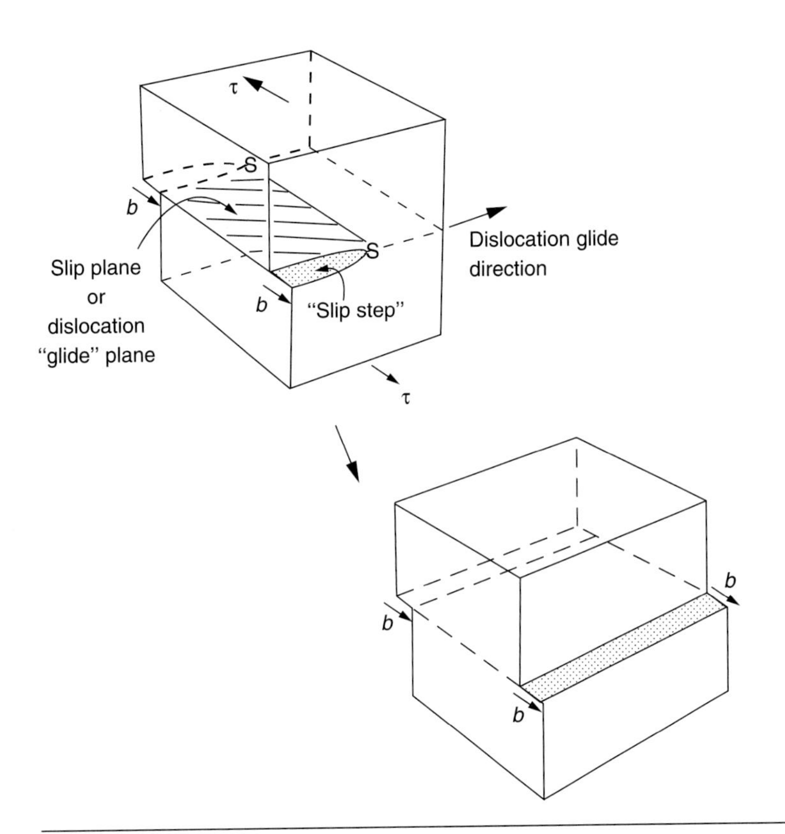

Figure 9.9 Screw-dislocation conventions.

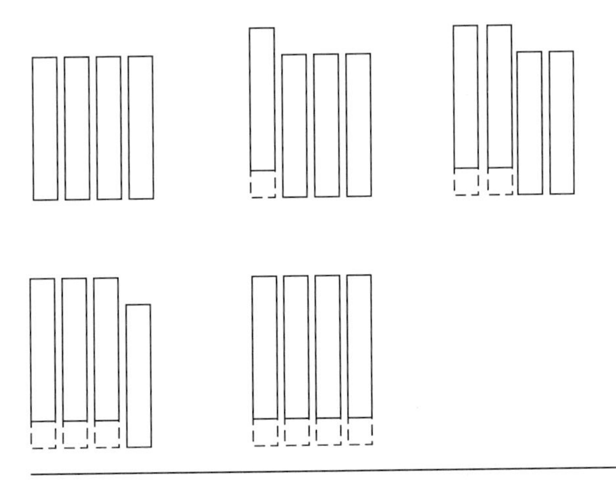

Figure 9.10 The "planking" analogy of the screw dislocation. Imagine four planks resting side by side on a factory floor. It is much easier to slide them across the floor one at a time than all at the same time.

Figure 9.11 An electron microscope picture of dislocation lines in stainless steel. The picture was taken by firing electrons through a very thin slice of steel about 100 nm thick. The dislocation lines here are only about 1000 atom diameters long because they have been "chopped off" where they meet the top and bottom surfaces of the thin slice. But a sugar-cube-sized piece of any engineering alloy contains about 10^5 km of dislocation line.

of as little steps of each. Dislocations can be seen by electron microscopy. Figure 9.11 shows an example.

9.4 The force acting on a dislocation

A shear stress (τ) exerts a force on a dislocation, pushing it through the crystal. For yielding to take place, this force must be great enough to overcome the *resistance* to the motion of the dislocation. This resistance is due to intrinsic friction opposing dislocation motion, plus contributions from alloying or work-hardening; they are discussed in detail in the next chapter. Here we show that the magnitude of the force is τb per unit length of dislocation.

We prove this by a virtual work calculation. We equate the work done by the applied stress when the dislocation moves completely through the crystal to the work done against the force f opposing its motion (Figure 9.12). The upper part is displaced relative to the lower by the distance b, and the applied stress does work $(\tau l_1 l_2) \times b$. In moving through the crystal, the dislocation travels

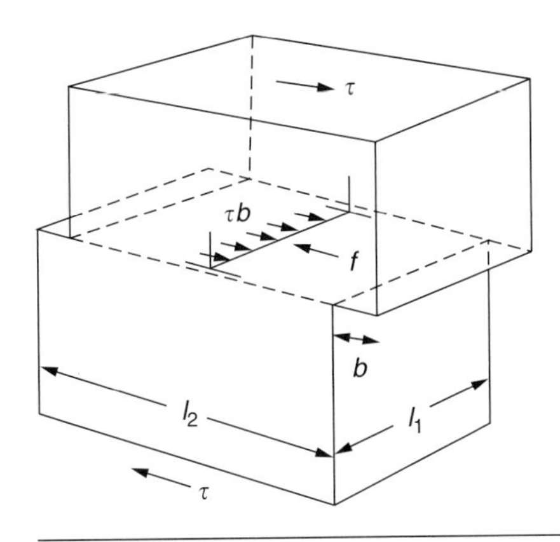

Figure 9.12 The force acting on a dislocation.

a distance l_2, doing work against the resistance, f per unit length, as it does so; this work is $fl_1\, l_2$. Equating the two gives

$$\tau b = f \tag{9.2}$$

This result holds for any dislocation—edge, screw or a mixture of both.

9.5 Other properties of dislocations

There are two remaining properties of dislocations that are important in understanding the plastic deformation of materials. These are:

(a) Dislocations always glide on crystallographic planes, as we might imagine from our earlier drawings of edge dislocation motion. In f.c.c. crystals, for example, the dislocations glide on {111} planes, and therefore plastic shearing takes place on {111} in f.c.c. crystals.

(b) The atoms near the core of a dislocation are displaced from their proper positions and thus have a higher energy. In order to keep the total energy as low as possible, the dislocation tries to be as short as possible—it behaves as if it had a *line tension*, T, like a rubber band. Very roughly, the strains at a dislocation core are of order 1/2; the stresses are therefore of order $G/2$ (Chapter 8) so the energy per unit volume of core is $G/8$. If we take the core radius to be equal to the atom size b, its volume, per unit length, is πb^2. The line tension is the energy per unit length (just as a surface tension is an energy per unit area), giving

$$T = \frac{\pi}{8} G b^2 \approx \frac{G b^2}{2} \tag{9.3}$$

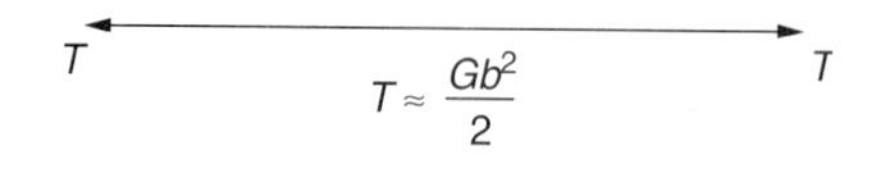

Figure 9.13 The line tension in a dislocation.

where G is the shear modulus. In absolute terms, T is small (we should need $\approx 10^8$ dislocations to hold an apple up) but it is large in relation to the size of a dislocation (Figure 9.13), and has an important bearing on the way in which obstacles obstruct the motion of dislocations.

We shall be looking in the next chapter at how we can use our knowledge of how dislocations work and how they behave in order to understand how materials deform plastically, and to help us design stronger materials.

Examples

9.1 Explain what is meant by the *ideal strength* of a material.

9.2 The energy per unit length of a dislocation is $\frac{1}{2}Gb^2$. Dislocations dilate a close-packed crystal structure very slightly, because at their cores the atoms are not close packed: the dilatation is $\frac{1}{4}b^2$ per unit length of dislocation. It is found that the density of a bar of copper changes from $8.9323\,\text{Mg}\,\text{m}^{-3}$ to $8.9321\,\text{Mg}\,\text{m}^{-3}$ when it is very heavily deformed. Calculate (a) the dislocation density introduced into the copper by the deformation and (b) the energy associated with that density. Compare this result with the latent heat of melting of copper ($1833\,\text{MJ}\,\text{m}^{-3}$) ($b$ for copper is $0.256\,\text{nm}$; G is $\frac{3}{8}E$).

Answers

(a) $1.4 \times 10^{15}\,\text{m}^{-2}$; (b) $2.1\,\text{MJ}\,\text{m}^{-3}$.

9.3 Explain briefly what is meant by a *dislocation*. Show with diagrams how the motion of (a) an edge dislocation and (b) a screw dislocation can lead to the plastic deformation of a crystal under an applied shear stress.

Chapter 10

Strengthening methods, and plasticity of polycrystals

Chapter contents

10.1 Introduction

We showed in the last chapter that:

(a) crystals contain dislocations;
(b) a shear stress τ, applied to the slip plane of a dislocation, exerts a force τb per unit length of the dislocation trying to push it forward;
(c) when dislocations move, the crystal deforms plastically—that is, it yields.

In this chapter, we examine ways of increasing the resistance to motion of a dislocation; it is this which determines the *dislocation yield strength* of a single isolated crystal, of a metal or a ceramic. But bulk engineering materials are aggregates of many crystals, or *grains*. To understand the plasticity of such an aggregate, we have to examine also how the individual crystals interact with each other. This lets us calculate the *polycrystal yield strength*—the quantity that enters engineering design.

10.2 Strengthening mechanisms

A crystal yields when the force τb (per unit length) exceeds f, the *resistance* (a force per unit length) opposing the motion of a dislocation. This defines the dislocation yield strength

$$\tau_y = \frac{f}{b} \qquad (10.1)$$

Most crystals have a certain *intrinsic* strength, caused by the bonds between the atoms which have to be broken and reformed as the dislocation moves. Covalent bonding, particularly, gives a very large *intrinsic lattice resistance*, f_i per unit length of dislocation. It is this that causes the enormous strength and hardness of diamond, and the carbides, oxides, nitrides, and silicates which are used for abrasives and cutting tools. But pure metals are very soft: they have a very low lattice resistance. Then it is useful to increase f by *solid solution strengthening*, by *precipitate* or *dispersion* strengthening, or by *work-hardening*, or by any combination of the three. Remember, however, that there is an upper limit to the yield strength: it can never exceed the ideal strength (Chapter 9). In practice, only a few materials have strengths that even approach it.

10.3 Solid solution hardening

A good way of hardening a metal is simply to make it impure. Impurities go into solution in a solid metal just as sugar dissolves in tea. A good example is the addition of zinc (Zn) to copper (Cu) to make the *alloy* called brass. The zinc

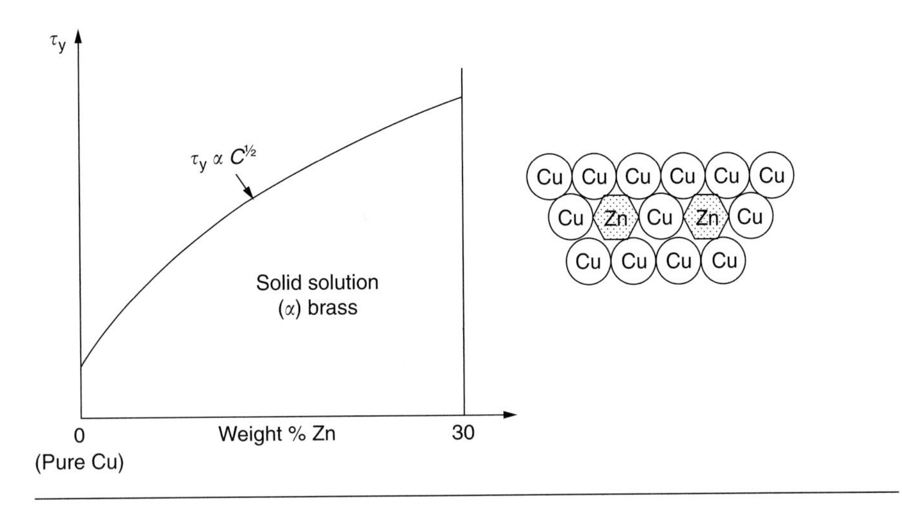

Figure 10.1 Solid solution hardening.

atoms replace copper atoms to form a *random substitutional solid solution*. At room temperature Cu will dissolve up to 30 percent Zn in this way. The Zn atoms are bigger than the Cu atoms, and, in squeezing into the Cu structure, generate stresses. These stresses "roughen" the slip plane, making it harder for dislocations to move; they increase the resistance f, and thereby increase the dislocation yield strength, τ_y (equation (10.1)). If the contribution to f given by the solid solution is f_{ss} then τ_y is increased by f_{ss}/b. In a solid solution of concentration C, the spacing of dissolved atoms on the slip plane (or on any other plane, for that matter) varies as $C^{-1/2}$; and the smaller the spacing, the "rougher" is the slip plane. As a result, τ_y increases about parabolically (i.e. as $C^{1/2}$) with solute concentration (Figure 10.1). Single-phase brass, bronze, and stainless steels, and many other metallic alloys, derive their strength in this way.

10.4 Precipitate and dispersion strengthening

If an impurity (copper, say) is dissolved in a metal or ceramic (aluminum, for instance) at a high temperature, and the alloy is cooled to room temperature, the impurity may *precipitate* as small particles, much as sugar will crystallize from a saturated solution when it is cooled. An alloy of Al containing 4 percent Cu ("Duralumin"), treated in this way, gives very small, closely spaced precipitates of the hard compound $CuAl_2$. Most steels are strengthened by precipitates of carbides, obtained in this way.*

* The optimum precipitate is obtained by a more elaborate *heat treatment:* the alloy is *solution heat-treated* (heated to dissolve the impurity), *quenched* (cooled fast to room temperature, usually by dropping it into oil or water) and finally *tempered* or *aged* for a controlled time and at a controlled temperature (to cause the precipitate to form).

Small particles can be introduced into metals or ceramics in other ways. The most obvious is to mix a dispersoid (such as an oxide) into a powdered metal (aluminum and lead are both treated in this way), and then compact and sinter the mixed powders.

Either approach distributes small, hard particles in the path of a moving dislocation. Figure 10.2 shows how they obstruct its motion. The stress τ has to push the dislocation between the obstacles. It is like blowing up a balloon in a bird cage: a very large pressure is needed to bulge the balloon between the bars, though once a large enough bulge is formed, it can easily expand further. The *critical configuration* is the semicircular one (Figure 10.2(c)): here the force τbL on one segment is just balanced by the force $2T$ due to the line tension, acting on either side of the bulge. The dislocation escapes (and yielding occurs) when

$$\tau_y = \frac{2T}{bL} \approx \frac{Gb}{L} \tag{10.2}$$

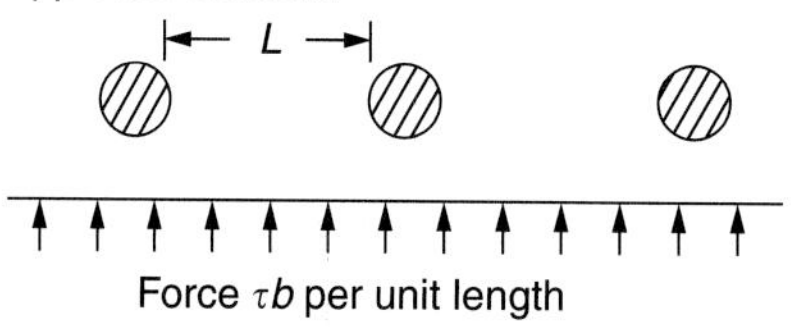

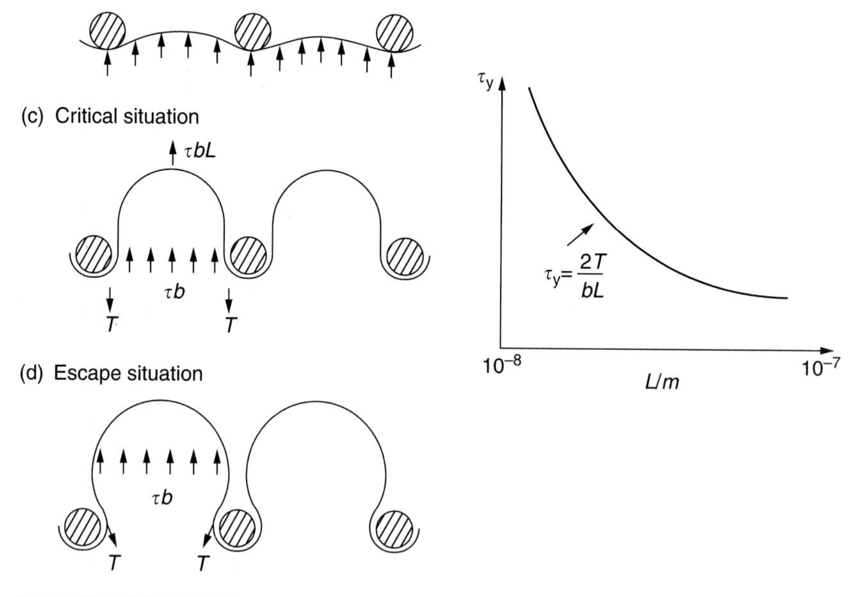

Figure 10.2 How dispersed precipitates help prevent the movement of dislocations, and help prevent plastic flow of materials.

The obstacles thus exert a resistance of $f_0 = 2T/L$. Obviously, the greatest hardening is produced by *strong, closely spaced* precipitates or dispersions (Figure 10.2).

10.5 Work-hardening

When crystals yield, dislocations move through them. Most crystals have several slip planes: the f.c.c. structure, which slips on {111} planes (Chapter 5), has four, for example. Dislocations on these intersecting planes interact, and obstruct each other, and accumulate in the material.

The result is *work-hardening*: the steeply rising stress–strain curve after yield, shown in Chapter 8. All metals and ceramics work-harden. It can be a nuisance: if you want to roll thin sheet, work-hardening quickly raises the yield strength so much that you have to stop and *anneal* the metal (heat it up to remove the accumulated dislocations) before you can go on. But it is also useful: it is a potent strengthening method, which can be added to the other methods to produce strong materials.

The analysis of work-hardening is difficult. Its contribution $f_{\omega h}$ to the total dislocation resistance f is considerable and increases with strain (Figure 10.3).

10.6 The dislocation yield strength

It is adequate to assume that the strengthening methods contribute in an additive way to the strength. Then

$$\tau_y = \frac{f_i}{b} + \frac{f_{ss}}{b} + \frac{f_o}{b} + \frac{f_{\omega h}}{b} \tag{10.3}$$

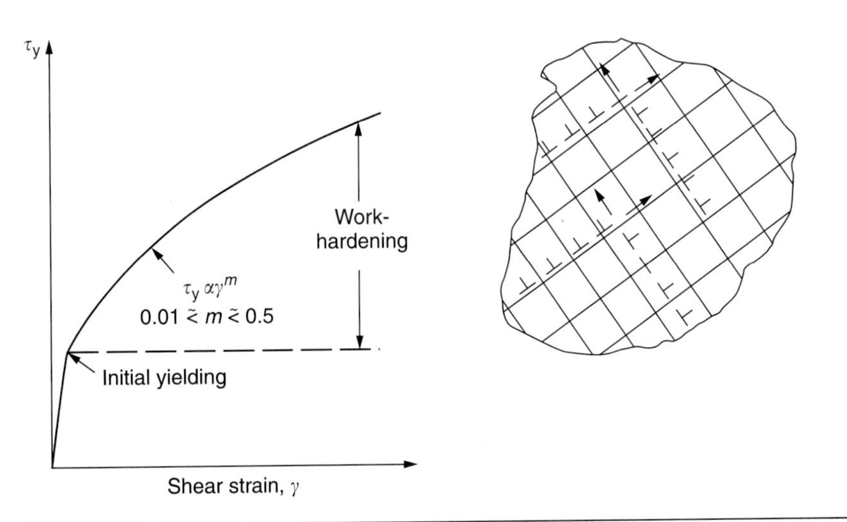

Figure 10.3 Collision of dislocations leads to work-hardening.

Strong materials either have a high intrinsic strength, f_i (like diamond), or they rely on the superposition of solid solution strengthening f_{ss}, obstacles f_o and work-hardening $f_{\omega h}$ (like high-tensile steels). But before we can use this information, one problem remains: we have calculated the yield strength of an *isolated crystal* in *shear*. We want the yield strength of a *polycrystalline aggregate* in *tension*.

10.7 Yield in polycrystals

The crystals, or *grains*, in a polycrystal fit together exactly but their crystal orientations differ (Figure 10.4). Where they meet, at *grain boundaries*, the

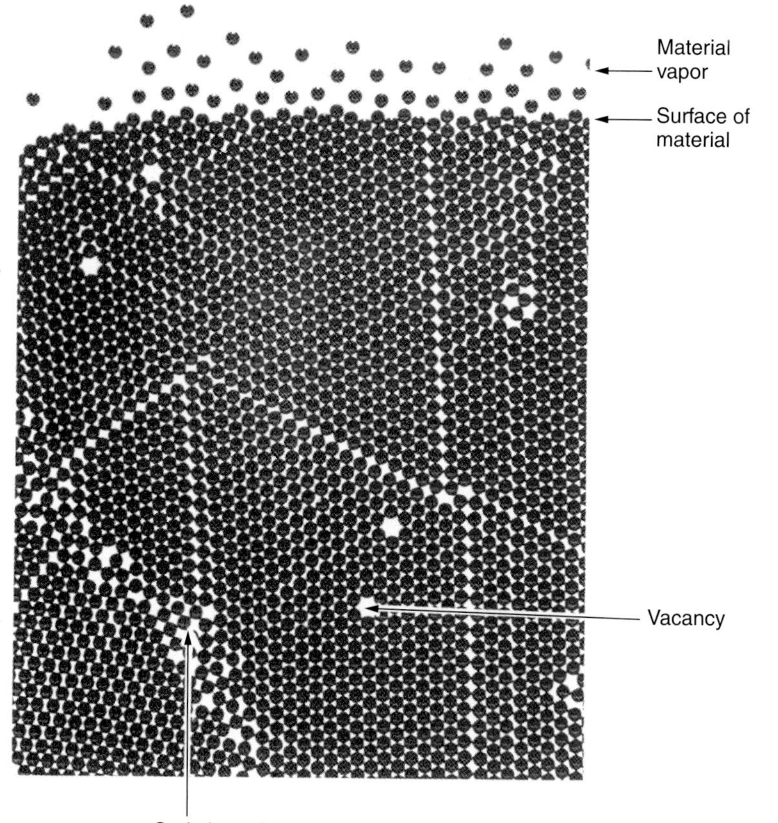

Figure 10.4 Ball bearings can be used to simulate how atoms are packed together in solids. Our photograph shows a ball-bearing model set up to show what the *grain boundaries* look like in a polycrystalline material. The model also shows up another type of defect — the vacancy — which is caused by a missing atom.

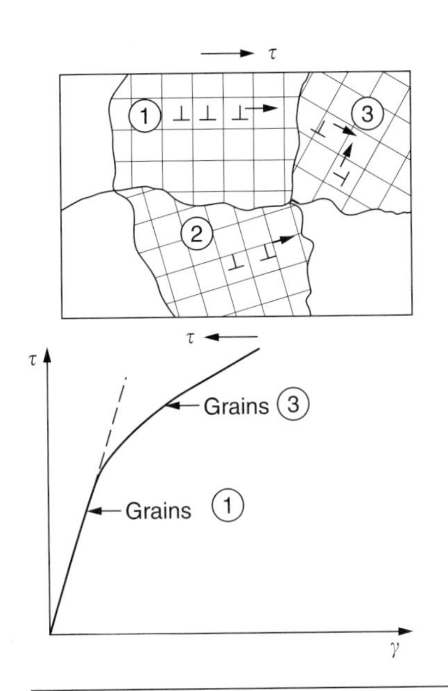

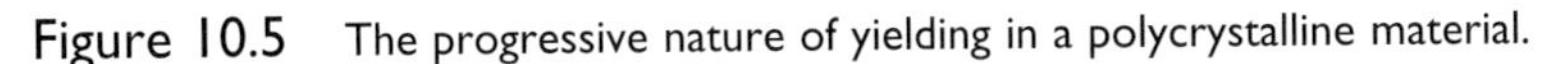

Figure 10.5 The progressive nature of yielding in a polycrystalline material.

crystal structure is disturbed, but the atomic bonds across the boundary are numerous and strong enough that the boundaries do not usually weaken the material.

Let us now look at what happens when a polycrystalline component begins to yield (Figure 10.5). Slip begins in grains where there are slip planes as nearly parallel to τ as possible, for example, grain (1). Slip later spreads to grains like (2) which are not so favorably oriented, and lastly to the worst oriented grains like (3). Yielding does not take place all at once, therefore, and there is no sharp polycrystalline yield point on the stress–strain curve. Further, gross (total) yielding does not occur at the dislocation yield strength τ_y, because not all the grains are oriented favorably for yielding. The gross yield strength is higher, by a factor called the Taylor factor, which is calculated (with difficulty) by averaging the stress over all possible slip planes; it is close to 1.5.

But we want the tensile yield strength, σ_y. A tensile stress σ creates a shear stress in the material that has a maximum value of $\tau = \sigma/2$. (We show this in Chapter 11 where we resolve the tensile stress onto planes within the material.) To calculate σ_y from τ_y, we combine the Taylor factor with the resolution factor to give

$$\sigma_y = 3\tau_y \tag{10.4}$$

σ_y is the quantity we want: the yield strength of bulk, polycrystalline solids. It is larger than the dislocation shear strength τ_y (by the factor 3) but is proportional to it. So all the statements we have made about increasing τ_y apply unchanged to σ_y.

Grain-boundary strengthening (Hall–Petch effect)

The presence of grain boundaries in a polycrystalline material has an additional consequence — they contribute to the yield strength because grain boundaries act as obstacles to dislocation movement. Because of this, dislocations pile up against grain boundaries, as shown in Figure 10.6.

The number of dislocations which can form in a pile up is given by

$$n = \frac{\alpha \tau d}{Gb}$$

where α is a constant and d is the grain size. If there were only one dislocation piled-up against the grain boundary, the force on it would be $f = \tau b$, as shown in equation (9.2). However, in a pile up of n dislocations, each dislocation exerts a force on the one in front, so the force on the leading (number 1) dislocation is $n\tau b$. If this force exceeds a critical value, F_{GB}, then the leading dislocation will move through the grain boundary, and yield will occur. The yield condition is given by

$$F_{GB} = n\tau b = \left(\frac{\alpha \tau d}{Gb}\right)\tau b = \frac{\alpha \tau^2 d}{G}$$

The contribution to the dislocation yield strength τ_y is therefore

$$\tau_y = \left(\frac{F_{GB} G}{\alpha}\right)^{1/2} d^{-1/2} = \beta d^{-1/2} \tag{10.5}$$

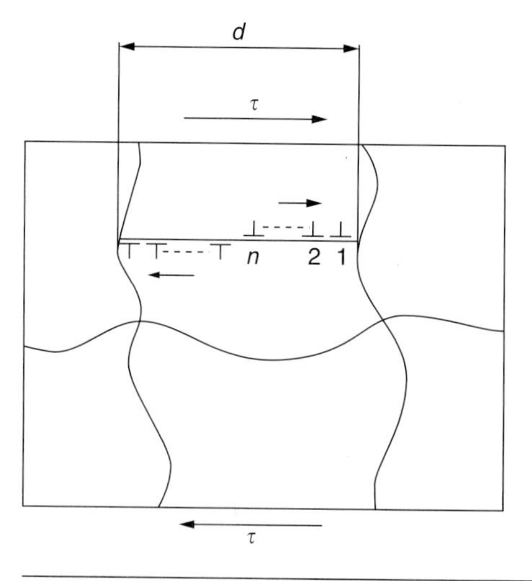

Figure 10.6 How dislocations pile up against grain boundaries.

where β is a constant. The interesting consequence of this result is that fine-grained materials (large $d^{-1/2}$) have a higher yield strength than coarse-grained materials (small $d^{-1/2}$). However, the value of β is specific to the material, and is less for some alloys than others.

10.8 Final remarks

A whole science of alloy design for high strength has grown up in which alloys are blended and heat-treated to achieve maximum τ_y. Important components that are strengthened in this way range from lathe tools ("high-speed" steels) to turbine blades ("Nimonic" alloys based on nickel). We shall have more to say about strong solids when we come to look at how materials are *selected* for a particular job. But first we must return to a discussion of plasticity at the non-atomistic, or continuum, level.

Examples

10.1 Show how dislocations can account for the following observations:

(a) cold working makes aluminum harder;
(b) an alloy of 20 percent Zn, 80 percent Cu is harder than pure copper;
(c) the hardness of nickel is increased by adding particles of thorium oxide.

10.2 Derive an expression for the shear stress τ needed to bow a dislocation line into a semicircle between small hard particles a distance L apart.

10.3 (a) A polycrystalline aluminum alloy contains a dispersion of hard particles of diameter 10^{-8} m and average center-to-center spacing of 6×10^{-8} m measured in the slip planes. Estimate their contribution to the tensile yield strength, σ_y, of the alloy.
(b) The alloy is used for the compressor blades of a small turbine. Adiabatic heating raises the blade temperature to 150°C, and causes the particles to coarsen slowly. After 1000 hours they have grown to a diameter of 3×10^{-8} m and are spaced 18×10^{-8} m apart. Estimate the drop in yield strength. (The shear modulus of aluminum is 26 GN m^{-2}, and $b = 0.286$ nm.)

Answers

(a) 450 MN m^{-2}; (b) 300 MN m^{-2}.

10.4 The yield stress of a sample of brass with a large grain size was 20 MN m^{-2}. The yield stress of an otherwise identical sample with a grain size of 4 μm was

120 $\mathrm{MN\,m^{-2}}$. Why did the yield stress increase in this way? What is the value of β in equation (10.5) for the brass?

Answer

0.2 $\mathrm{MN\,m^{-3/2}}$.

Chapter 11

Continuum aspects of plastic flow

Chapter contents

11.1 Introduction

Plastic flow occurs by shear. Dislocations move when the shear stress on the slip plane exceeds *the dislocation yield strength* τ_y of a single crystal. If this is averaged over all grain orientations and slip planes, it can be related to *the tensile yield strength* σ_y of a polycrystal by $\sigma_y = 3\tau_y$ (Chapter 10). But in solving problems of plasticity, it is more useful to define *the shear yield strength k* of a polycrystal. It is equal to $\sigma_y/2$, and differs from τ_y because it is an average shear resistance over all orientations of slip plane. When a structure is loaded, the planes on which shear will occur can often be identified or guessed, and the collapse load calculated approximately by requiring that the stress exceed k on these planes.

In this chapter, we show that $k = \sigma_y/2$, and use k to relate the hardness to the yield strength of a solid. We then examine tensile instabilities which appear in the drawing of metals and polymers.

11.2 The onset of yielding and the shear yield strength, *k*

A tensile stress applied to a piece of material will create a shear stress at an angle to the tensile stress. Let us examine the stresses in more detail. Resolving forces in Figure 11.1 gives the shearing force as

$$F \sin\theta$$

The area over which this force acts in shear is

$$\frac{A}{\cos\theta}$$

and thus the shear stress, τ is

$$\tau = \frac{F\sin\theta}{A/\cos\theta} = \frac{F}{A}\sin\theta\cos\theta$$
$$= \sigma\sin\theta\cos\theta \qquad (11.1)$$

If we plot this against θ as in Figure 11.2 we find a maximum τ at $\theta = 45°$ to the tensile axis. This means that the *highest value of the shear stress is found at 45° to the tensile axis, and has a value of $\sigma/2$.*

Now, from what we have said in Chapters 9 and 10, if we are dealing with a single crystal, the crystal will *not* in fact slip on the 45° plane—it will slip on the nearest lattice plane to the 45° plane on which dislocations can glide (Figure 11.3). In a polycrystal, neighboring grains each yield on their nearest-to-45° slip planes. On a microscopic scale, slip occurs on a zigzag path; but the *average* slip path is at 45° to the tensile axis. The shear stress on this plane

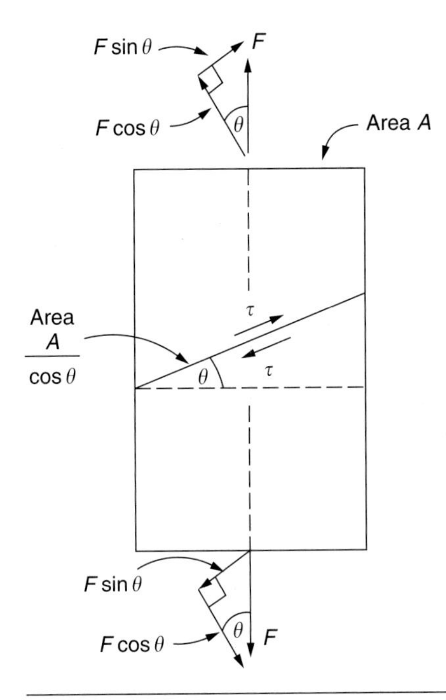

Figure 11.1 A tensile stress, F/A, produces a shear stress, τ, on an inclined plane in the stressed material.

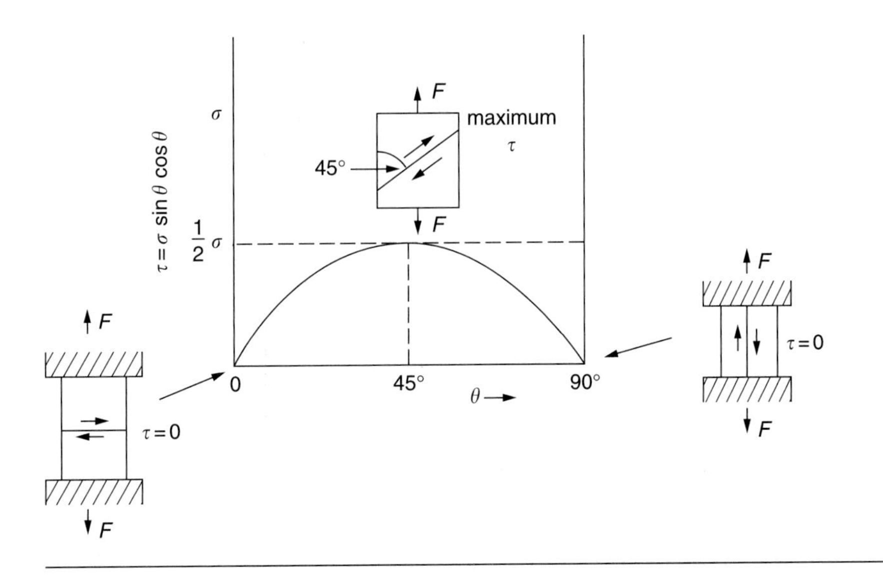

Figure 11.2 Shear stresses in a material have their maximum value on planes at 45° to the tensile axis.

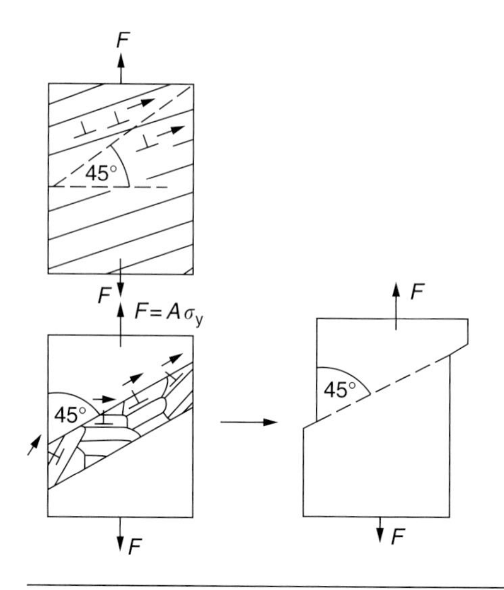

Figure 11.3 In a polycrystalline material the *average* slip path is at 45° to the tensile axis.

when yielding occurs is therefore $\tau = \sigma_y/2$, and we define this as the shear yield strength k:

$$k = \sigma_y/2 \tag{11.2}$$

11.3 Analyzing the hardness test

The concept of shear yielding — where we ignore the details of the grains in our polycrystal and treat the material as a *continuum* — is useful in many respects. For example, we can use it to calculate the loads that would make our material yield for all sorts of quite complicated geometries.

A good example is the problem of the *hardness indenter* that we referred to in the hardness test in Chapter 8. Then, we stated that the hardness

$$H = \frac{F}{A} = 3\sigma_y$$

(with a correction factor for materials that work-harden appreciably — most do). For simplicity, let us assume that our material does not work-harden; so that as the indenter is pushed into the material, the yield strength does not change. Again, for simplicity, we will consider a two-dimensional model. (A real indenter, of course, is three-dimensional, but the result is, for practical purposes, the same.)

As we press a flat indenter into the material, shear takes place on the 45° planes of maximum shear stress shown in Figure 11.4, at a value of shear stress

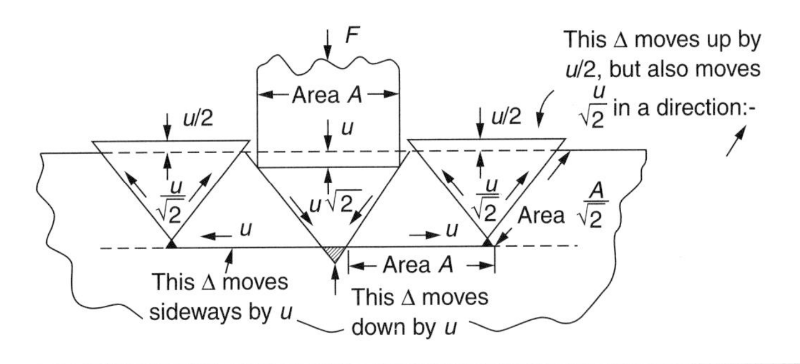

Figure 11.4 The plastic flow of material under a hardness indenter—a simplified two-dimensional visualization.

equal to k. By equating the work done by the force F as the indenter sinks a distance u to the work done against k on the shear planes, we get:

$$Fu = 2 \times \frac{Ak}{\sqrt{2}} \times u\sqrt{2} + 2 \times Ak \times u + 4 \times \frac{Ak}{\sqrt{2}} \times \frac{u}{\sqrt{2}}$$

This simplifies to

$$F = 6Ak$$

from which

$$\frac{F}{A} = 6k = 3\sigma_y$$

But F/A is the hardness, H; so

$$H = 3\sigma_y \tag{11.3}$$

(Strictly, shear occurs not just on the shear planes we have drawn, but on a myriad of 45° planes near the indenter. If our assumed geometry for slip is wrong it can be shown rigorously by a theorem called the *upper-bound* theorem that the value we get for F at yield—the so-called "limit" load—is always on the high side.)

Similar treatments can be used for all sorts of two-dimensional problems: for calculating the plastic collapse load of structures of complex shape, and for analyzing metal-working processes like forging, rolling, and sheet drawing.

11.4 Plastic instability: necking in tensile loading

We now turn to the other end of the stress–strain curve and explain why, in tensile straining, materials eventually start to *neck*, a name for *plastic instability*. It means that flow becomes localized across one section of the specimen or component, as shown in Figure 11.5, and (if straining continues)

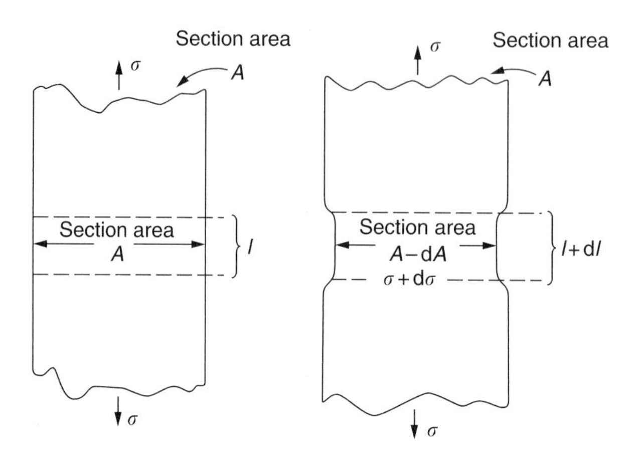

Figure 11.5 The formation of a neck in a bar of material which is being deformed plastically.

the material fractures there. Plasticine necks readily; chewing gum is very resistant to necking.

We analyze the instability by noting that if a force F is applied to the end of the specimen of Figure 11.5, then any section must carry this load. But is it capable of doing so? Suppose one section deforms a little more than the rest, as the figure shows. Its section is less, and the stress in it is therefore larger than elsewhere. If work-hardening has raised the yield strength enough, the reduced section can still carry the force F; but if it has not, plastic flow will become localized at the neck and the specimen will fail there. Any section of the specimen can carry a force $A\sigma$, where A is its area, and σ its current strength. If $A\sigma$ increases with strain, the specimen is stable. If it decreases, it is unstable and will neck. The critical condition for the start of necking is that

$$A\sigma = F = \text{constant}$$

Then

$$A\,\mathrm{d}\sigma + \sigma\,\mathrm{d}A = 0$$

or

$$\frac{\mathrm{d}\sigma}{\sigma} = -\frac{\mathrm{d}A}{A}$$

But volume is conserved during plastic flow, so

$$-\frac{\mathrm{d}A}{A} = \frac{\mathrm{d}l}{l} = \mathrm{d}\epsilon$$

(prove this by differentiating $Al = \text{constant}$). So

$$\frac{\mathrm{d}\sigma}{\sigma} = \mathrm{d}\epsilon$$

or

$$\frac{d\sigma}{d\epsilon} = \sigma \tag{11.4}$$

This equation is given in terms of true stress and true strain. As we said in Chapter 8, tensile data are usually given in terms of nominal stress and strain. From Chapter 8:

$$\sigma = \sigma_n(1 + \epsilon_n)$$
$$\epsilon = \ln(1 + \epsilon_n)$$

If these are differentiated and substituted into the necking equation we get

$$\frac{d\sigma_n}{d\epsilon_n} = 0 \tag{11.5}$$

In other words, on the point of instability, the *nominal* stress–strain curve is at its maximum as we know experimentally from Chapter 8.

To see what is going on physically, it is easier to return to our first condition. At low stress, if we make a little neck, the material in the neck will work-harden and will be able to carry the extra stress it has to stand because of its smaller area; load will therefore be continuous, and the material will be stable. At high stress, the *rate of work-hardening* is less as the true stress–true strain curve shows: that is, the slope of the σ/ϵ curve is less. Eventually, we reach a point at which, when we make a neck, the work-hardening is only *just* enough to stand the extra stress. This is the point of necking (Figure 11.6) with

$$\frac{d\sigma}{d\epsilon} = \sigma$$

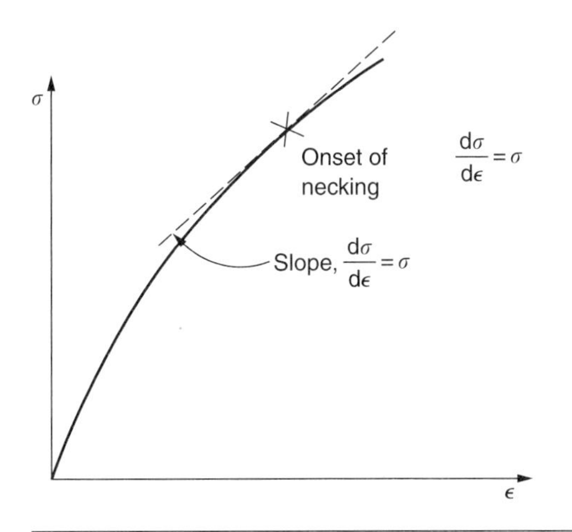

Figure 11.6 The condition for necking.

At still higher *true* stress $d\sigma/d\epsilon$, the rate of work-hardening, decreases further, becoming insufficient to maintain stability — the extra stress in the neck can no longer be accommodated by the work-hardening produced by making the neck, and the neck grows faster and faster, until final fracture takes place.

Consequences of plastic instability

Plastic instability is very important in processes like deep drawing sheet metal to form car bodies, cans, etc. Obviously we must ensure that the materials and press designs are chosen carefully to *avoid* instability.

Mild steel is a good material for deep drawing in the sense that it flows a great deal before necking starts. It can therefore be drawn very deeply without breaking (Figure 11.7).

Aluminum alloy is much less good (Figure 11.8) — it can only be drawn a little before instabilities form. Pure aluminum is not nearly as bad, but is much too *soft* to use for most applications.

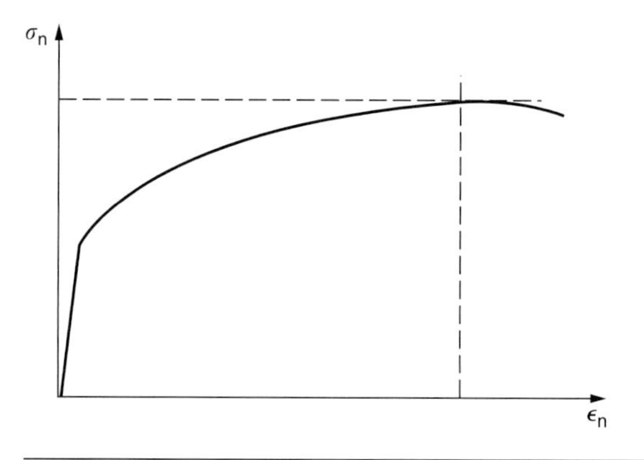

Figure 11.7 Mild steel can be drawn out a lot before it fails by necking.

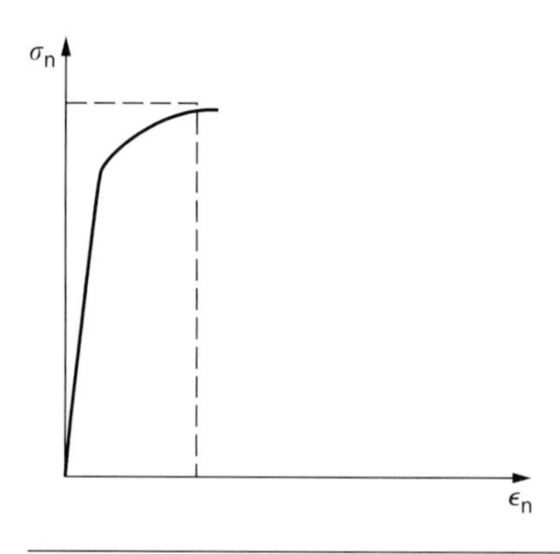

Figure 11.8 Aluminum alloy quickly necks when it is drawn out.

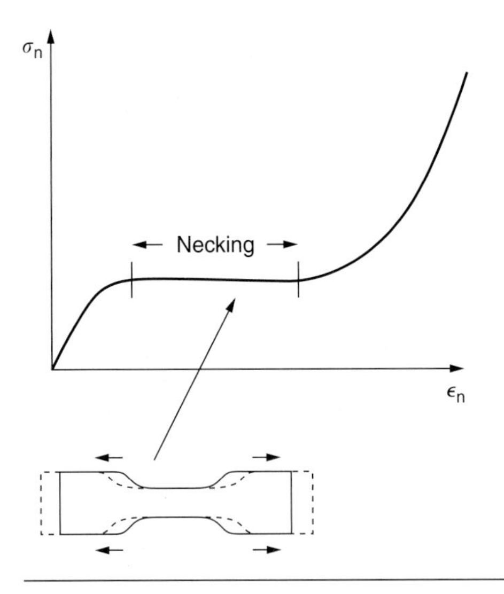

Figure 11.9 Polythene forms a *stable* neck when it is drawn out; drawn polythene is very strong.

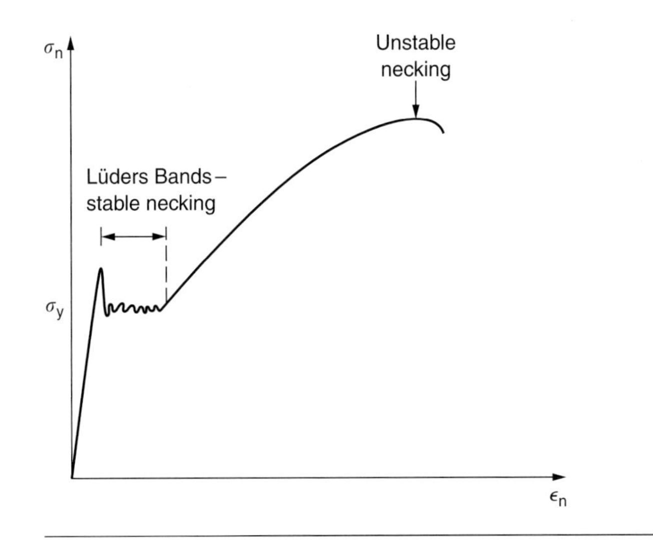

Figure 11.10 Mild steel often shows both stable and unstable necks.

Polythene shows a kind of necking that does *not* lead to fracture. Figure 11.9 shows its σ_n/ϵ_n curve. At quite low, stress

$$\frac{d\sigma_n}{d\epsilon_n}$$

becomes zero and necking begins. However, the neck never becomes *unstable*—it simply grows in length—because at high strain the material work-hardens considerably, and is able to support the increased stress at the reduced cross-section of the neck. This odd behavior is caused by the lining up of the polymer chains in the neck along the direction of the neck — and for this sort of reason *drawn* (i.e. "fully necked") polymers can be made to be very strong indeed — much stronger than the undrawn polymers.

Finally, mild steel can sometimes show an instability like that of polythene. If the steel is annealed, the stress/strain curve looks like that in Figure 11.10. A stable neck, called a Lüders Band, forms, and propagates (as it did in polythene) without causing fracture because the strong work-hardening of the later part of the stress/strain curve prevents this. Lüders Bands are a problem when sheet steel is pressed because they give lower precision and disfigure the pressing.

Examples

11.1 By calculating the plastic work done in each process, determine whether the bolt passing through the plate will fail, when loaded in tension, by yielding of the shaft or shearing-off of the head. (Assume no work-hardening.)

Answer

The bolt will fail by shearing-off of the head.

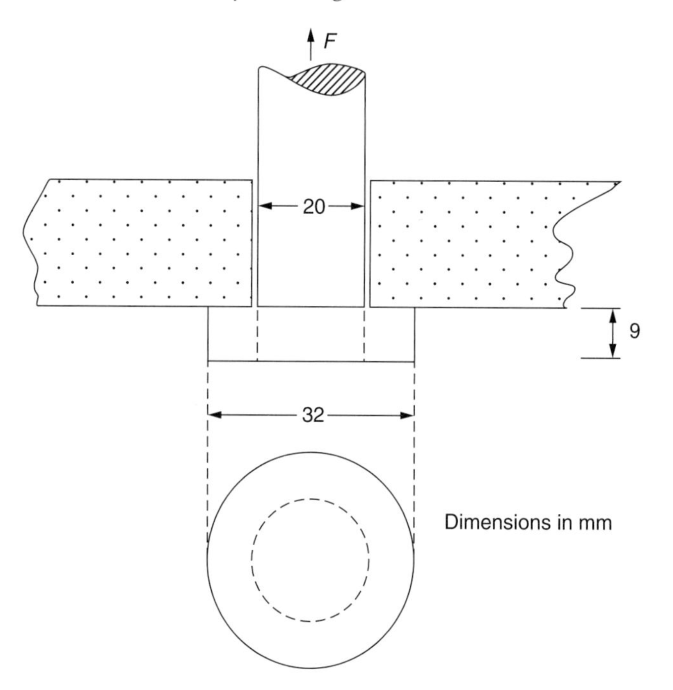

11.2 A metal bar of width w is compressed between two hard anvils as shown in the diagram. The third dimension of the bar, L, is much greater than w. Plastic deformation takes place as a result of shearing along planes, defined by the dashed lines in the diagram, at a shear stress k. Find an upper bound for the load F when (a) there is no friction between anvils and bar, and (b) there is sufficient friction to effectively weld the anvils to the bar. Show that the solution to case (b) satisfies the general formula

$$F \leq 2wLK\left(1 + \frac{w}{4d}\right)$$

which defines upper bounds for all integral values of $w/2d$.

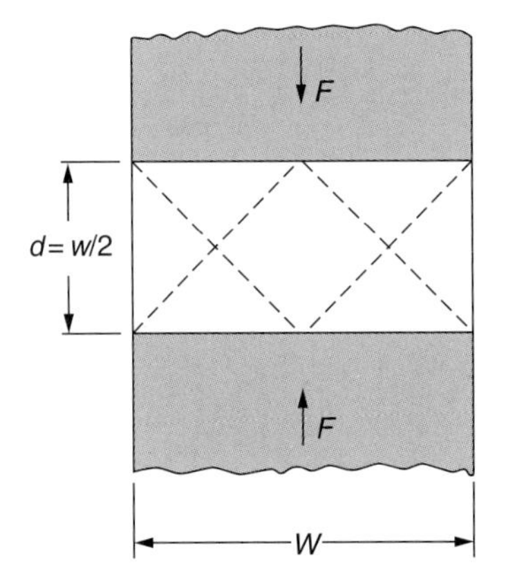

Answers

(a) $2wLk$, (b) $3wLk$.

11.3 A composite material used for rock-drilling bits consists of an assemblage of tungsten carbide cubes (each 20 μm in size) stuck together with a thin layer of cobalt. The material is required to withstand compressive stresses of $4000\,\mathrm{MN\,m^{-2}}$ in service. Use the equation in example 11.2 to estimate an upper limit for the thickness of the cobalt layer. You may assume that the compressive yield stress of tungsten carbide is well above $4000\,\mathrm{MN\,m^{-2}}$, and that the cobalt yields in shear at $k = 175\,\mathrm{MN\,m^{-2}}$.

Answer

0.48 μm.

11.4 (a) Discuss the assumption that, when a piece of metal is deformed at constant temperature, its volume is unchanged.

(b) A ductile metal wire of uniform cross section is loaded in tension until it just begins to neck. Assuming that volume is conserved, derive a differential expression relating the *true* stress to the *true* strain at the onset of necking.

11.5 The curve of true stress against true strain for the metal wire approximates to

$$\sigma = 350\epsilon^{0.4}\ \mathrm{MN\,m^{-2}}$$

Estimate the tensile strength of the wire and the work required to take $1\ \mathrm{m^3}$ of the wire to the point of necking.

Answers

$163\ \mathrm{MN\,m^{-2}}$, 69.3 MJ.

11.6 (a) If the *true* stress–*true* strain curve for a material is defined by $\sigma = A\epsilon^n$, where A and n are constants, find the tensile strength σ_{TS}.

(b) For a nickel alloy, $n = 0.2$, and $A = 800\ \mathrm{MN\,m^{-2}}$. Evaluate the tensile strength of the alloy. Evaluate the true stress in an alloy specimen loaded to σ_{TS}.

Answers

$$\text{(a)}\ \sigma_{TS} = \frac{An^n}{e^n}, \quad \text{(b)}\ \sigma_{TS} = 475\ \mathrm{MN\,m^{-2}},\ \sigma = 580\ \mathrm{MN\,m^{-2}}$$

11.7 The indentation hardness, H, is given by $H \approx 3\sigma_y$ where σ_y is the true yield stress at a nominal plastic strain of 8%. If the true stress–strain curve of a material is given by

$$\sigma = A\epsilon^n$$

and $n = 0.2$, calculate the tensile strength of a material for which the indentation hardness is $600\ \mathrm{MN\,m^{-2}}$. You may assume that $\sigma_{TS} = An^n/e^n$.

Answer

$\sigma_{TS} = 198\ \mathrm{MN\,m^{-2}}$.

Chapter 12

Case studies in yield-limited design

Chapter contents

12.1 Introduction

We now examine three applications of our understanding of plasticity. The first (material selection for a spring) requires that there be *no plasticity whatever*. The second (material selection for a pressure vessel) typifies plastic design of a large structure. It is unrealistic to expect no plasticity: there will always be some, at bolt holes, loading points, or changes of section. The important thing is that yielding should not spread entirely through any section of the structure—that is, that *plasticity must not become general*. Finally, we examine an instance (the rolling of metal strip) in which yielding is deliberately induced, to give *large-strain plasticity*.

12.2 Case study 1: elastic design-materials for springs

Springs come in many shapes and have many purposes. One thinks of axial springs (a rubber band, for example), leaf springs, helical springs, spiral springs, torsion bars. Regardless of their shape or use, the best material for a spring of minimum volume is that with the greatest value of σ_y^2/E. Here E is Young's modulus and σ_y the failure strength of the material of the spring: its yield strength if ductile, its fracture strength or modulus of rupture if brittle. Some materials with high values of this quantity are listed in Table 12.1.

The argument, at its simplest, is as follows. The primary function of a spring is that of storing elastic energy and—when required—releasing it again. The elastic energy stored per unit volume in a block of material stressed uniformly to a stress σ is:

$$U^{el} = \frac{\sigma^2}{2E}$$

It is this that we wish to maximize. The spring will be damaged if the stress σ exceeds the yield stress or failure stress σ_y; the constraint is $\sigma \leq \sigma_y$. So the

Table 12.1 Materials for springs

	E (GNm^{-2})	σ_y (MNm^{-2})	σ_y^2/E (MJm^{-3})	σ_y/E
Brass (cold-rolled)	120	638	3.38	5.32×10^{-3}
Bronze (cold-rolled)		640	3.41	5.33×10^{-3}
Phosphor bronze		770	4.94	6.43×10^{-3}
Beryllium copper		1380	15.9	11.5×10^{-3}
Spring steel	200	1300	8.45	6.5×10^{-3}
Stainless steel (cold-rolled)		1000	5.0	5.0×10^{-3}
Nimonic (high-temp. spring)		614	1.9	3.08×10^{-3}

maximum energy density is

$$U^{\mathrm{el}} = \frac{\sigma_y^2}{2E}$$

Torsion bars and leaf springs are less efficient than axial springs because some of the material is not fully loaded: the material at the neutral axis, for instance, is not loaded at all. Consider—since we will need the equations in a moment—the case of a leaf spring.

The leaf spring

Even leaf springs can take many different forms, but all of them are basically elastic beams loaded in bending. A rectangular section elastic beam, simply supported at both ends, loaded centrally with a force F, deflects by an amount

$$\delta = \frac{Fl^3}{4Ebt^3} \tag{12.1}$$

ignoring self weight (Figure 12.1). Here l is the length of the beam, t its thickness, b its width, and E is the modulus of the material of which it is made. The elastic energy stored in the spring, per unit volume, is

$$U^{\mathrm{el}} = \frac{1}{2}\frac{F\delta}{btl} = \frac{F^2 l^2}{8Eb^2 t^4} \tag{12.2}$$

Figure 12.2 shows that the stress in the beam is zero along the neutral axis at its center, and is a maximum at the surface, at the midpoint of the beam

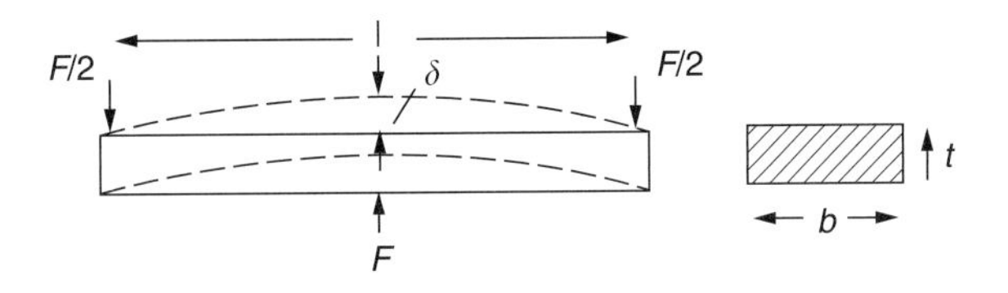

Figure 12.1 A leaf spring under load.

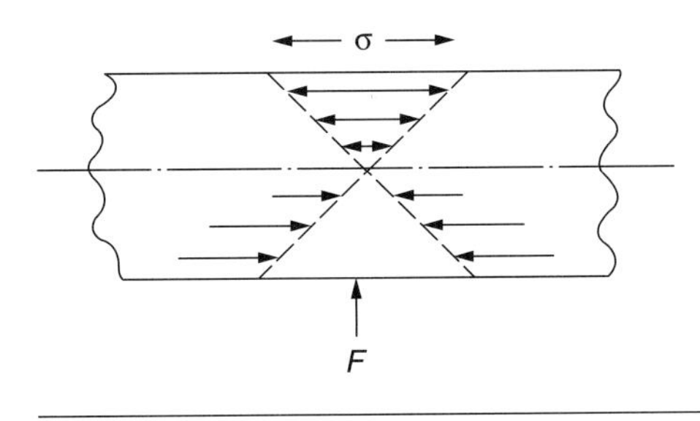

Figure 12.2 Stresses inside a leaf spring.

(because the bending moment is biggest there). The maximum surface stress is given by

$$\sigma = \frac{3Fl}{2bt^2} \tag{12.3}$$

Now, to be successful, a spring must not undergo a permanent set during use: it must always "spring" back. The condition for this is that the maximum stress (equation (12.3)) always be less than the yield stress:

$$\frac{3Fl}{2bt^2} < \sigma_y \tag{12.4}$$

Eliminating t between this and equation (12.2) gives

$$U^{el} = \frac{1}{18}\left(\frac{\sigma_y^2}{E}\right)$$

This equation says: if in service a spring has to undergo a given deflection δ under a force F, the ratio of σ_y^2/E must be high enough to avoid a permanent set. This is why we have listed values of σ_y^2/E in Table 12.1: the best springs are made of materials with high values of this quantity. For this reason spring materials are heavily strengthened (see Chapter 10): by solid solution strengthening plus work-hardening (cold-rolled, single-phase brass and bronze), solid solution and precipitate strengthening (spring steel), and so on. Annealing any spring material removes the work-hardening, and may cause the precipitate to coarsen (increasing the particle spacing), reducing σ_y and making the material useless as a spring.

Example: *Springs for a centrifugal clutch.* Suppose that you are asked to select a material for a spring with the following application in mind. A spring-controlled clutch like that shown in Figure 12.3 is designed to transmit 20 horsepower at 800 rpm; the clutch is to begin to pick up load at 600 rpm. The blocks are lined with Ferodo or some other friction material. When properly adjusted, the maximum deflection of the springs is to be 6.35 mm (but the friction pads may wear, and larger deflections may occur; this is a standard problem with springs—almost always, they must withstand occasional extra deflections without losing their sets).

Mechanics

The force on the spring is

$$F = Mr\omega^2 \tag{12.5}$$

where M is the mass of the block, r the distance of the center of gravity of the block from the center of rotation, and ω the angular velocity. The *net* force the block exerts on the clutch rim at full speed is

$$Mr\left(\omega_2^2 - \omega_1^2\right) \tag{12.6}$$

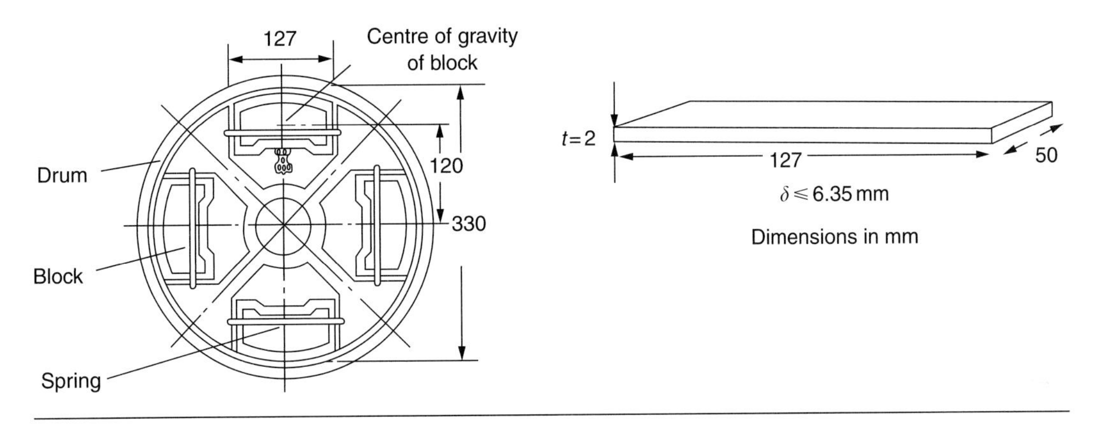

Figure 12.3 Leaf springs in a centrifugal clutch.

where ω_2 and ω_1 correspond to the angular velocities at 800 and 600 rpm (the *net* force must be zero for $\omega_2 = \omega_1$, at 600 rpm). The full power transmitted is given by 4 μ_s Mr $(\omega_2^2 - \omega_1^2)$ × distance moved per second by inner rim of clutch at full speed, that is

$$\text{power} = 4\mu_s Mr(\omega_2^2 - \omega_1^2) \times \omega_2 r \tag{12.7}$$

μ_s is the coefficient of static friction, r is specified by the design (the clutch cannot be too big), and μ_s is a constant (partly a property of the clutch-lining material). Both the power and ω_2 and ω_1 are specified in equation (12.7), so M is specified also; and finally the maximum force on the spring, too, is determined by the design from $F = Mr\omega_1^2$ The requirement that this force deflect the beam by only 6.35 mm with the linings just in contact is what determines the thickness, t, of the spring via equation (12.1) (l and b are fixed by the design).

Metallic materials for the clutch springs

Given the spring dimensions ($t = 2$ mm, $b = 50$ mm, $l = 127$ mm) and given $\delta \leq 6.35$ mm, all specified by design, which material should we use? Eliminating F between equations (12.1) and (12.4) gives

$$\frac{\sigma_y}{E} > \frac{6\delta t}{l^2} = \frac{6 \times 6.35 \times 2}{127 \times 127} = 4.7 \times 10^{-3} \tag{12.8}$$

As well as seeking materials with high values of $\sigma_y{}^2/E$, we must also ensure that the material we choose — if it is to have the dimensions specified above and also deflect through 6.35 mm without yielding — meets the criterion of equation (12.8).

Table 12.1 shows that spring steel, the cheapest material listed, is adequate for this purpose, but has a worryingly small safety factor to allow for wear of the linings. Only the expensive beryllium–copper alloy, of all the metals shown, would give a significant safety factor ($\sigma_y / E = 11.5 \times 10^{-3}$).

In many designs, the mechanical requirements are such that single springs of the type considered so far would yield even if made from beryllium copper. This commonly arises in the case of suspension springs for vehicles, etc., where both large δ ("soft" suspensions) and large F (good load-bearing capacity) are required. The solution then can be to use multi-leaf springs (Figure 12.4). t can be made *small* to give *large* δ without yield according to

$$\left(\frac{\sigma_y}{E}\right) > \frac{6\delta t}{l^2} \tag{12.9}$$

whilst the lost load-carrying capacity resulting from small t can be made up by having several leaves combining to support the load.

Nonmetallic materials

Finally, materials other than the metals originally listed in Table 12.1 can make good springs. Glass, or fused silica, with σ_y/E as large as 58×10^{-3} is excellent, *provided* it operates under protected conditions where it cannot be scratched or suffer impact loading (it was used for galvanometer suspensions). Nylon is good—provided the forces are low—having $\sigma_y/E \approx 22 \times 10^{-3}$, and it is widely used in household appliances and children's toys (you probably brushed your teeth with little nylon springs this morning). Leaf springs for heavy trucks are now being made of CFRP: the value of σ_y/E (6×10^{-3}) is similar to that of spring steel, and the weight saving compensates for the higher cost. CFRP is always worth examining where an innovative use of materials might offer advantages.

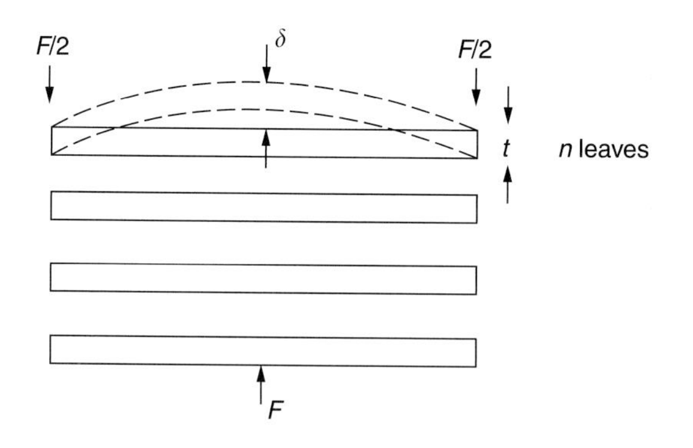

Figure 12.4 Multi-leaved springs (schematic).

12.3 Case study 2: plastic design-materials for a pressure vessel

We shall now examine material selection for a pressure vessel able to contain a gas at pressure p, first minimizing the *weight*, and then the *cost*. We shall seek a design that will not fail by plastic collapse (i.e. general yield). But we must be cautious: structures can also fail by *fast fracture*, by *fatigue*, and by *corrosion* superimposed on these other modes of failure. We shall discuss these in Chapters 13, 17 and 26. Here we shall assume that plastic collapse is our only problem.

Pressure vessel of minimum weight

The body of an aircraft, the hull of a spacecraft, the fuel tank of a rocket: these are examples of pressure vessels which must be as light as possible.

The stress in the vessel wall (Figure 12.5) is:

$$\sigma = \frac{pr}{2t} \tag{12.10}$$

r, the radius of the pressure vessel, is fixed by the design. For safety, $\sigma \le \sigma_y / S$, where S is the safety factor. The vessel mass is

$$M = 4\pi r^2 t\rho \tag{12.11}$$

so that

$$t = \frac{M}{4\pi r^2 \rho} \tag{12.12}$$

Substituting for t in equation (12.8) we find that

$$\frac{\sigma_y}{S} \ge \frac{pr}{2}\frac{4\pi r^2 \rho}{M} = \frac{2\pi p r^3 \rho}{M} \tag{12.13}$$

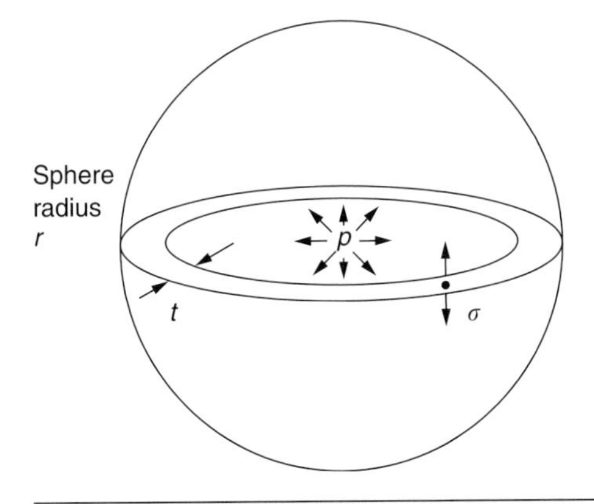

Figure 12.5 Thin-walled spherical pressure vessel.

From equation (12.11) we have, for the mass,

$$M = S2\pi p r^3 \left(\frac{\rho}{\sigma_y}\right) \tag{12.14}$$

so that for the lightest vessel we require the smallest value of (ρ/σ_y). Table 12.2 gives values of (ρ/σ_y) for candidate materials.

By far the lightest pressure vessel is that made of CFRP. Aluminum alloy and pressure-vessel steel come next. Reinforced concrete or mild steel results in a very heavy vessel.

Pressure vessel of minimum cost

If the relative cost of the material is $\tilde{p}$ then the relative cost of the vessel is

$$\tilde{p}M = \text{constant } \tilde{p}\left(\frac{\rho}{\sigma_y}\right) \tag{12.15}$$

Thus relative costs are minimized by minimizing $\tilde{p}(\rho/\sigma_y)$. Data are given in Table 12.2.

The proper choice of material is now a quite different one. Reinforced concrete is now the best choice—that is why many water towers, and pressure vessels for nuclear reactors, are made of reinforced concrete. After that comes pressure-vessel steel—it offers the best compromise of both price and weight. CFRP is very expensive.

12.4 Case study 3: large-strain plasticity—rolling of metals

Forging, *sheet drawing*, and *rolling* are metal-forming processes by which the section of a billet or slab is reduced by compressive plastic deformation. When a slab is rolled (Figure 12.6) the section is reduced from t_1 to t_2 over a length l as it passes through the rolls. At first sight, it might appear that there would be no sliding (and thus no friction) between the slab and the rolls, since these move with the slab. But the metal is elongated in the rolling direction, so it

Table 12.2 Materials for pressure vessels

Material	σ_y (MNm^{-2})	ρ (Mgm^{-3})	$\tilde{p}$	$\frac{\rho}{\sigma_y} \times 10^3$	$\frac{\tilde{p}\rho}{\sigma_y} \times 10^6$
Reinforced concrete	200	2.5	50	13	0.65
Alloy steel (pressure-vessel steel)	1000	7.8	200	7.8	1.6
Mild steel	220	7.8	100	36	3.6
Aluminum alloy	400	2.7	400	6.8	2.7
Fiberglass	200	1.8	1000	9.0	9
CFRP	600	1.5	20,000	2.5	50

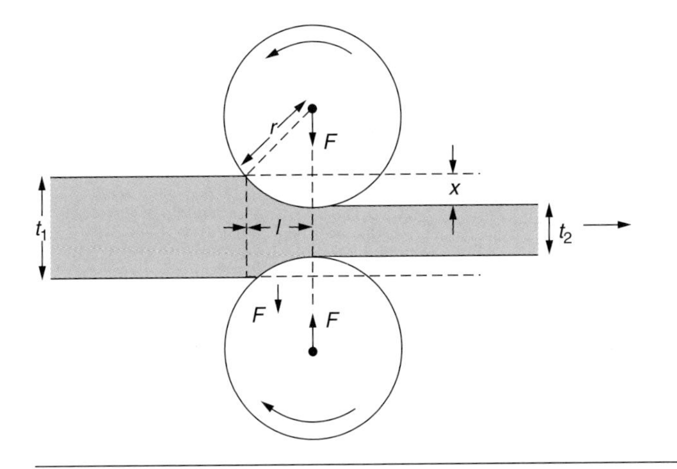

Figure 12.6 The rolling of metal sheet.

speeds up as it passes through the rolls, and some slipping is inevitable. If the rolls are polished and lubricated (as they are for precision and cold-rolling) the frictional losses are small. We shall ignore them here (though all detailed treatments of rolling include them) and calculate the *rolling torque* for perfectly lubricated rolls.

From the geometry of Figure 12.6

$$l^2 + (r - x)^2 = r^2$$

or, if $x = \frac{1}{2}(t_1 - t_2)$ is small (as it almost always is),

$$l = \sqrt{r(t_1 - t_2)}$$

The rolling force F must cause the metal to yield over the length l and width w (normal to Figure 12.6). Thus

$$F = \sigma_y w l$$

If the reaction on the rolls appears halfway along the length marked l, as shown on the lower roll, the torque is

$$T = \frac{Fl}{2} = \frac{\sigma_y w l^2}{2}$$

giving

$$T = \frac{\sigma_y w r(t_1 - t_2)}{2}. \tag{12.16}$$

The torque required to drive the rolls increases with yield strength σ_y, so hot-rolling (when σ_y is low—see Chapter 17) takes less power than cold-rolling. It obviously increases with the reduction in section $(t_1 - t_2)$. And it increases with roll diameter $2r$; this is one of the reasons why small-diameter rolls, often backed by two or more rolls of larger diameter (simply to stop them bending), are used.

Rolling can be analyzed in much more detail to include important aspects which we have ignored: friction, the elastic deformation of the rolls, and the constraint of plane strain imposed by the rolling geometry. But this case study gives an idea of why an understanding of plasticity, and the yield strength, is important in forming operations, both for metals and polymers.

Examples

12.1 You have been asked to prepare an outline design for the pressure hull of a deep-sea submersible vehicle capable of descending to the bottom of the Mariana Trench in the Pacific Ocean. The external pressure at this depth is approximately $100\,\text{MN}\,\text{m}^{-2}$, and the design pressure is to be taken as $200\,\text{MN}\,\text{m}^{-2}$. The pressure hull is to have the form of a thin-walled sphere with a specified radius r of 1 m and a uniform thickness t. The sphere can fail by yield or compressive failure at a pressure p_f given by

$$p_f = 2\sigma_f\left(\frac{t}{r}\right)$$

where σ_f is the yield stress or the compressive failure stress as appropriate.

The basic design requirement is that the pressure hull shall have the minimum possible mass compatible with surviving the design pressure.

By eliminating t from the equations, show that the minimum mass of the hull is given by the expression

$$M_f = 2\pi r^3 p_f\left(\frac{\rho}{\sigma_f}\right)$$

Hence obtain a merit index to meet the design requirement for the failure mechanism. [You may assume that the surface area of the sphere is $4\pi r^2$.]

Answer

σ_f/ρ

12.2 Consider the pressure hull of examples 7.3 and 12.1. For each material listed in the following table, calculate the minimum mass and wall thickness of the pressure hull for both failure mechanisms at the design pressure.

Material	E (GNm^{-2})	σ_f (MNm^{-2})	ρ (kg m^{-3})
Alumina	390	5000	3900
Glass	70	2000	2600
Alloy steel	210	2000	7800
Titanium alloy	120	1200	4700
Aluminum alloy	70	500	2700

Hence determine the limiting failure mechanism for each material. [Hint: this is the failure mechanism which gives the larger of the two values of t.]

What is the optimum material for the pressure hull? What are the mass, wall thickness, and limiting failure mechanism of the optimum pressure hull?

Answers

Material	M_b (tonne)	t_b (mm)	M_f (tonne)	t_f (mm)	*Limiting failure mechanism*
Alumina	2.02	41	0.98	20	Buckling
Glass	3.18	97	1.63	50	Buckling
Alloy steel	5.51	56	4.90	50	Buckling
Titanium alloy	4.39	74	4.92	83	Yielding
Aluminum alloy	3.30	97	6.79	200	Yielding

The optimum material is alumina, with a mass of 2.02 tonne, a wall thickness of 41 mm and a limiting failure mechanism of external-pressure buckling.

12.3 The drawing shows a bolted flanged connection in a tubular supporting pier for an old bridge. You are required to determine the bending moment M needed to make the connection fail by yielding of the wrought-iron bolts. Because the data for the bridge are old, the units for the dimensions in the drawing are inches. The units for force and stress are tons and tons/square inch (tsi). Assume that the yield strength of wrought iron is 11 tsi. [Hint: find the yield load for one bolt, then sum the moments generated by all the bolts about the axis X-X].

Answer

126 ton foot (1 foot = 12 inches).

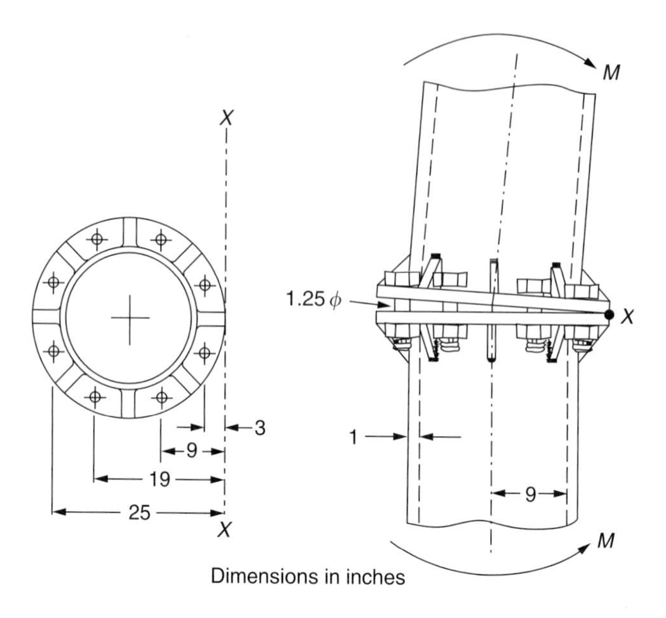

12.4 The diagram gives the dimensions of a steel bicycle chain. The chain is driven by a chain wheel which has pitch diameter of 190 mm. The chain wheel is connected to the pedals by a pair of cranks set at 180° in the usual way.

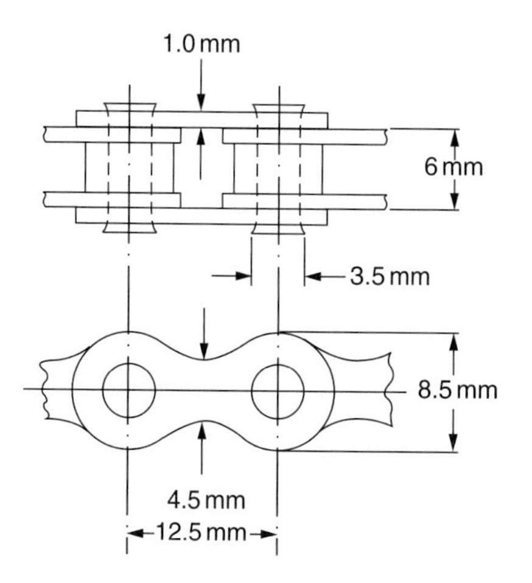

The center of each pedal is 170 mm away from the center of the chain wheel. If the cyclist weighs 90 kg, estimate the factor of safety of the chain. You may assume that a link would fail in simple tension at the position of minimum cross-sectional area and that a pin would fail in double shear. The yield strength of the steel is $1500\,\mathrm{N\,m^{-2}}$.

Answer

8.5.

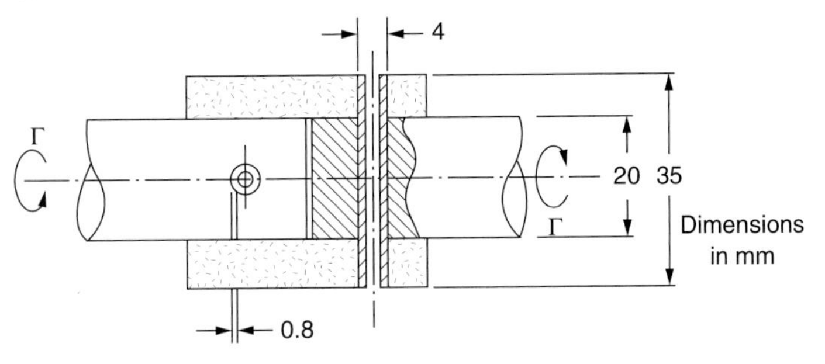

12.5 The diagram shows a coupling between two rotating shafts designed to transmit power from a low-speed hydraulic motor to a gearbox. The coupling sleeve was a sliding fit on the shafts and the torque was taken by the two Bissell pins as shown in the diagram. Owing to a malfunction in the gearbox one of the pins sheared, disconnecting the drive. Assuming that the shear yield stress k is $750\,\mathrm{MN\,m^{-2}}$, estimate the failure torque.

Answer

12 kgf m.

Part D

Fast fracture, brittle fracture and toughness

Chapter 13

Fast fracture and toughness

Chapter contents

13.1 Introduction

Sometimes, structures which were properly designed to avoid both excessive elastic deflection and plastic yielding fail in a catastrophic way by *fast fracture*. Common to these failures—of things like welded ships, welded bridges, and gas pipelines and pressure vessels under large internal pressures—is the presence of cracks, often the result of imperfect welding. Fast fracture is caused by the growth—at the speed of sound in the material—of existing cracks that suddenly became unstable. Why do they do this?

13.2 Energy criterion for fast fracture

If you blow up a balloon, energy is stored in it. There is the energy of the compressed gas in the balloon, and there is the elastic energy stored in the rubber membrane itself. As you increase the pressure, the total amount of elastic energy in the system increases.

If we then introduce a flaw into the system, by poking a pin into the inflated balloon, the balloon will explode, and all this energy will be released. The membrane fails by fast fracture, *even though well below its yield strength*. But if we introduce a flaw of the same dimensions into a system with *less* energy in it, as when we poke our pin into a *partially* inflated balloon, the flaw is stable and fast fracture does not occur. Finally, if we blow up the punctured balloon progressively, we eventually reach a pressure at which it suddenly bursts. In other words, we have arrived at a *critical* balloon *pressure* at which our pin-sized flaw is just unstable, and fast fracture *just* occurs. Why is this?

To make the flaw grow, say by 1 mm, we have to tear the rubber to create 1 mm of new crack surface, and this consumes energy: the tear energy of the rubber per unit area × the area of surface torn. If the work done by the gas pressure inside the balloon, plus the release of elastic energy from the membrane itself, is less than this energy the tearing simply cannot take place—it would infringe the laws of thermodynamics.

We can, of course, increase the energy in the system by blowing the balloon up a bit more. The crack or flaw will remain stable (i.e. it will not grow) until the system (balloon plus compressed gas) has stored in it enough energy that, if the crack advances, *more energy is released than is absorbed*. There is, then, a *critical pressure* for fast fracture of a pressure vessel containing a crack or flaw of a given *size*.

All sorts of accidents (the sudden collapsing of bridges, sudden explosion of steam boilers) have occurred—and still do—due to this effect. In all cases, the critical stress—above which enough energy is available to provide the tearing energy needed to make the crack advance—was exceeded, taking the designer completely by surprise. But how do we calculate this critical stress?

From what we have said already, we can write down an energy balance which must be met if the crack is to advance, and fast fracture is to occur.

Suppose a crack of length a in a material of thickness t advances by δa, then we require that: work done by loads $\geq$ change of elastic energy+energy absorbed at the crack tip, that is

$$\delta W \geq \delta U^{\text{el}} + G_c t \delta a \tag{13.1}$$

where G_c is the energy absorbed per unit area of *crack* (*not* unit area of new surface), and $t\delta a$ is the crack area.

G_c is a material property—it is the energy absorbed in making unit area of crack, and we call it the *toughness* (or, sometimes, the "critical strain energy release rate"). Its units are energy m^{-2} or J m^{-2}. A high toughness means that it is hard to make a crack propagate (as in copper, for which $G_c \approx 10^6\,\text{J m}^{-2}$). Glass, on the other hand, cracks very easily; G_c for glass is only $\approx 10\,\text{J m}^{-2}$.

This same quantity G_c measures the strength of adhesives. You can measure it for the adhesive used on sticky tape (like Sellotape) by hanging a weight on a partly peeled length while supporting the roll so that it can freely rotate (hang it on a pencil) as shown in Figure 13.1. Increase the load to the value M that just causes rapid peeling (= fast fracture). For this geometry, the quantity δU^{el} is small compared to the work done by M (the tape has comparatively little "give") and it can be neglected. Then, from our energy formula,

$$\delta W = G_c t \delta a$$

for fast fracture. In our case,

$$Mg\delta a = G_c t \delta a$$

$$Mg = G_c t$$

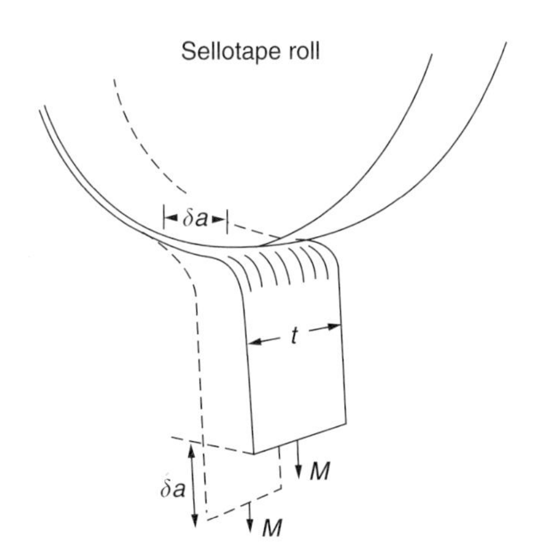

Figure 13.1 How to determine G_c for Sellotape adhesive.

and therefore,

$$G_c = \frac{Mg}{t}$$

Typically, $t = 2\,\text{cm}$, $M = 0.15\,\text{kg}$ and $g \approx 10\,\text{m}\,\text{s}^{-2}$, giving

$$G_c \approx 75\,\text{J}\,\text{m}^{-2}$$

This is a reasonable value for adhesives, and a value bracketed by the values of G_c for many polymers.

Naturally, in most cases, we cannot neglect δU^{el}, and must derive more general relationships. Let us first consider a cracked plate of material loaded so that the displacements at the boundary of the plate are fixed. This is a common mode of loading a material — it occurs frequently in welds between large pieces of steel, for example — and is one which allows us to calculate δU^{el} quite easily.

Fast fracture at fixed displacements

The plate shown in Figure 13.2 is clamped under tension so that its upper and lower ends are fixed. Since the ends cannot move, the forces acting on them can do no work, and $\delta W = 0$. Accordingly, our energy formula gives, for the onset of fast fracture,

$$-\delta U^{\text{el}} = G_c t \delta a \tag{13.2}$$

Now, as the crack grows into the plate, it allows the material of the plate to *relax*, so that it becomes less highly stressed, and *loses* elastic energy. δU^{el} is thus *negative*, so that $-\delta U^{\text{el}}$ is *positive*, as it must be since G_c is defined positive. We can estimate δU^{el} in the way shown in Figure 13.3.

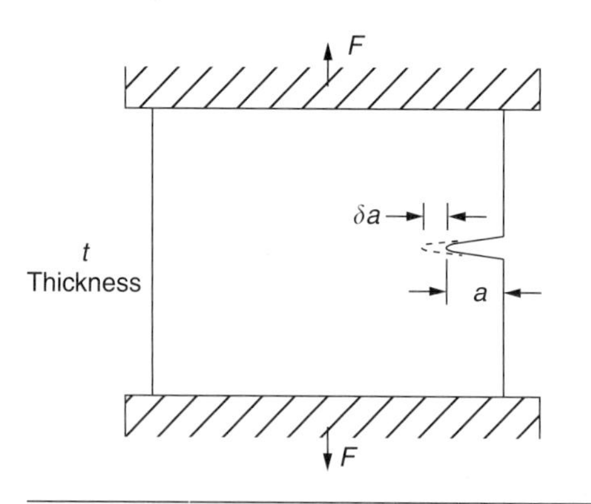

Figure 13.2 Fast fracture in a fixed plate.

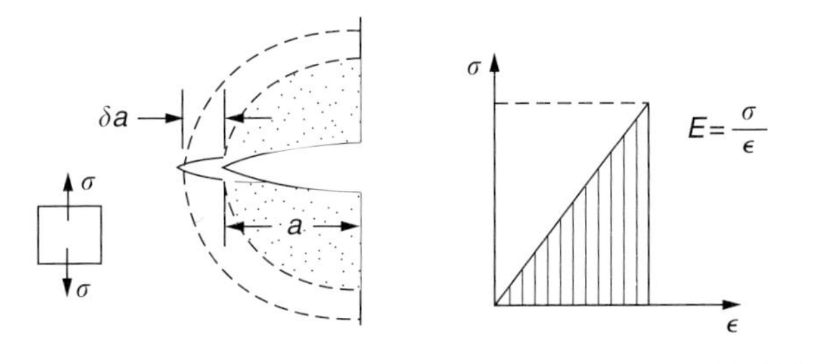

Figure 13.3 The release of stored strain energy as a crack grows.

Let us examine a small cube of material of unit volume inside our plate. Due to the load F this cube is subjected to a stress σ, producing a strain ϵ. Each unit cube therefore has strain energy U^{el} of $\frac{1}{2}\sigma\epsilon$, or $\sigma^2/2E$. If we now introduce a crack of length a, we can consider that the material in the dotted region relaxes (to zero stress) so as to lose all its strain energy. The energy change is

$$U^{\mathrm{el}} = -\frac{\sigma^2}{2E}\frac{\pi a^2 t}{2}.$$

As the crack spreads by length δa, we can calculate the appropriate δU^{el} as

$$\delta U^{\mathrm{el}} = \frac{\mathrm{d}U^{\mathrm{el}}}{\mathrm{d}a}\delta a = -\frac{\sigma^2}{2E}\frac{2\pi a t}{2}\delta a$$

The critical condition (equation (13.2)) then gives

$$\frac{\sigma^2 \pi a}{2E} = G_c$$

at onset of fast fracture.

Actually, our assumption about the way in which the plate material relaxes is obviously rather crude, and a rigorous mathematical solution of the elastic stresses and strains indicates that our estimate of δU^{el} is too low by exactly a factor of 2. Thus, correctly, we have

$$\frac{\sigma^2 \pi a}{E} = G_c$$

which reduces to

$$\sigma\sqrt{\pi a} = \sqrt{EG_c} \tag{13.3}$$

at fast fracture.

Fast fracture at fixed loads

Another, obviously very common way of loading a plate of material, or any other component for that matter, is simply to hang weights on it (fixed loads)

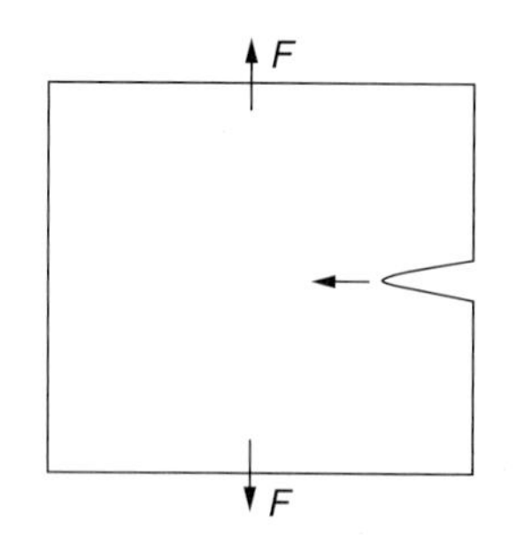

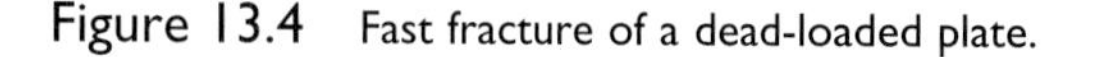

Figure 13.4 Fast fracture of a dead-loaded plate.

(Figure 13.4). Here the situation is a little more complicated than it was in the case of fixed displacements. As the crack grows, the plate becomes less *stiff*, and relaxes so that the applied forces move and do work. δW is therefore finite and positive. However, δU^{el} is now positive also (it turns out that some of δW goes into increasing the strain energy of the plate) and our final result for fast fracture is in fact found to be unchanged.

The fast-fracture condition

Let us now return to our condition for the onset of fast fracture, knowing it to be general* for engineering structures

$$\sigma\sqrt{\pi a} = \sqrt{EG_c}$$

The left-hand side of our equation says that *fast fracture will occur when, in a material subjected to a stress* σ, *a crack reaches some critical size a:* or, alternatively, when *material containing cracks of size a is subjected to some critical stress* σ. The right-hand side of our result *depends on material properties only; E* is obviously a material constant, and G_c, the energy required to generate unit area of crack, again must depend only on the basic properties of our material. Thus, the important point about the equation is that *the critical combination of stress and crack length at which fast fracture commences is a material constant.*

The term $\sigma\sqrt{\pi a}$ crops up so frequently in discussing fast fracture that it is usually abbreviated to a single symbol, K, having units $MN\,m^{-3/2}$; it is called, somewhat unclearly, the *stress intensity factor*. Fast fracture therefore occurs when

$$K = K_c$$

* But see note at end of this chapter.

where $K_c(=\sqrt{EG_c})$ is the *critical* stress intensity factor, more usually called the *fracture toughness*.

To summarize:

G_c = *toughness* (sometimes, critical strain energy release rate). Usual units: $kJ\,m^{-2}$;

$K_c=\sqrt{EG_c}$ = *fracture toughness* (sometimes: critical stress intensity factor). Usual units: $MN\,m^{-3/2}$;

$K=\sigma\sqrt{\pi a}$ = stress intensity factor*. Usual units: $MN\,m^{-3/2}$.

Fast fracture occurs when $K=K_c$.

13.3 Data for G_c and K_c

K_c can be determined experimentally for any material by inserting a crack of known length a into a piece of the material and loading until fast fracture occurs. G_c can be derived from the data for K_c using the relation $K_c=\sqrt{EG_c}$. Figures 13.5 and 13.6 and Table 13.1 show experimental data for K_c and G_c for a wide range of metals, polymers, ceramics, and composites. The values of K_c and G_c range considerably, from the least tough materials, like ice and ceramics, to the toughest, like ductile metals; polymers have intermediate toughness, G_c, but low fracture toughness, K_c (because their *moduli* are low). However, reinforcing polymers to make *composites* produces materials having good fracture toughnesses. Finally, although most metals are tough at or above room temperature, when many (e.g. b.c.c. metals like steels, or c.p.h. metals) are cooled sufficiently, they become quite brittle as the data show.

Obviously these figures for toughness and fracture toughness are extremely important—ignorance of such data has led, and can continue to lead, to engineering disasters of the sort we mentioned at the beginning of this chapter. But just how do these large variations between various materials arise? Why *is* glass so brittle and annealed copper so tough? We shall explain why in Chapter 14.

A note on the stress intensity, K

In Section 13.2 we showed that

$$K=\sigma\sqrt{\pi a}=\sqrt{EG_c}$$

* But see note at end of this chapter.

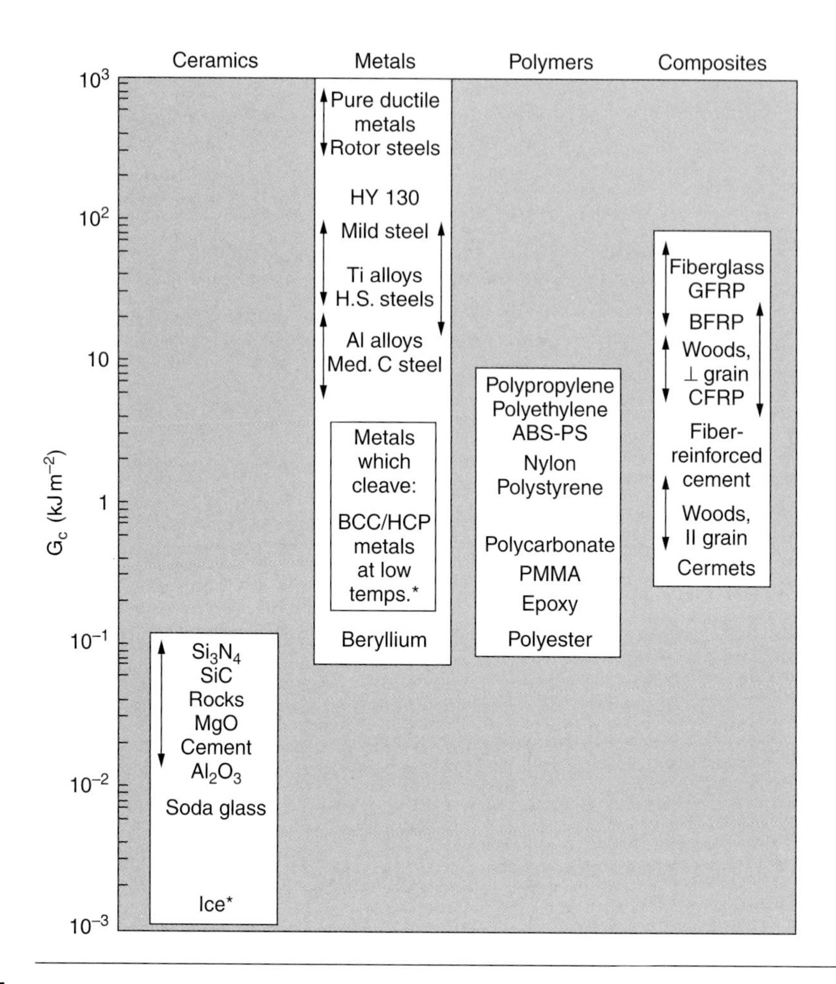

Figure 13.5 Toughness, G_c (values at room temperature unless starred).

at onset of fast fracture. Strictly speaking, this result is valid only for a crack through the center of a wide plate of material. In practice, the problems we encounter seldom satisfy this geometry, and a numerical correction to $\sigma\sqrt{\pi a}$ is required to get the strain energy calculation right. In general we write:

$$K = Y\sigma\sqrt{\pi a},$$

where Y is the numerical correction factor. Values of Y can be found from tables in standard reference books such as those listed in the References at the end of the book. However, provided the crack length a is small

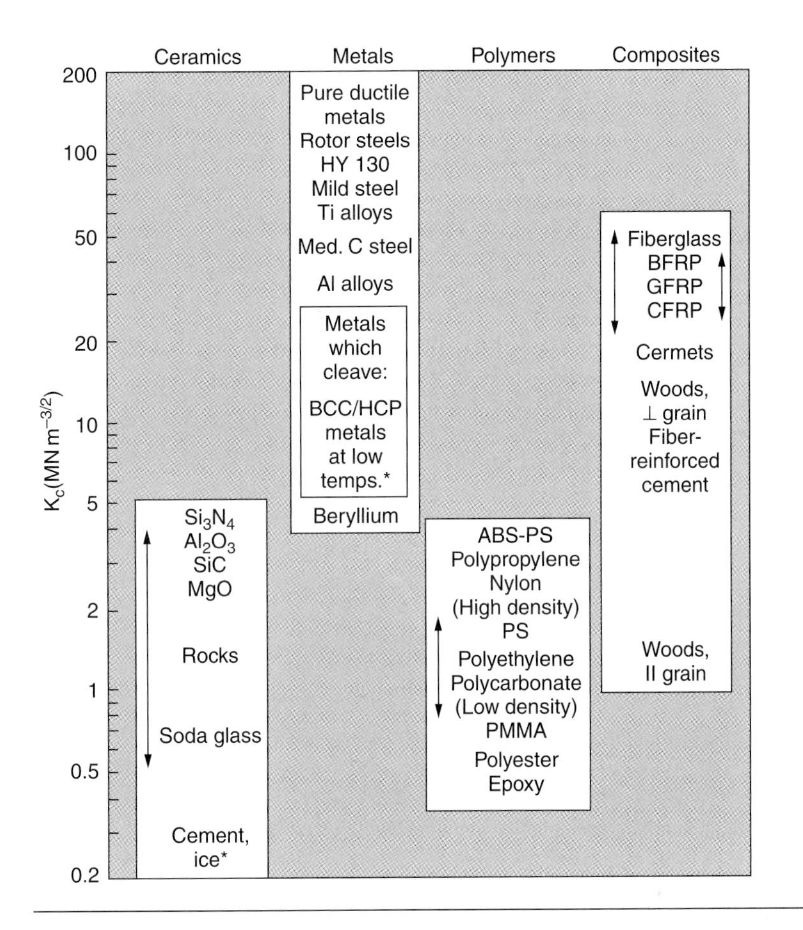

Figure 13.6 Fracture toughness, K_c (values at room temperature unless starred).

compared to the width of the plate W, it is usually safe to assume that $Y \approx 1$.

Examples

13.1 Two wooden beams are butt-jointed using an epoxy adhesive as shown in the diagram. The adhesive was stirred before application, entraining air bubbles which, under pressure in forming the joint, deform to flat, penny-shaped discs of diameter $2a = 2$ mm. If the beam has the dimensions shown, and epoxy has a fracture toughness of $0.5\,\mathrm{MN\,m^{-3/2}}$, calculate the maximum load F that the beam can support. Assume $K = \sigma\sqrt{\pi a}$ for the disc-shaped bubbles.

Table 13.1 Toughness, G_c, and fracture toughness, K_c

Material	G_c (kJ m^{-2})	K_c (MN $m^{-3/2}$)
Pure ductile metals (e.g. Cu, Ni, Ag, Al)	100–1000	100–350
Rotor steels (A533; Discalloy)	220–240	204–214
Pressure-vessel steels (HY130)	150	170
High-strength steels (HSS)	15–118	50–154
Mild steel	100	140
Titanium alloys (Ti6Al4V)	26–114	55–115
GFRPs	10–100	20–60
Fiberglass (glassfiber epoxy)	40–100	42–60
Aluminum alloys (high strength–low strength)	8–30	23–45
CFRPs	5–30	32–45
Common woods, crack ⊥ to grain	8–20	11–13
Boron-fiber epoxy	17	46
Medium-carbon steel	13	51
Polypropylene	8	3
Polyethylene (low density)	6–7	1
Polyethylene (high density)	6–7	2
ABS Polystyrene	5	4
Nylon	2–4	3
Steel-reinforced cement	0.2–4	10–15
Cast iron	0.2–3	6–20
Polystyrene	2	2
Common woods, crack || to grain	0.5–2	0.5–1
Polycarbonate	0.4–1	1.0–2.6
Cobalt/tungsten carbide cermets	0.3–0.5	14–16
PMMA	0.3–0.4	0.9–1.4
Epoxy	0.1–0.3	0.3–0.5
Granite (Westerly Granite)	0.1	3
Polyester	0.1	0.5
Silicon nitride, Si_3N_4	0.1	4–5
Beryllium	0.08	4
Silicon carbide SiC	0.05	3
Magnesia, MgO	0.04	3
Cement/concrete, unreinforced	0.03	0.2
Calcite (marble, limestone)	0.02	0.9
Alumina, Al_2O_3	0.02	3–5
Shale (oilshale)	0.02	0.6
Soda glass	0.01	0.7–0.8
Electrical porcelain	0.01	1
Ice	0.003	0.2*

* Values at room temperature unless starred.

Answer

2.97 kN.

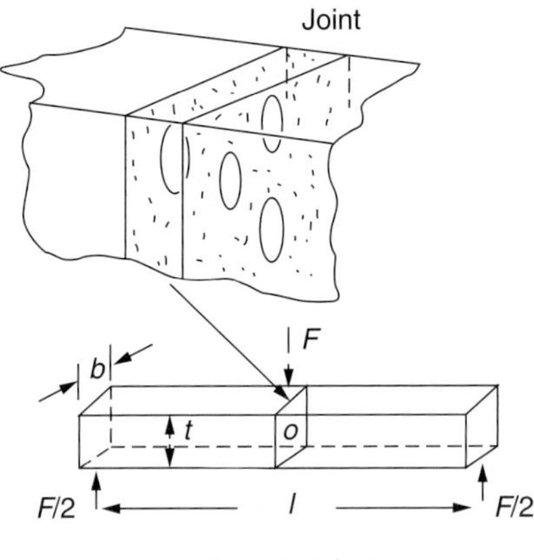

$t = b = 0.1$ m; $l = 2$ m

13.2 A large thick plate of steel is examined by X-ray methods, and found to contain no detectable cracks. The equipment can detect a single edge-crack of depth $a = 5$ mm or greater. The steel has a fracture toughness K_c of 40 MN m$^{-3/2}$ and a yield strength of 500 MN m^{-2}. Assuming that the plate contains cracks on the limit of detection, determine whether the plate will undergo general yield or will fail by fast fracture before general yielding occurs. What is the stress at which fast fracture would occur?

Answer

Failure by fast fracture at 319 MN m^{-2}.

13.3 The fuselage of a passenger aircraft can be considered to be an internally pressurized thin walled tube of diameter 7 m and wall thickness 3 mm. It is made from aluminum alloy plate with a fracture toughness K_c of 100 MN m$^{-3/2}$. At cruising altitude, the internal gauge pressure is 0.06 MN m^{-2}. Multiple fatigue cracks initiated at a horizontal row of rivet holes, and linked to form a single long axial through thickness crack in the fuselage. Estimate the critical length at which this crack will run, resulting in the break-up of the fuselage.

Answer

0.65 m.

Chapter 14

Micromechanisms of fast fracture

Chapter contents

14.1 Introduction

In Chapter 13 we showed that, if a material contains a crack, and is sufficiently stressed, the crack becomes unstable and grows—at up to the speed of sound in the material—to cause catastrophically rapid fracture, or *fast fracture* at a stress less than the yield stress. We were able to quantify this phenomenon and obtained a relationship for the onset of fast fracture

$$\sigma\sqrt{\pi a} = \sqrt{EG_c}$$

or, in more succinct notation,

$$K = K_c \text{ for fast fracture.}$$

It is helpful to compare this with other, similar, "failure" criteria:

$$\sigma = \sigma_y \text{ for yielding,}$$

$$M = M_p \text{ for plastic collapse,}$$

$$P/A = H \text{ for indentation.}$$

(Here M is the moment and M_p the fully-plastic moment of, for instance, a beam; P/A is the indentation pressure and H the hardness of, for example, armor plating.) The left-hand side of each of these equations describes the *loading conditions*; the right-hand side is a *material property*. When the left-hand side (which increases with load) equals the right-hand side (which is fixed), failure occurs.

Some materials, like glass, have low G_c and K_c, and crack easily; ductile metals have high G_c and K_c and are very resistant to fast-fracture; polymers have intermediate G_c, but can be made tougher by making them into composites; and (finally) many metals, when cold, become brittle—that is, G_c and K_c fall with temperature. How can we explain these important observations?

14.2 Mechanisms of crack propagation, 1: ductile tearing

Let us first of all look at what happens when we load a cracked piece of a *ductile* metal—in other words, a metal that can flow readily to give large plastic deformations (like pure copper; or mild steel at, or above, room temperature). If we load the material sufficiently, we can get fracture to take place starting from the crack. If you examine the surfaces of the metal after it has fractured (Figure 14.1) you see that the fracture surface is extremely rough, indicating that a great deal of plastic work has taken place. Let us explain this observation. Whenever a crack is present in a material, the stress close to the crack, σ_{local}, is greater than the average stress σ applied to the piece of material;

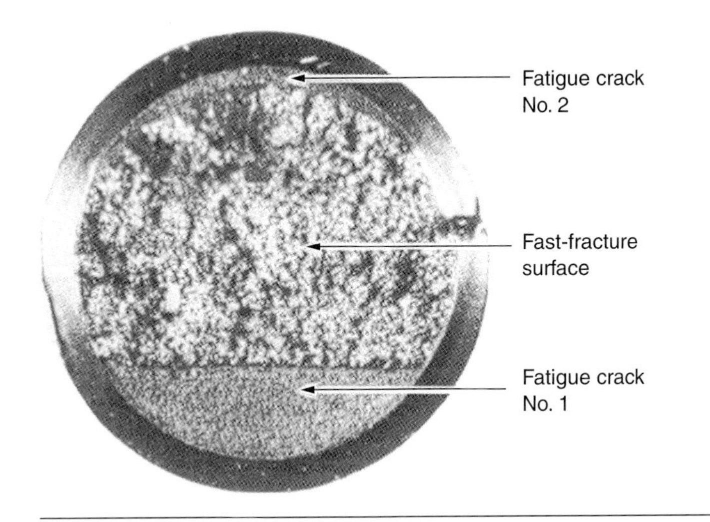

Figure 14.1 Before it broke, this steel bolt held a seat onto its mounting at Milan airport. Whenever someone sat down, the lower part of the cross-section went into tension, causing a crack to grow there by *metal fatigue* (Chapter 17; crack No. 1). When someone got up again, the upper part went into tension, causing fatigue crack No. 2 to grow. Eventually the bolt failed by fast fracture from the larger of the two fatigue cracks. The victim was able to escape with the fractured bolt!

the crack has the effect of *concentrating* the stress. Mathematical analysis shows that the local stress ahead of a *sharp* crack in an elastic material is

$$\sigma_{\text{local}} = \sigma + \sigma\sqrt{\frac{a}{2r}} \tag{14.1}$$

The closer one approaches to the tip of the crack, the higher the local stress becomes, until at some distance r_y from the tip of the crack the stress reaches the yield stress, σ_y of the material, and plastic flow occurs (Figure 14.2). The distance r_y is easily calculated by setting $\sigma_{\text{local}} = \sigma_y$ in equation (14.1). Assuming r_y to be small compared to the crack length, a, the result is

$$\begin{aligned} r_y &= \frac{\sigma^2 a}{2\sigma_y^2} \\ &= \frac{K^2}{2\pi\sigma_y^2} \end{aligned} \tag{14.2}$$

The crack propagates when K is equal to K_c; the width of the *plastic zone*, r_y, is then given by equation (14.2) with K replaced by K_c. Note that the zone of plasticity shrinks rapidly as σ_y increases: cracks in soft metals have a large plastic zone; cracks in hard ceramics have a small zone, or none at all.

Even when nominally pure, most metals contain tiny *inclusions* (or particles) of chemical compounds formed by reaction between the metal and impurity

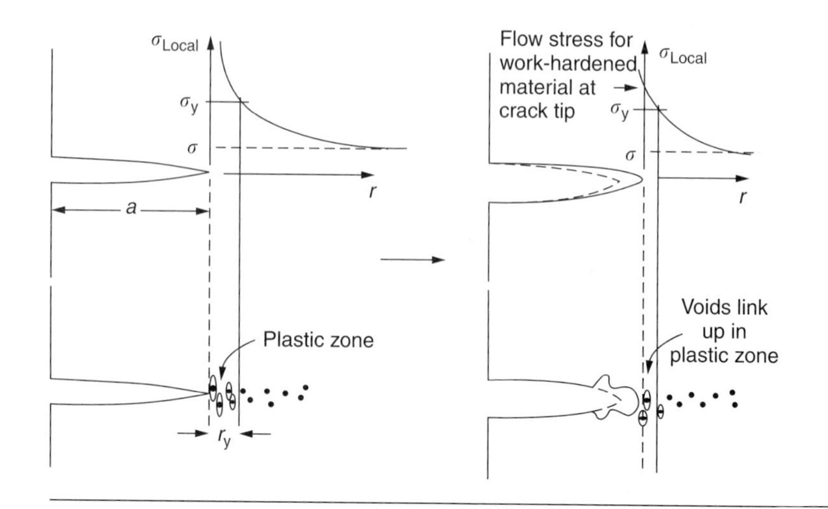

Figure 14.2 Crack propagation by ductile tearing.

atoms. Within the plastic zone, plastic flow takes place around these inclusions, leading to elongated cavities, as shown in Figure 14.2. As plastic flow progresses, these cavities link up, and the crack advances by means of this *ductile tearing*. The plastic flow at the crack tip naturally turns our initially sharp crack into a *blunt* crack, and it turns out from the stress mathematics that this *crack blunting* decreases σ_{local} so that, at the crack tip itself, σ_{local} is just sufficient to keep on plastically deforming the work-hardened material there, as the figure shows.

The important thing about crack growth by ductile tearing is that *it consumes a lot of energy by plastic flow;* the bigger the plastic zone, the more energy is absorbed. High energy absorption means that G_c is high, and so is K_c. This is why ductile metals are so tough. Other materials, too, owe their toughness to this behavior—plasticine is one, and some polymers also exhibit toughening by processes similar to ductile tearing.

14.3 Mechanisms of crack propagation, 2: cleavage

If you now examine the fracture surface of something like a ceramic, or a glass, you see a very different state of affairs. Instead of a very rough surface, indicating massive local plastic deformation, you see a rather featureless, flat surface suggesting little or no plastic deformation. How is it that cracks in ceramics or glasses can spread without plastic flow taking place? Well, the local stress ahead of the crack tip, given by our formula

$$\sigma_{\text{local}} = \sigma + \sigma\sqrt{\frac{a}{2r}}$$

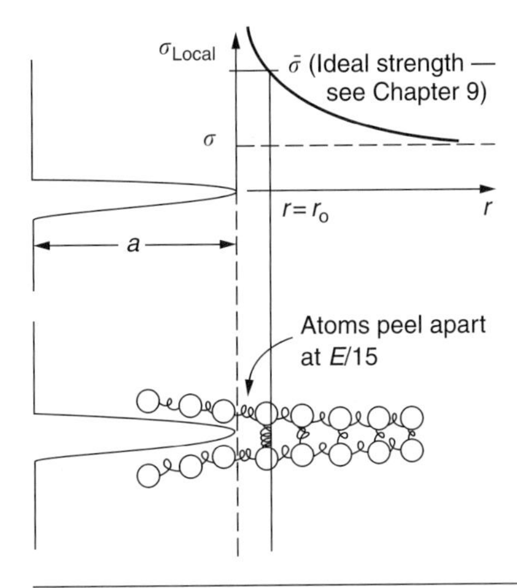

Figure 14.3 Crack propagation by cleavage.

can clearly approach very high values very near to the crack tip *provided that blunting of our sharp crack tip does not occur.* As we showed in Chapter 8, ceramics and glasses have very high yield strengths, and thus very little plastic deformation takes place at crack tips in these materials. Even allowing for a small degree of crack blunting, the local stress at the crack tip is still in excess of the ideal strength and is thus large enough to literally break apart the interatomic bonds there; the crack then spreads between a pair of atomic planes giving rise to an atomically flat surface by *cleavage* (Figure 14.3). The energy required simply to break the interatomic bonds is *much* less than that absorbed by ductile tearing in a tough material, and this is why materials like ceramics and glasses are so brittle. It is also why some steels become brittle and fail like glass, at low temperatures—as we shall now explain.

At low temperatures metals having b.c.c. and c.p.h. structures become brittle and fail by cleavage, even though they may be tough at or above room temperature. In fact, only those metals with an f.c.c. structure (like copper, lead, aluminum) remain unaffected by temperature in this way. In metals not having an f.c.c. structure, the motion of dislocations is assisted by the *thermal agitation* of the atoms (we shall talk in more detail about *thermally activated* processes in Chapter 21). At lower temperatures this thermal agitation is less, and the dislocations cannot move as easily as they can at room temperature in response to a stress—the intrinsic lattice resistance (Chapter 10) increases. The result is that the yield strength rises, and the plastic zone at the crack tip shrinks until it becomes so small that the fracture mechanism changes from ductile tearing to cleavage. This effect is called the *ductile-to-brittle* transition; for steels it can be as high as $\approx 0^{\circ}$C, depending on the composition of the steel;

steel structures like ships, bridges, and oil rigs are much more likely to fail in winter than in summer.

A somewhat similar thing happens in many polymers at the *glass–rubber transition* that we mentioned in Chapter 6. Below the transition these polymers are much more brittle than above it, as you can easily demonstrate by cooling a piece of rubber or polyethylene in liquid nitrogen. (Many other polymers, like epoxy resins, have low G_c values at *all* temperatures simply because they are heavily cross-linked at all temperatures by *covalent* bonds and the material does not flow at the crack tip to cause blunting.)

14.4 Composites, including wood

As Figures 13.5 and 13.6 show, composites are tougher than ordinary polymers. The low toughness of materials like epoxy resins, or polyester resins, can be enormously increased by reinforcing them with carbon fiber or glass fiber. But why is it that putting a second, equally (or more) brittle material like graphite or glass into a brittle polymer makes a tough composite? The reason is the *fibers act as crack stoppers* (Figure 14.4).

The sequence in the figure shows what happens when a crack runs through the brittle matrix towards a fiber. As the crack reaches the fiber, the stress field just ahead of the crack separates the matrix from the fiber over a small region (a process called *debonding*) and the crack is blunted so much that its motion is *arrested*. Naturally, this only works if the crack is running normal to the fibers: wood is very tough across the grain, but can be split easily (meaning that G_c is low) along it. One of the reasons why fiber composites are so useful in engineering design—in addition to their high *stiffnesses* that we talked about in Chapter 6—is their high *toughness* produced in this way. Of course, there are other ways of making polymers tough. The addition of small particles ("fillers") of various sorts to polymers can modify their properties considerably. Rubber-toughened polymers (like ABS), for example, derive their toughness from the small rubber particles they contain. A crack intersects and stretches them as shown in Figure 14.5. The particles act as little springs, clamping the crack shut, and thereby increasing the load needed to make it propagate.

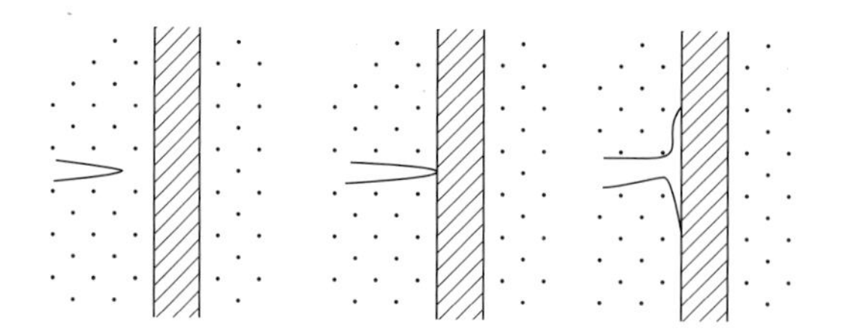

Figure 14.4 Crack stopping in composites.

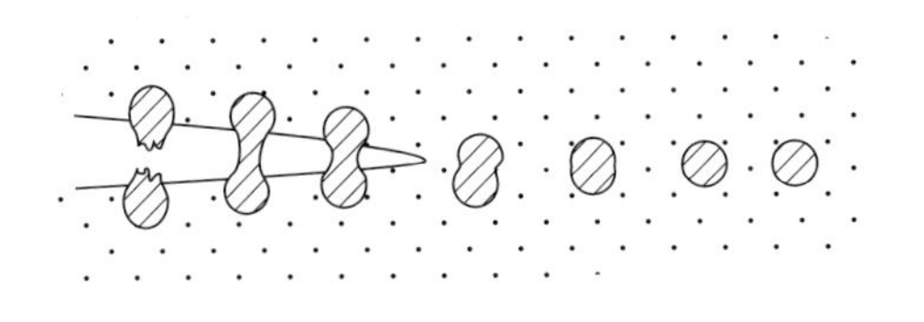

Figure 14.5 Rubber-toughened polymers.

14.5 Avoiding brittle alloys

Let us finally return to the toughnesses of metals and alloys, as these are by far the most important class of materials for highly stressed applications. Even at, or above, room temperature, when nearly all common pure metals are tough, alloying of these metals with other metals or elements (e.g. with carbon to produce steels) can reduce the toughness. This is because alloying increases the resistance to dislocation motion (Chapter 10), raising the yield strength and causing the plastic zone to shrink. A more marked decrease in toughness can occur if enough impurities are added to make precipitates of chemical *compounds* formed between the metal and the impurities. These compounds can often be very brittle and, if they are present in the shape of extended plates (e.g. sigma-phase in stainless steel; graphite in cast iron), cracks can spread along the plates, leading to brittle fracture. Finally, heat-treatments of alloys like steels can produce different *crystal structures* having great hardness (but also therefore great brittleness because crack blunting cannot occur). A good example of such a material is high-carbon steel after quenching into water from bright red heat: it becomes as brittle as glass. Proper heat treatment, following suppliers' specifications, is essential if materials are to have the properties you want.

Examples

14.1 A large furnace flue operating at 440°C was made from a low-alloy steel. After 2 years in service, specimens were removed from the flue and fracture toughness tests were carried out at room temperature. The average value of K_c was only about 30 $\text{MN m}^{-3/2}$ compared to a value when new of about 80. The loss in toughness was due to temper embrittlement caused by the impurity phosphorus.

The skin of the flue was made from plate 10 mm thick. Owing to the self-weight of the flue the plate has to withstand primary membrane stresses of up to 60 MN m^{-2}. Estimate the length of through-thickness crack that will lead to the fast fracture of the flue when the plant is shut down.

Answer

160 mm.

14.2 A ceramic bidet failed catastrophically in normal use, causing serious lacerations to the person concerned. The fracture initiated at a pre-existing through-thickness crack of length $2a = 20$ mm located at a circular flush hole. Strain gauge tests on an identical bidet gave live load stress $= 0.4$ to $1.0\,\text{MN m}^{-2}$ (tensile) and residual stress (from shrinkage during manufacture) $= 4\,\text{MN m}^{-2}$ (tensile). Specimen tests gave $K_c = 1.3\,\text{MN m}^{-3/2}$. Assuming that the hole introduces a local stress concentration factor of 3, account for the failure using fracture mechanics.

14.3 Account for the following observations.

(a) Ductile metals have high toughnesses whereas ceramics, glasses, and rigid polymers have low toughnesses.
(b) Aligned fiber composites are much tougher when the crack propagates perpendicular to the fibers than parallel to them.

14.4 The photograph shows a small part of a ductile tearing fracture surface taken in the scanning electron microscope (SEM). The material is aluminum alloy.

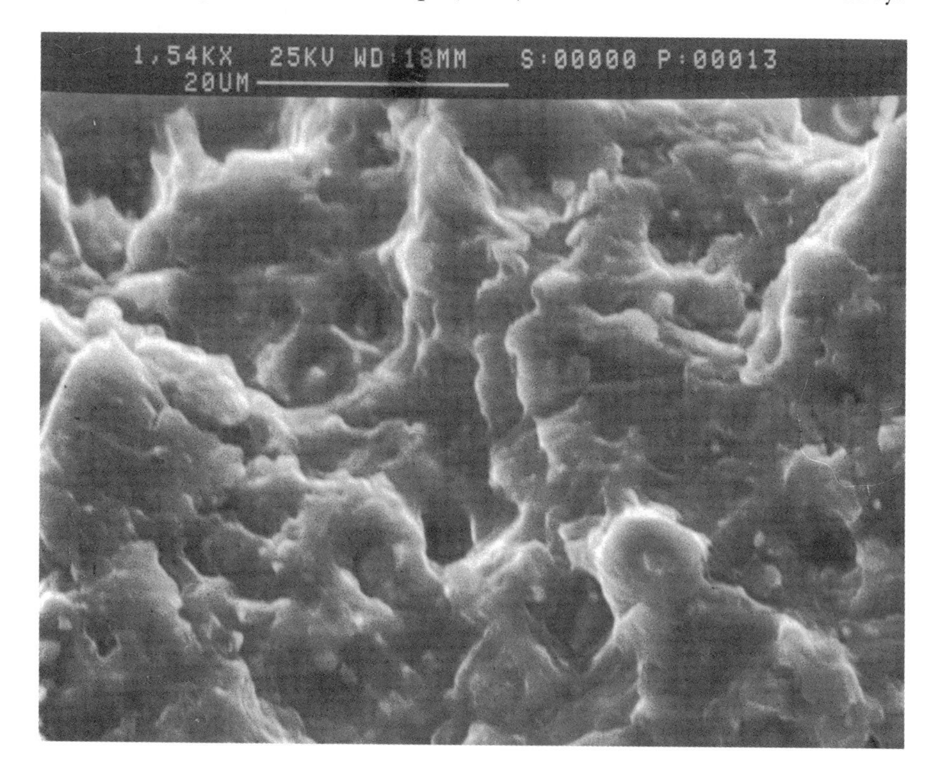

Comment on the features you observe, and relate them to the description in the text. [Note: this fracture surface comes from the fast fracture region of the broken roller arm in Case Study 1 of Chapter 19.]

14.5 The photograph shows a small part of a cleavage fracture surface taken in the scanning electron microscope (SEM). The material is low-alloy steel. Comment

on the features you observe, and relate them to the description in the text. [Note: this fracture surface comes from the fast fracture region of the broken bolt in Case Study 2 of Chapter 19.] The white marker is 100 μm long.

Chapter 15

Case studies in fast fracture

Chapter contents

15.1 Introduction

In this chapter we look at four real situations where failure occurred because of the catastrophic growth of a crack by fast fracture: a steel ammonia tank which exploded because of weld cracks; a perspex pressure window which exploded during hydrostatic testing; a polyurethane foam jacket on a liquid methane tank which cracked during cooling; and a wooden balcony railing which fell over because the end posts split.

15.2 Case study 1: fast fracture of an ammonia tank

Figure 15.1 shows part of a steel tank which came from a road tank vehicle. The tank consisted of a cylindrical shell about 6 m long. A hemispherical cap was welded to each end of the shell with a circumferential weld. The tank was used to transport liquid ammonia. In order to contain the liquid ammonia the pressure had to be equal to the saturation pressure (the pressure at which a mixture of liquid and vapor is in equilibrium). The saturation pressure increases rapidly with temperature: at 20°C the absolute pressure is 8.57 bar; at 50°C it is 20.33 bar. The *gauge* pressure at 50°C is 19.33 bar, or 1.9 MN m^{-2}. Because of this the tank had to function as a pressure vessel. The maximum operating pressure was 2.07 MN m^{-2} gauge. This allowed the tank to be used safely to 50°C, above the maxmimum temperature expected in even a hot climate.

While liquid was being unloaded from the tank a fast fracture occurred in one of the circumferential welds and the cap was blown off the end of the shell. In order to decant the liquid the space above the liquid had been pressurized with ammonia gas using a compressor. The normal operating pressure of the

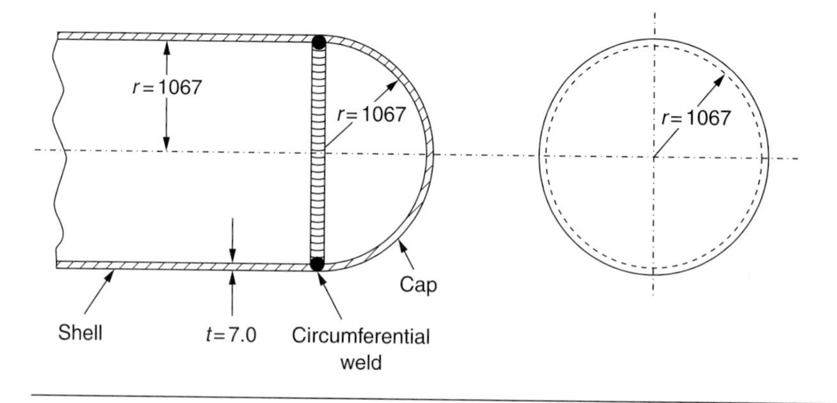

Figure 15.1 The weld between the shell and the end cap of the pressure vessel. Dimensions in mm.

compressor was $1.83\,\mathrm{MN\,m^{-2}}$; the maximum pressure (set by a safety valve) was $2.07\,\mathrm{MN\,m^{-2}}$. One can imagine the effect on nearby people of this explosive discharge of a large volume of highly toxic vapor.

Details of the failure

The geometry of the failure is shown in Figure 15.2. The initial crack, 2.5 mm deep, had formed in the heat-affected zone between the shell and the circumferential weld. The defect went some way around the circumference of the vessel. The cracking was intergranular, and had occurred by a process called stress corrosion cracking (see Chapter 26). The final fast fracture occurred by transgranular cleavage (see Chapter 14). This indicates that the heat-affected zone must have had a very low fracture toughness. In this case

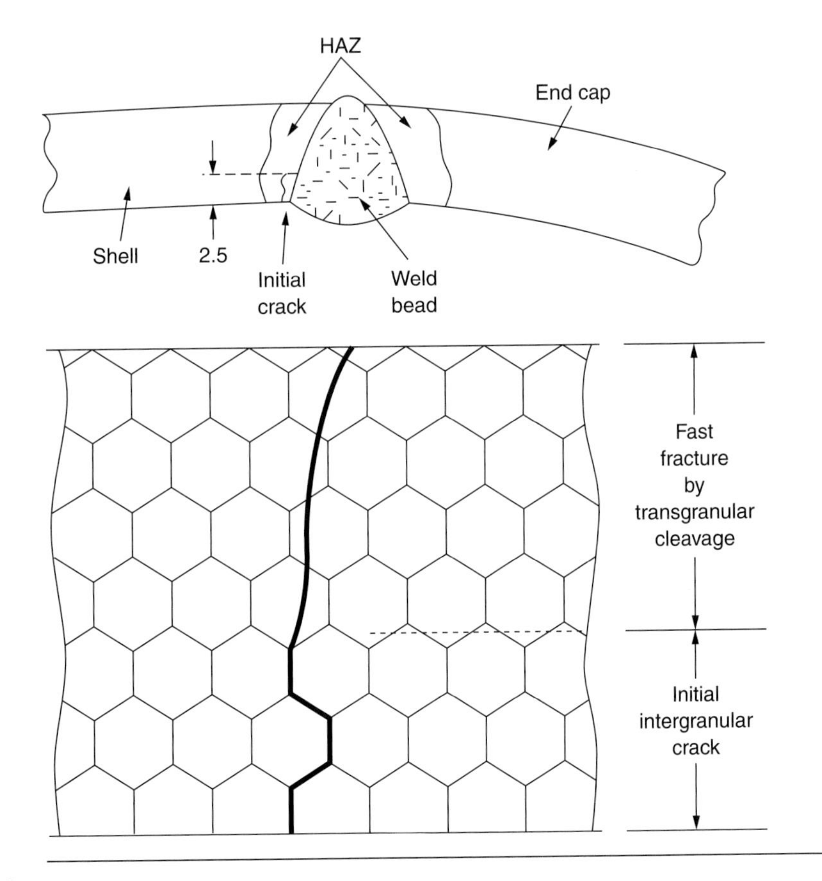

Figure 15.2 The geometry of the failure. Dimensions in mm.

study we predict the critical crack size for fast fracture using the fast fracture equation.

Material properties

The tank was made from high-strength low-alloy steel with a yield strength of $712\,\text{MN}\,\text{m}^{-2}$ and a fracture toughness of $80\,\text{MN}\,\text{m}^{-3/2}$. The heat from the welding process had altered the structure of the steel in the heat-affected zone to give a much greater yield strength ($940\,\text{MN}\,\text{m}^{-2}$) but a much lower fracture toughness ($39\,\text{MN}\,\text{m}^{-3/2}$).

Calculation of critical stress for fast fracture

The longitudinal stress σ in the wall of a cylindrical pressure vessel containing gas at pressure p is given by

$$\sigma = \frac{pr}{2t}$$

provided that the wall is thin ($t \ll r$). $p = 1.83\,\text{MN}\,\text{m}^{-2}$, $r = 1067\,\text{mm}$ and $t = 7\,\text{mm}$, so $\sigma = 140\,\text{MN}\,\text{m}^{-2}$. The fast fracture equation is

$$Y\sigma\sqrt{\pi a} = K_c$$

Because the crack penetrates a long way into the wall of the vessel, it is necessary to take into account the correction factor Y (see Chapter 13). Figure 15.3 shows that $Y = 1.92$ for our crack. The critical stress for fast fracture is given by

$$\sigma = \frac{K_c}{Y\sqrt{\pi a}} = \frac{39}{1.92\sqrt{\pi .0.0025}} = 229\ MN\ \text{m}^{-2}$$

The critical stress is 64 percent greater than the longitudinal stress. However, the change in section from a cylinder to a sphere produces something akin to a stress concentration; when this is taken into account the failure is accurately predicted.

Conclusions and recommendations

This case study provides a good example of the consequences of having an inadequate fracture toughness. However, even if the heat-affected zone had a high toughness, the crack would have continued to grow through the wall of the tank by stress-corrosion cracking until fast fracture occurred. The critical crack

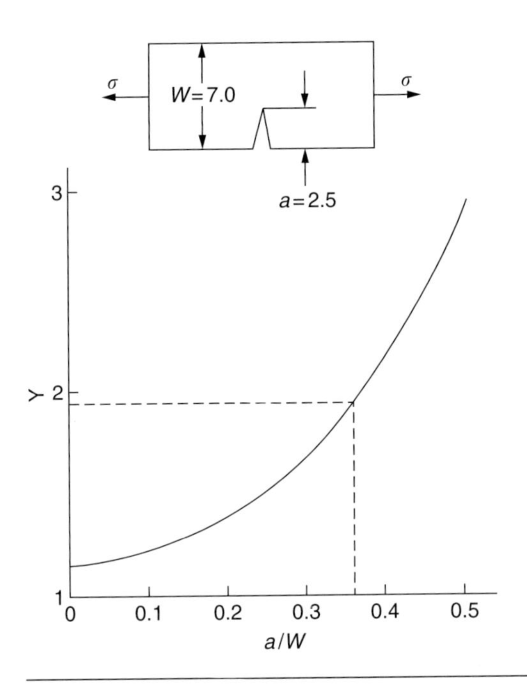

Figure 15.3 *Y* value for the crack. Dimensions in mm.

size would have been greater, but failure would have occurred eventually. The only way of avoiding failures of this type is to prevent stress corrosion cracking in the first place.

15.3 Case study 2: explosion of a perspex pressure window during hydrostatic testing

Figure 15.4 shows the general arrangement drawing for an experimental rig, which is designed for studying the propagation of buckling in externally pressurized tubes. A long open-ended tubular specimen is placed on the horizontal axis of the rig with the ends emerging through pressure seals. The rig is partially filled with water and the space above the water is filled with nitrogen. The nitrogen is pressurized until buckling propagates along the length of the specimen. The volumes of water and nitrogen in the rig can be adjusted to give stable buckling propagation. Halfway along the rig is a flanged perspex connector which allows the propagation of the buckle to be observed directly using a high-speed camera. Unfortunately, this perspex window exploded during the first hydrostatic test at a pressure of only one-half of the specified hydrostatic test pressure.

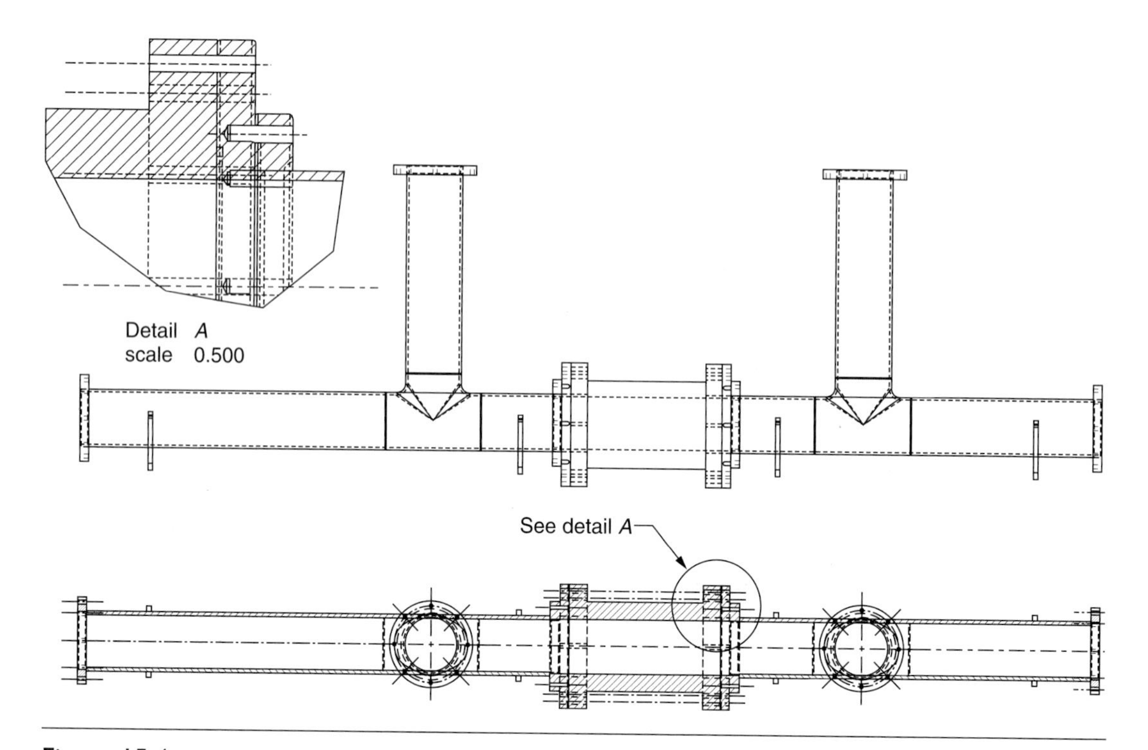

Figure 15.4 General arrangement drawing of the experimental rig.

Design data

Relevant design data for the perspex connector are given as follows.

Internal diameter of cylindrical portion $2B = 154\,\text{mm}$.
External diameter of cylindrical portion $2A = 255\,\text{mm}$.
Forming process: casting.
Tensile strength, $\sigma_f \approx 62\,\text{MN}\,\text{m}^{-2}$ (minimum), $77\,\text{MN}\,\text{m}^{-2}$ (average).
Fracture toughness $K_c \approx 0.8$ to $1.75\,\text{MN}\,\text{m}^{-3/2}$.
Working pressure $= 7\,\text{MN}\,\text{m}^{-2}$ gauge.
Hydraulic test pressure $= 8.6\,\text{MN}\,\text{m}^{-2}$ gauge.
Failure pressure $= 4.8\,\text{MN}\,\text{m}^{-2}$ gauge.

Failure analysis

Figure 15.5 is a photograph of the perspex connector taken after the explosion. Detailed visual inspection of the fracture surface indicated that the fracture

Figure 15.5 Photograph of the perspex connector taken after the explosion.

initiated as a hoop stress tensile failure in the cylindrical portion and subsequently propagated towards each flange.

The hoop stress σ in the cylindrical portion can be calculated from the standard result for thick-walled tubes

$$\sigma = p\left(\frac{B}{r}\right)^2 \frac{A^2 + r^2}{A^2 - B^2},$$

where p is the internal gauge pressure and r is the radius at which the stress is calculated. The hoop stress is a maximum at the bore of the tube, with $r = B$ and

$$\sigma = p\frac{A^2 + B^2}{A^2 - B^2} = 2.13p.$$

The hoop stress is a minimum at the external surface of the tube, with $r = A$ and

$$\sigma = p\frac{2B^2}{A^2 - B^2} = 1.13p.$$

We can see that the most probable site for failure initiation is the bore of the tube: the hoop stress here is calculated to be 10 MN m^{-2} at the failure pressure.

This is only *one-sixth* of the minimum tensile strength. Using the fast fracture equation

$$K_c = \sigma\sqrt{\pi a}$$

with a fracture toughness of 1 $\mathrm{MN\,m^{-3/2}}$ and a hoop stress of 10 $\mathrm{MN\,m^{-2}}$ gives a critical defect size a of 3.2 mm at the failure pressure. At the operating pressure the critical defect size would only be 1.5 mm. A defect of this size would be difficult to find under production conditions in such a large volume. In addition, it would be easy to introduce longitudinal scratches in the bore of the connector during routine handling and use.

Conclusions

The most probable explanation for the failure is that a critical defect was present in the wall of the perspex connector. The connector was a standard item manufactured for flow visualization in pressurized systems. The designers had clearly used a stress-based rather than a fracture-mechanics based approach with entirely predictable consequences.

15.4 Case study 3: cracking of a polyurethane foam jacket on a liquid methane tank

Figure 15.6 is a schematic half section through a tank used for storing liquid methane at atmospheric pressure. Because methane boils at −162°C, the tank is made from an aluminum alloy in order to avoid any risk of brittle failure. Even so, it is considered necessary to have a second line of defense should the tank spring a leak. This is achieved by placing the tank into a leakproof mild-steel jacket, and inserting a layer of thermal insulation into the space between the two. The jacket is thereby protected from the cooling effect of the methane, and the temperature of the steel is kept above the ductile-to-brittle transition temperature. But what happens if the tank springs a leak? If the insulation is porous (like fiberglass matting) then the liquid methane will flow through the insulation to the wall of the jacket and will boil off. As a result the jacket will cool down to −162°C and may fail by brittle fracture. To avoid this possibility the inner wall of the jacket is coated with a layer of closed-cell foam made from rigid polyurethane (PUR). The tank is then lowered into the jacket and the assembly gap is filled with fiberglass matting. The theory is that if the tank leaks, the flow of methane will be arrested by the closed-cell structure of the PUR and the jacket will be protected. This system has been used in ships designed for the bulk carriage of liquid methane. In such applications, the mild-steel jacket surrounding the tank is the single hull of the ship itself.

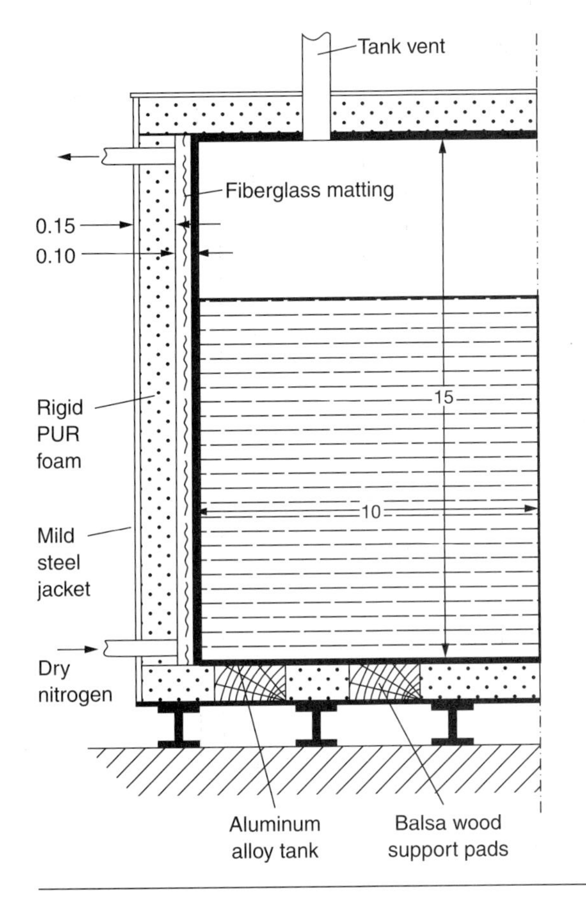

Figure 15.6 Schematic half-section through a typical liquid methane storage tank using closed-cell polyurethane foam for thermal insulation/secondary containment. Typical dimensions in m.

It is vital in order to protect the structural integrity of the hull that the PUR provides effective containment. However, incidents have occurred where the PUR has cracked, compromising the integrity of the hull. In one reported instance, three brand-new bulk carriers had to be written off as a result of multiple cracking of the PUR foam layer.

Thermal stresses in the foam

Under normal operating conditions, the temperature of the foam decreases linearly with distance through the layer, as shown in Figure 15.7. The foam wants to contract as it gets cold, but is prevented from doing so by the rigid

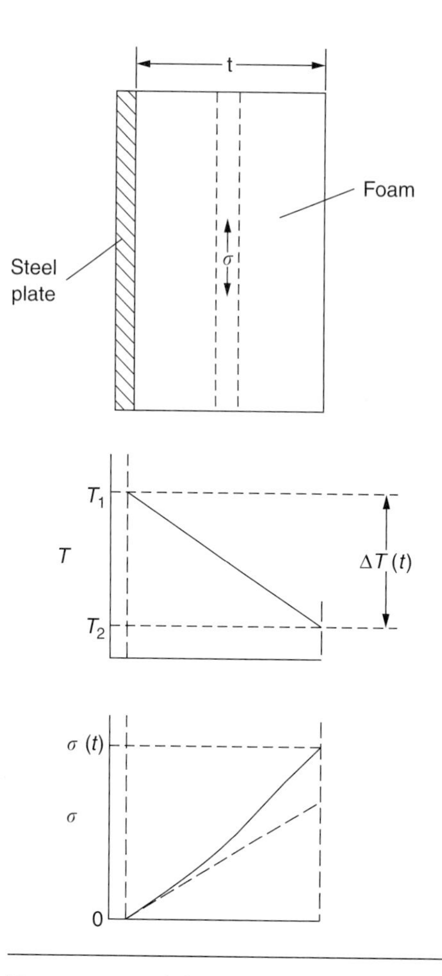

Figure 15.7 Temperature and thermal stress in the foam layer.

steel wall of the jacket to which it is stuck. The temperature differential ΔT generates a biaxial tensile stress σ in the plane of the layer which is given by

$$\sigma = \frac{\alpha \Delta T E}{(1 - \upsilon)}$$

α, E and υ are respectively the coefficient of thermal expansion, Young's modulus and Poisson's ratio of the foam. Figure 15.7 also shows the variation of stress with distance through the layer. It is nonlinear because Young's modulus for polyurethane (PU) increases as the temperature decreases. The thermal stress is a maximum at the inner surface of the PUR layer.

We saw in Chapter 6 that polymers behave as elastic-brittle solids provided they are colder than the glass transition temperature, T_G. For PUR,

$T_G \approx 100°C$, or 373 K. Presumably the foam failed by brittle cracking when the maximum thermal stress reached the fracture stress of the foam. In order to check this hypothesis we first list the relevant data for a typical PUR foam as used in cryogenic applications.

Cell size $\approx$ 0.5 mm.
Thermal expansion coefficient $\alpha \approx 10^{-4}°C^{-1}$.
Poisson's ratio $v \approx 0.3$.
Young's modulus $E \approx 34\,MN\,m^{-2}$ at $-100°C$.
Fracture stress $\sigma_f \approx 1.4\,MN\,m^{-2}$ at $-100°C$.
Fracture toughness $K_c \approx 0.05\,MN\,m^{-3/2}$ at $-100°C$.

Referring to Figure 15.7, reasonable estimates of temperature are $T_1 \approx 0°C$, $T_2 \approx -100°C$ and $\Delta T(t) \approx 100°C$. For a temperature differential of this magnitude, $\sigma \approx 0.5\,MN\,m^{-2}$. However, this value for the maximum stress is substantially less than the fracture stress of approximately $1.4\,MN\,m^{-2}$ expected at $-100°C$. This means that, on a simple basis, the foam should not have fractured in service. In order to understand why the foam did in fact break it is necessary to analyze the problem using fracture mechanics.

Fracture mechanics analysis

It is a simple matter to estimate the size of a critical defect in the foam layer by using the fast fracture equation

$$K_c = \sigma\sqrt{\pi a}$$

Since $K_c \approx 0.05\,MN\,m^{-3/2}$ and $\sigma(t) \approx 0.5\,MN\,m^{-2}$, then $a \approx 3$ mm. This is a small defect and is only of the order of six times the typical cell size (0.5 mm).

It is important to note that the method of applying the foam to the interior of the hull was conducive to the presence of defects. The foam was delivered through a hand-held nozzle as a gas-blown froth and was applied in layers about 25 mm thick. In practice this method is likely to result in considerable variation in cell size and the introduction of defects.

The most reasonable explanation for the formation of cracks in the foam layer is therefore that small defects (introduced into the foam at the time of application) became critical as the temperature of the foam fell during charging with liquid methane. The most probable location for crack initiation would have been the inner face of the foam where the thermal stress was a maximum (and where surface defects of size a as opposed to buried defects of size $2a$ would have been critical). In the incidents which were reported, cracks had propagated through the full thickness of the layer.

Conclusions

Considerable financial loss has resulted from the cracking of PUR foam in liquid methane insulation/containment systems. This has been caused by a combination of the very low fracture toughness of PUR foam at low temperature and the likelihood that the method of applying the foam will introduce defects exceeding the critical size for crack propagation. An elementary knowledge of fracture mechanics would have suggested that the use of plain (un-reinforced) PUR in the present situation was fundamentally unsound.

15.5 Case study 4: collapse of a wooden balcony railing

A student residence built approximately 30 years ago has eight balconies which are accessed from public areas in the building. Each balcony is flanked by a pair of brick walls between which is secured a balustrade fabricated from wooden components. Unfortunately, while a number of people were gathered on a second-floor balcony, the balustrade gave way. As a consequence, five people fell to the ground below and were injured. The failure was caused by the propagation of pre-existing cracks in the balustrade while people were leaning against the top rail.

Design and construction of the balustrade

Figures 15.8 and 15.9 are reproductions of scale drawings of the balustrade prepared during the failure investigation. The balustrades had originally been dimensioned in inches and these units are retained here (1 inch = 25.40 mm). The numbers on the drawings are identified as follows: 1, end post (31 inch long); 2, baluster; 3, top rail (105 inch long); 4, top stringer; 5, bottom stringer; 6, kick board; 7, rebate groove; 8, blind mortise; 9, tenon; 10, 2-inch nail (head at outer face of baluster); 11, 2-inch nail (head at inner face of baluster); 12, galvanized steel fixing bracket cemented into brick wall; 13, steel coach screw ($\frac{1}{4}$ inch diameter).

Figure 15.10 shows that the application of a horizontal force to the top rail will tend to initiate a longitudinal crack in the end post at the root of the tenon. If the force is maintained, such a crack would run along the grain of the end post from top to bottom. This in turn would open-out the mortises for the stringers. The tenon at the end of each stringer is retained by a single nail (and possibly glue) and should detach relatively easily. Full separation of the balustrade from the retained part of the end post would occur by torsional fracture of the joint between end post and kick board.

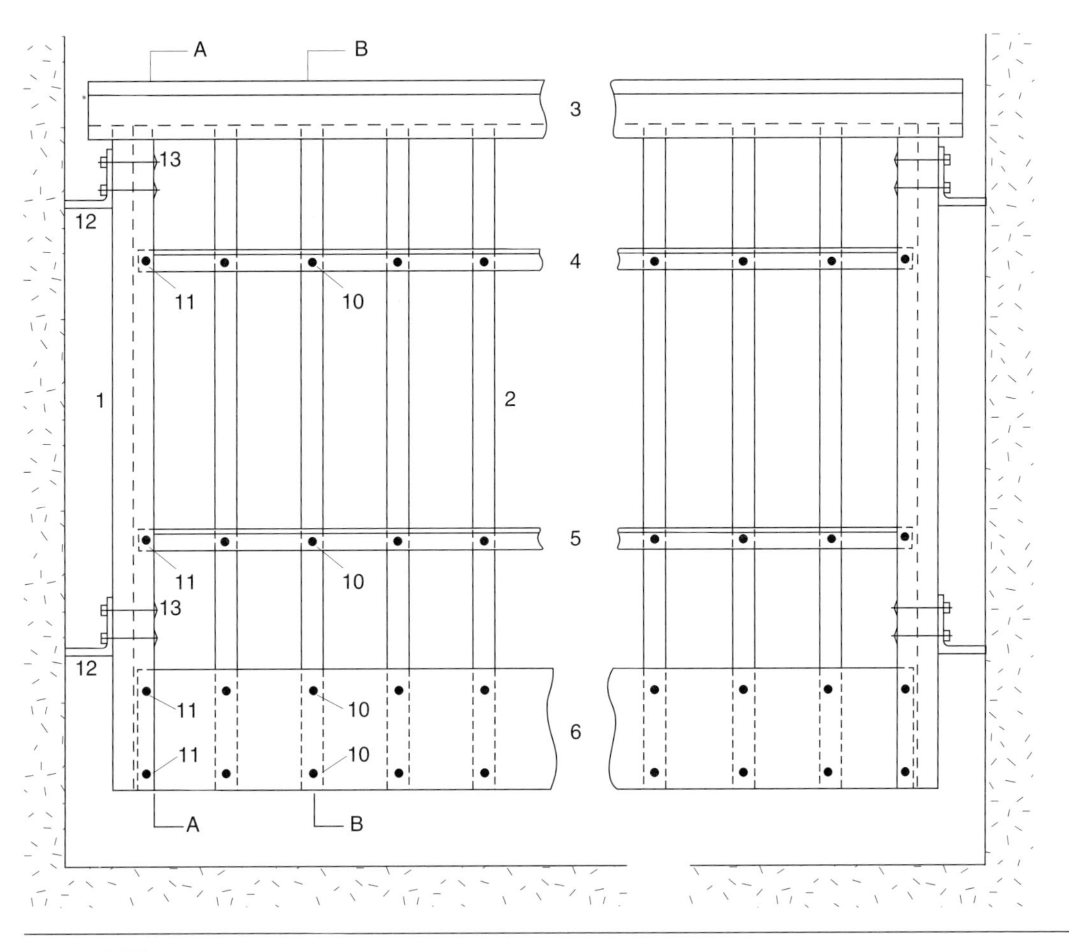

Figure 15.8 Elevation of the balustrade as seen by an observer standing on the balcony and looking outwards. The top and bottom of the balustrade are respectively 39 and 6 inches above the balcony floor. There is a total of 24 balusters along the length of the balustrade (not all shown).

Because the end post provides the connection between the top rail and the balcony walls, it is a safety-critical component. However, for aesthetic reasons, the post was given an L-shaped cross section which halved the width of the tenon and also halved the load required to initiate and propagate a longitudinal crack.

Details of the failure

In spite of these design errors, it appears unlikely that cracks would initiate from the root of the tenon under normal service loads. However, the presence

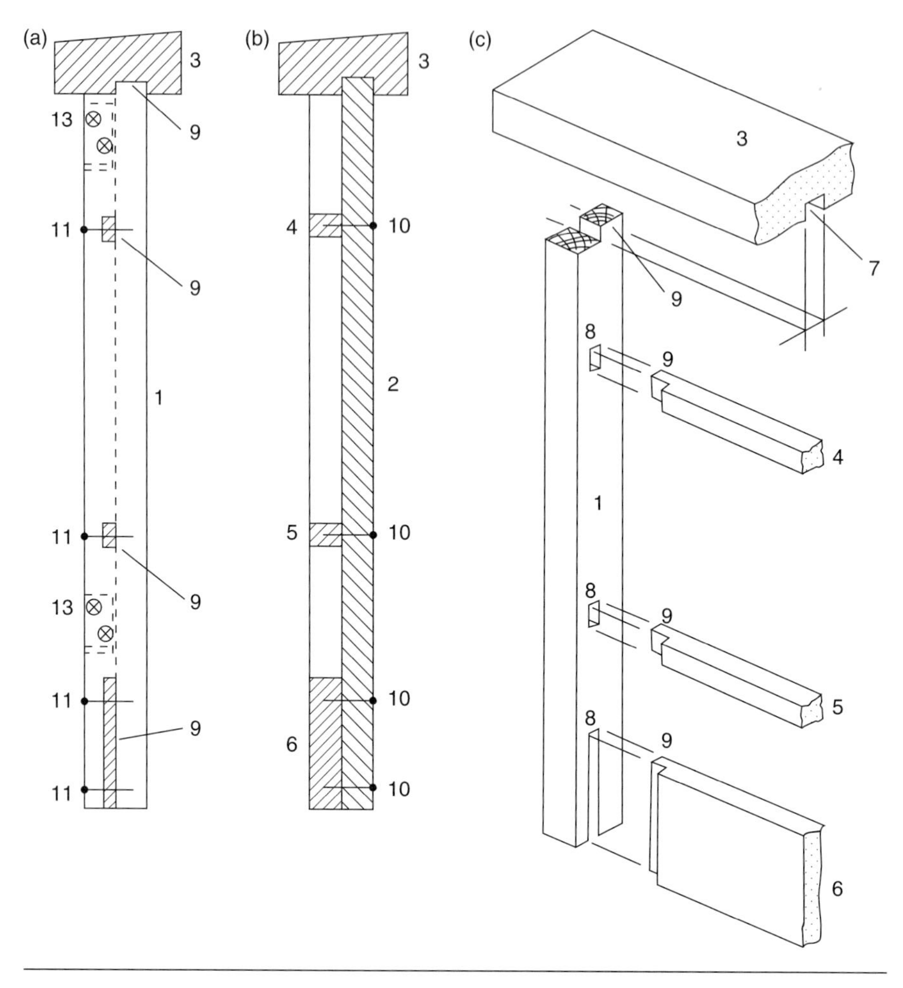

Figure 15.9 (a) Cross section drawn on A-A (see Figure 15.8) looking at the left-hand end of the balustrade. (b) Cross section drawn on B-B (see Figure 15.8) looking at the left-hand end of the balustrade. (c) Exploded view showing joints.

of wood stain on the fracture surfaces of the failed end posts indicates that the balustrade collapsed because of pre-existing cracks. These cracks were at least 8 inches long at the time of the most recent timber treatment. Cracks presumably developed over a period of time through exposure to the weather. This hypothesis was confirmed by a high incidence of cracks in the other balustrades. Figure 15.11 shows an 18-inch crack observed in one of the other balustrades. The crack path is identical to that observed in the failed end posts, as shown in Figure 15.12.

In order to add weight to the failure scenario, fracture mechanics tests were carried out on new end posts. The measured crack propagation forces for

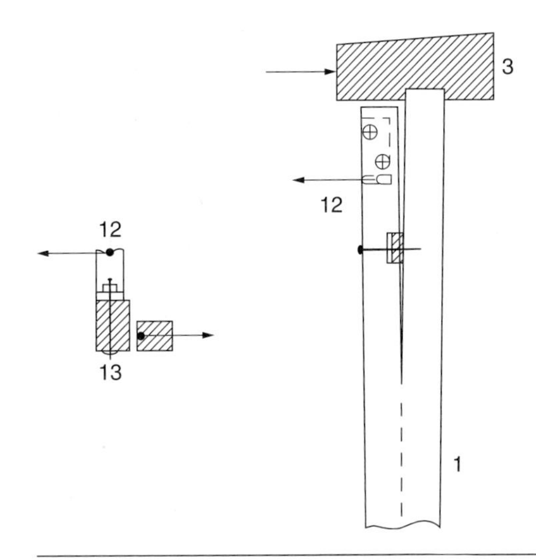

Figure 15.10 A horizontal outward force applied to the top rail tends to split the end post.

representative crack lengths were comparable to the combined lateral force delivered by five people leaning on the top rail.

Conclusions

The collapse of the railing was caused by the propagation of pre-existing cracks in the end posts, which are safety-critical components. The collapse was caused by a lack of awareness by both architect and maintenance personnel of the poor along-grain fracture toughness of wood, the potential sources of crack-opening forces in the design, the propensity of externally exposed timber to develop cracks and the significance of cracks for the integrity of the structure.

Examples

15.1 A cylindrical pressure vessel is rolled up from flat steel plate 10 mm thick, and the edges are welded together to produce a longitudinal welded seam. Calculate the stress intensity factor K at a hoop stress of $100\,\mathrm{MN\,m^{-2}}$ for two possible types of weld defect:

(a) a crack running along the whole length of the weld, and extending from the inner surface of the vessel to a depth of 5 mm;
(b) a through-thickness crack of length $2a = 40\,\mathrm{mm}$.

Figure 15.11 An 18-inch crack in another balustrade.

Which is the more dangerous crack, and why?
[Hint: make sure you use the correct values for Y].

Answers

(a) 37.6 MN m$^{-3/2}$; (b) 25.1 MN m$^{-3/2}$; crack (a), because it has the larger K, and also because it does not reveal itself by creating a leak path-crack (b) does create a leak path.

15.2 With reference to Case study 2, why was perspex an unsuitable material for the pressure window?

15.3 With reference to Case study 3, how would you redesign the insulation system to prevent cracking in a new installation?

Figure 15.12 The crack path in a failed end post.

15.4 With reference to Case study 4, how would you modify the design to prevent collapse in a new installation?

15.5 Give examples from your own experience of the many *advantages* of wood having a low fracture toughness along the grain.

Chapter 16

Probabilistic fracture of brittle materials

Chapter contents

16.1 Introduction

We saw in Chapter 13 that the fracture toughness K_c of ceramics and rigid polymers was very low compared to that of metals and composites. Cement and ice have the lowest K_c, at $\approx 0.2\,\text{MN}\,\text{m}^{-3/2}$. Traditional manufactured ceramics (brick, pottery, china, porcelain) and natural stone or rock are better, at ≈ 0.5–$2\,\text{MN}\,\text{m}^{-3/2}$. But even the "high-tech" engineering ceramics like silicon nitride, alumina, and silicon carbide only reach $\approx 4\,\text{MN}\,\text{m}^{-3/2}$. Rigid polymers (thermoplastics below the glass transition temperature, or heavily cross-linked thermosets such as epoxy) have $K_c \approx 0.5$–$4\,\text{MN}\,\text{m}^{-3/2}$. This low fracture toughness makes ceramics and rigid polymers very vulnerable to the presence of crack-like defects. They are *defect-sensitive* materials, liable to fail by fast fracture from defects well before they can yield.

Unfortunately, many of these materials tend to be awash with small-scale crack-like defects. Manufactured ceramics almost always contain cracks and flaws left by the production process (e.g. the voids left between the particles from which the ceramic was fabricated). The defects are even worse in cement, because of the rather crude nature of the mixing and setting process. Ice usually contains small bubbles of trapped air (and, in the case of sea ice, concentrated brine). Moulded polymer components often contain small voids. And all but the hardest brittle materials accumulate additional defects when they are handled or exposed to an abrasive environment.

The design strength of a brittle material in tension is therefore determined by its low fracture toughness in combination with the lengths of the crack-like defects it contains. If the longest microcrack in a given sample has length $2a_m$, then the tensile strength is simply given by

$$\sigma_{TS} = \frac{K_c}{\sqrt{\pi a_m}} \tag{16.1}$$

from the fast fracture equation. Some engineering ceramics have tensile strengths about half that of steel—around $200\,\text{MN}\,\text{m}^{-2}$. Taking a typical fracture toughness of $2\,\text{MN}\,\text{m}^{-3/2}$, the largest microcrack has a size of $60\,\mu\text{m}$, which is of the same order as the original particle size. Pottery, brick, and stone generally have tensile strengths which are much lower than this—around $20\,\text{MN}\,\text{m}^{-2}$. This indicates defects of the order of 2 mm for a typical fracture toughness of $1\,\text{MN}\,\text{m}^{-3/2}$. The tensile strength of cement and concrete is even lower—$2\,\text{MN}\,\text{m}^{-2}$ in large sections—implying the presence of at least one crack 6 mm or more in length for a fracture toughness of $0.2\,\text{MN}\,\text{m}^{-3/2}$.

Measuring the tensile strength of brittle materials

In Chapter 8 we saw that properties such as the yield and tensile strengths could be measured easily using a long cylindrical specimen loaded in simple tension. But it is difficult to perform tensile tests on brittle materials—the specimens tend to break where they are gripped by the testing machine. This is because the local contact stresses exceed the fracture strength, and premature failure occurs at the grips. It is much easier to measure the force required to break a beam in bending (Figure 16.1). The maximum tensile stress in the surface of the beam when it breaks is called the modulus of rupture, σ_r; for an elastic beam it is related to the maximum moment in the beam, M_r, by

$$\sigma_r = \frac{6M_r}{bd^2} \tag{16.2}$$

where d is the depth and b the width of the beam. You might think that σ_r should be equal to the tensile strength σ_{TS}. But it is actually a little larger (typically 1.7 times larger), for reasons which we will get to when we discuss the statistics of strength later in this chapter.

The third test shown in Figure 16.1 is the compression test. For metals (or any plastic solid) the strength measured in compression is the same as that measured in tension. But for brittle solids this is not so; for these, the compressive strength is roughly 15 times larger, with

$$\sigma_c \approx 15\sigma_{TS} \tag{16.3}$$

The reason for this is explained by Figure 16.2. Cracks in compression propagate *stably*, and twist out of their original orientation to propagate *parallel to the compression axis*. Fracture is not caused by the rapid unstable propagation of one crack, but the slow extension of many cracks to form a crushed zone. It is not the size of the largest crack (a_m) that counts, but that of the average $\bar{a}$. The compressive strength is still given by a formula 1 - like equation (16.1), with

$$\sigma_c = C\frac{K_c}{\sqrt{\pi\bar{a}}} \tag{16.4}$$

but the constant C is about 15, instead of 1.

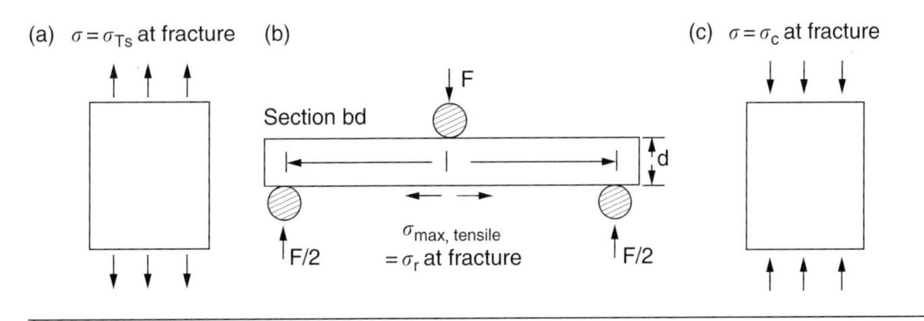

Figure 16.1 Tests which measure the fracture strengths of brittle materials (a) The tensile test measures the tensile strength, σ_{TS}. (b) The bend test measures the modulus of rupture, σ_r, typically $1.73 \times \sigma_{TS}$. (c) The compression test measures the crushing strength, σ_c, typically $15 \times \sigma_{TS}$.

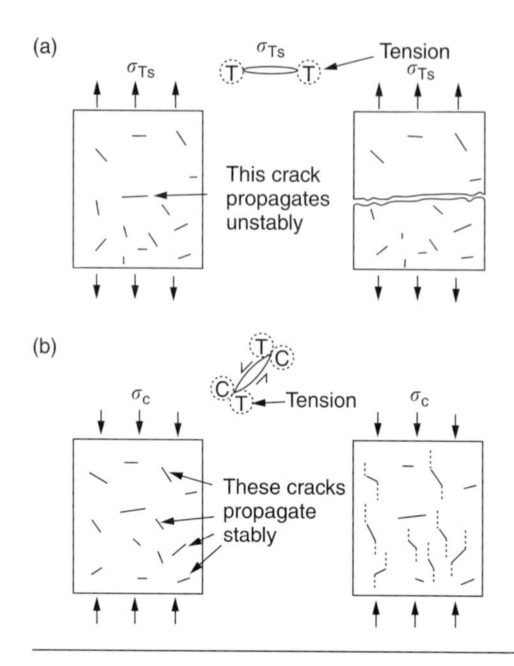

Figure 16.2 (a) In tension the largest flaw propagates unstably. (b) In compression, many flaws propagate stably to give general crushing.

16.2 The statistics of strength and the Weibull distribution

The chalk with which I write on the blackboard when I teach is a brittle solid. Some sticks of chalk are weaker than others. On average, I find (to my slight irritation), that about 3 out of 10 sticks break as soon as I start to write with them; the other 7 survive. The failure probability, P_f, for this chalk, loaded in bending under my (standard) writing load is 3/10, that is,

$$P_f = 0.3$$

When you write on a blackboard with chalk, you are not unduly incovenienced if 3 pieces in 10 break while you are using it; but if 1 in 2 broke, you might seek an alternative supplier. So the failure probability, P_f, of 0.3 is acceptable (just barely). If the component were a ceramic cutting tool, a failure probability of 1 in 100 ($P_f = 10^{-2}$) might be acceptable, because a tool is easily replaced. But if it were the window of a vacuum system, the failure of which can cause injury, one might aim for a P_f of 10^{-6}; and for a ceramic protective tile on the re-entry vehicle of a space shuttle, when one failure in any one of 10,000 tiles could be fatal, you might calculate that a P_f of 10^{-8} was needed.

When using a brittle solid under load, it is not possible to be certain that a component will not fail. But if an acceptable risk (the failure probability) can be assigned to the function filled by the component, then it *is* possible to design so that this acceptable risk is met. This chapter explains why ceramics have this dispersion of strength; and shows how to design components so they have a given probability of survival. The method is an interesting one, with application beyond ceramics to the malfunctioning of any complex system in which the breakdown of one component will cause the entire system to fail.

Chalk is a porous ceramic. It has a fracture toughness of $0.9\,\text{MN}\,\text{m}^{-3/2}$ and, being poorly consolidated, is full of cracks and angular holes. The average tensile strength of a piece of chalk is $15\,\text{MN}\,\text{m}^{-2}$, implying an average length for the longest crack of about 1 mm (calculated from equation (16.1)). But the chalk itself contains a distribution of crack lengths. Two nominally identical pieces of chalk can have tensile strengths that differ greatly—by a factor of 3 or more. This is because one was cut so that, by chance, all the cracks in it are small, whereas the other was cut so that it includes one of the longer flaws of the distribution. Figure 16.3 illustrates this: if the block of chalk is cut into pieces, piece A will be weaker than piece B because it contains a larger flaw. It is inherent in the strength of brittle materials that there will be a statistical variation in strength. There is no single "tensile strength"; but there is a certain, definable, *probability* that a given sample will have a given strength.

The distribution of crack lengths has other consequences. A large sample will fail at a lower stress than a small one, on average, because it is more likely that it will contain one of the larger flaws (Figure 16.3). So there is a *volume dependence* of the strength. For the same reason, a brittle rod is stronger in bending than in simple tension: in tension the entire sample carries the tensile stress, while in bending only a thin layer close to one surface (and thus a

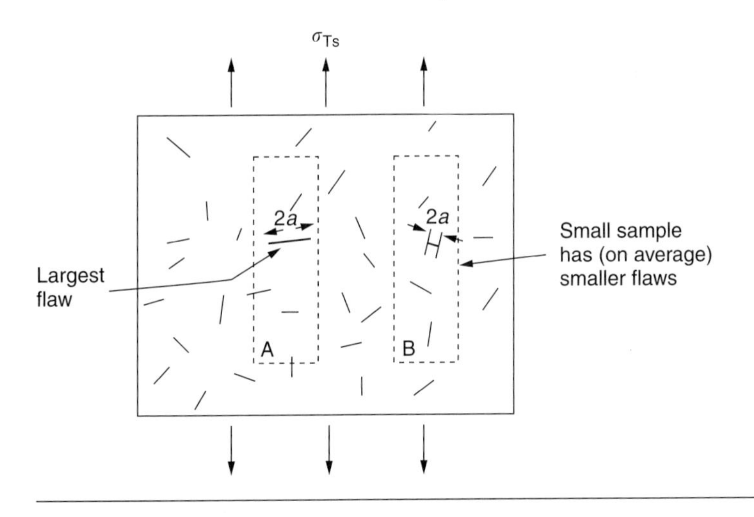

Figure 16.3 If small samples are cut from a large block of a brittle material, they will show a dispersion of strengths because of the dispersion of flaw sizes. The average strength of the small samples is greater than that of the large sample.

relatively smaller volume) carries the peak tensile stress (Figure 16.4). That is why the modulus of rupture is larger than the tensile strength.

The Swedish engineer, Weibull, invented the following way of handling the statistics of strength. He defined the *survival probability* $P_s(V_0)$ as the fraction of identical samples, each of volume V_0, which survive loading to a tensile stress σ. He then proposed that

$$P_s(V_0) = \exp\left\{-\left(\frac{\sigma}{\sigma_0}\right)^m\right\} \tag{16.5}$$

where σ_0 and m are constants. This equation is plotted in Figure 16.5(a). When $\sigma = 0$ all the samples survive, of course, and $P_s(V_0) = 1$. As σ increases, more and more samples fail, and $P_s(V_0)$ decreases. Large stresses cause virtually all the samples to break, so $P_s(V_0) \rightarrow 0$ and $\sigma \rightarrow \infty$.

If we set $\sigma = \sigma_0$ in equation (16.5) we find that $P_s(V_0) = 1/e$ $(= 0.37)$. So σ_0 is simply the tensile stress that allows 37 percent of the samples to survive. The constant m tells us how rapidly the strength falls as we approach σ_0 (see Figure 16.5(b)). It is called the *Weibull modulus*. The lower m, the greater the *variability* of strength. For ordinary chalk, m is about 5, and the variability is great. Brick, pottery, and cement are like this too. The engineering ceramics (e.g. SiC, Al_2O_3, and Si_3N_4) have values of m of about 10, for these, the strength varies rather less. Even steel shows some variation in strength, but it is small: it can be described by a Weibull modulus of about 100. Figure 16.5(b) shows that, for $m \approx 100$, a material can be treated as having a single, well-defined failure stress.

σ_0 and m can be found from experiment. A batch of samples, each of volume V_0, is tested at a stress σ_1 and the fraction $P_{s1}(V_0)$ that survive is determined. Another batch is tested at σ_2 and so on. The points are then plotted on Figure 16.5(b). It is easy to determine σ_0 from the graph, but m has to be found by curve-fitting.

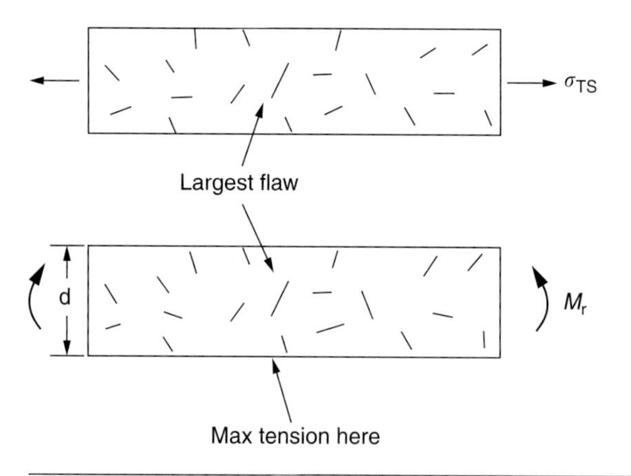

Figure 16.4 Brittle materials appear to be stronger in bending than in tension because the largest flaw may not be near the surface.

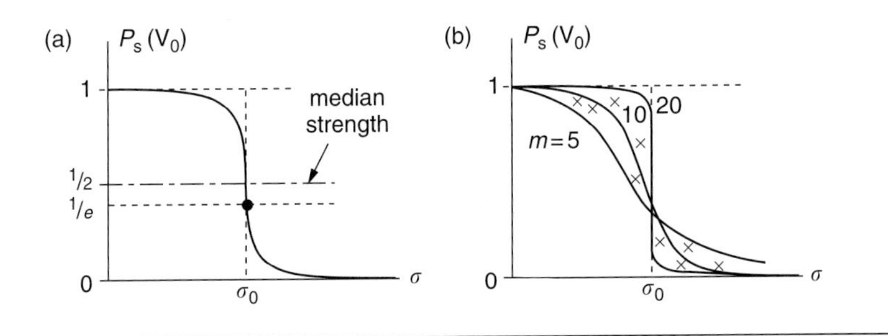

Figure 16.5 (a) The Weibull distribution function. (b) When the modulus, m, changes, the survival probability changes as shown.

So much for the stress dependence of P_s. But what of its volume dependence? We have already seen that the probability of one sample surviving a stress σ is $P_s(V_0)$. The probability that a batch of n such samples all survive the stress is just $\{P_s(V_0)\}^n$. If these n samples were stuck together to give a single sample of volume $V = nV_0$ then its survival probability would still be $\{P_s(V_0)\}^n$. So

$$P_s(V) = \{P_s(V_0)\}^n = \{P_s(V_0)\}^{V/V_0}$$

This is equivalent to

$$\ln P_s(V) = \frac{V}{V_0} \ln P_s(V_0)$$

or

$$P_s(V) = \exp\left\{\frac{V}{V_0} \ln P_s(V_0)\right\} \tag{16.6}$$

The Weibull distribution (equation (16.5)) can be rewritten as

$$\ln P_s(V_0) = -\left(\frac{\sigma}{\sigma_0}\right)^m$$

If we insert this result into equation (16.6) we get

$$P_s(V) = \exp\left\{-\frac{V}{V_0}\left(\frac{\sigma}{\sigma_0}\right)^m\right\} \tag{16.7}$$

or

$$\ln P_s(V) = -\frac{V}{V_0}\left(\frac{\sigma}{\sigma_0}\right)^m$$

This, then, is our final design equation. It shows how the survival probability depends on both the stress σ and the volume V of the component. In using it, the first step is to fix on an acceptable failure probability, P_f: 0.3 for chalk,

10^{-2} for the cutting tool, 10^{-6} for the vacuum-chamber window. The survival probability is then given by $P_s = 1 - P_f$. (It is useful to remember that, for small P_f, $\ln P_s = \ln(1 - P_f) \approx -P_f$.) We can then substitute suitable values of σ_0, m, and V/V_0 into the equation to calculate the design stress.

Note that equation (16.7) assumes that the component is subjected to a *uniform* tensile stress σ. In many applications, σ is not constant, but instead varies with position throughout the component. Then, we rewrite equation (16.7) as

$$P_s(V) = \exp\left\{-\frac{1}{\sigma_0^m V_0}\int_V \sigma^m \, dV\right\} \tag{16.8}$$

16.3 Case study: cracking of a polyurethane foam jacket on a liquid methane tank

In Chapter 15, we looked at the case of a polyurethane foam jacket on a liquid methane tank, which had cracked because of thermal stress. The cracking was explained by using the fast fracture equation

$$K_c = \sigma\sqrt{\pi a}$$

to estimate the size of the critical defect a, and then showing that a was comparable to the size of the defects introduced into the foam during the application process. This explanation is certainly adequate for solving a failure problem, but for the purposes of predictive design a more accurate assessment of the conditions for fracture is desirable. This is a good example of how the Weibull equation can be used to solve a design problem.

Referring to Chapter 15, we recall that the maximum stress σ_{max} in the foam jacket was 0.5 MN m^{-2}, and the thickness t of the jacket was 0.15 m. Tensile tests on laboratory samples of volume $V_0 \approx 5 \times 10^{-5}$ m^3 gave a median fracture strength σ_f of 1.4 MN m^{-2} and a Weibull modulus m of 8.

The situation in the tensile tests is described by equation (16.5), with appropriate substitutions to give

$$0.5 = \exp\left\{-\left(\frac{\sigma_f}{\sigma_0}\right)^m\right\} \tag{16.9}$$

The tensile stress in the jacket varies with distance through the thickness, being zero where the jacket is stuck to the steel plating and being a maximum at the free surface. The Weibull equation for a varying stress field is equation (16.8). As shown in Figure 16.6, we define a volume element $dV = a^2\,dx$, and approximate the stress at the position x as

$$\sigma = \sigma_{max}\left(\frac{x}{t}\right)$$

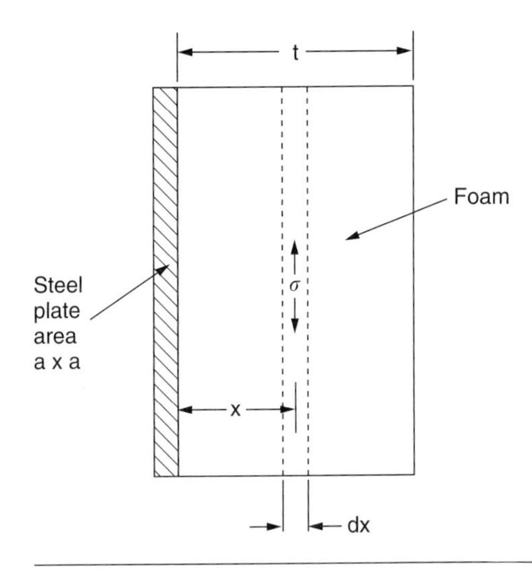

Figure 16.6 Volume element with constant tensile stress in the foam jacket.

Equation (16.8) can then be written as

$$0.5 = \exp\left\{-\frac{1}{\sigma_0^m V_0}\int_0^t \sigma_{\max}^m \left(\frac{x}{t}\right)^m a^2\, dx\right\} \tag{16.10}$$

Combining equations (16.9) and (16.10) we get

$$\left(\frac{\sigma_f}{\sigma_0}\right)^m = \left(\frac{\sigma_{\max}}{\sigma_0}\right)^m \frac{a^2}{V_0 t^m}\int_0^t x^m\, dx = \frac{1}{(m+1)}\left(\frac{\sigma_{\max}}{\sigma_0}\right)^m \frac{a^2 t}{V_0}$$

This gives

$$a^2 t = (m+1)\left(\frac{\sigma_f}{\sigma_{\max}}\right)^m V_0$$

and

$$a^2 t = 9\left(\frac{1.4}{0.5}\right)^8 5\times 10^{-5}\,\mathrm{m}^3 = 1.70\,\mathrm{m}^3 = 3.4\times 3.4\times 0.15\,\mathrm{m}^3$$

This means that there is a 50 percent chance that cracks will occur in the foam jacket at intervals of approximately 3.4 m. The probability of survival could be increased by reducing the thickness of the foam layer (increasing the thickness of the fiberglass matting so as to keep the overall thickness of the insulation constant). This would reduce the thermal stresses and also the volume V. However, the safest way to avoid cracking would be to reinforce the foam with fibers of glass or Kevlar. These will bridge any cracks, and prevent crack propagation towards the outer parts of the foam. Possibilities include mixing short chopped fibers into the foam while it is being sprayed on, or incorporating layers of fiber mesh as the layers of foam are built up.

Examples

16.1 Why are ceramics usually much stronger in compression than in tension? Al_2O_3 has a fracture toughness K_c of about $3\,MN\,m^{-3/2}$. A batch of Al_2O_3 samples is found to contain surface flaws about $30\,\mu m$ deep. Estimate (a) the tensile strength and (b) the compressive strength of the samples.

Answers

(a) $309\,MN\,m^{-2}$, (b) $4635\,MN\,m^{-2}$.

16.2 Modulus-of-rupture tests are carried out using the arrangement shown in Figure 16.1. The specimens break at a load F of about 330 N. Find the modulus of rupture, given that $l = 50$ mm, and that $b = d = 5$ mm.

Answer

$198\,MN\,m^{-2}$.

16.3 In order to test the strength of a ceramic, cylindrical specimens of length 25 mm and diameter 5 mm are put into axial tension. The tensile stress σ which causes 50 percent of the specimens to break is $120\,MN\,m^{-2}$. Cylindrical ceramic components of length 50 mm and diameter 11 mm are required to withstand an axial tensile stress σ^1 with a survival probability of 99 percent. Given that $m = 5$, use equation (16.7) to determine σ^1.

Answer

$32.6\,MN\,m^{-2}$.

16.4 Modulus-of-rupture tests were carried out on samples of silicon carbide using the three-point bend test geometry shown in Figure 16.1. The samples were 100 mm long and had a 10 mm by 10 mm square cross section. The median value of the modulus of rupture was $400\,MN\,m^{-2}$. Tensile tests were also carried out using samples of identical material and dimensions, but loaded in tension along their lengths. The median value of the tensile strength was only $230\,MN\,m^{-2}$. Account in a qualitative way for the difference between the two measures of strength.

Answer

In the tensile test, the whole volume of the sample is subjected to a tensile stress of 230 MN m^{-2}. In the bend test, only the lower half of the sample is subjected to a tensile stress. Furthermore, the average value of this tensile stress is considerably less than the peak value of 400 MN m^{-2} (which is only reached at the underside of the sample beneath the central loading point). The probability of finding a fracture-initiating defect in the small volume subjected to the highest stresses is small.

16.5 Modulus-of-rupture tests were done on samples of ceramic with dimensions $l = 100$ mm, $b = d = 10$ mm. The median value of σ_r (i.e. σ_r for $P_s = 0.5$) was 300 MN m^{-2}. The ceramic is to be used for components with dimensions $l = 50$ mm, $b = d = 5$ mm loaded in simple tension along their length. Calculate the tensile stress σ that will give a probability of failure, P_f, of 10^{-6}. Assume that $m = 10$. Note that, for $m = 10$, $\sigma_{TS} = \sigma_r/1.73$.

Answer

55.7 MN m^{-2}.

16.6 The diagram is a schematic of a stalactite, a cone-shaped mineral deposit hanging downwards from the roof of a cave. Its failure due to self weight loading is to be modelled using Weibull statistics. The geometry of the stalactite is idealized as a cone of length L and semiangle α. The cone angle is assumed small so that the base radius equals αL. The stalactite density is ρ.

(a) Show that the variation of tensile stress σ with height x is given by $\sigma = \frac{1}{3}\rho g x$. You may assume that the volume of a cone is given by $(\pi/3) \times (\text{base radius})^2 \times$ height.

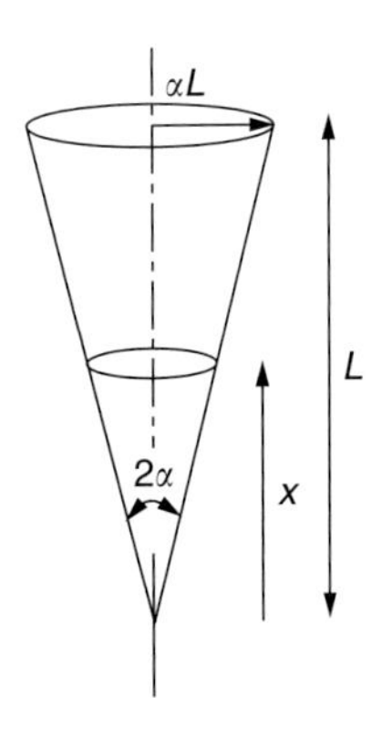

(b) Use equation (16.8) for Weibull statistics with a varying stress to show that the probability of survival $P_s(L)$ for a stalactite of length L is given by

$$P_s(L) = \exp\left\{ -\left(\frac{\rho g}{3\sigma_0}\right)^m \frac{\pi \alpha^2 L^{m+3}}{(m+3)V_0} \right\}$$

Explain why there is a dependency on cone angle α, even though the stress variation up the stalactite is independent of α.

16.7 (a) Cylindrical samples of radius r and length l taken from stalactites are tested in uniform tension. If the stress level which gives a 50 percent failure rate for specimens is equal to σ_t, derive an expression for the length of stalactite which can be expected to have a 50 percent survival rate.

(b) Comment on possible practical difficulties with the sample tests.

Answer

$$L = \left\{ \frac{lr^2(m+3)}{\alpha^2} \left(\frac{3\sigma_t}{\rho g}\right)^m \right\}^{1/(m+3)}$$

Part E

Fatigue failure

Chapter 17

Fatigue failure

Chapter contents

17.1 Introduction

In Chapters 13 and 14, we examined the conditions under which a crack was stable, and would not grow, and the condition

$$K = K_c$$

under which it would propagate catastrophically by fast fracture. If we know the maximum size of crack in the structure we can then choose a working load at which fast fracture will not occur.

But cracks can form, and grow slowly, at loads lower than this, if either the stress is cycled or if the environment surrounding the structure is corrosive (most are). The first process of slow crack growth—*fatigue*—is the subject of this chapter. The second—*corrosion*—is discussed later, in Chapters 26 and 27.

More formally: *if a component or structure is subjected to repeated stress cycles*, like the loading on the connecting rod of a petrol engine or on the wings of an aircraft—*it may fail at stresses well below the tensile strength*, σ_{TS}, *and often below the yield strength*, σ_y, of the material. The processes leading to this failure are termed "fatigue". When the clip of your pen breaks, when the pedals fall off your bicycle, when the handle of the refrigerator comes away in your hand, it is usually fatigue which is responsible.

We distinguish three categories of fatigue (Table 17.1).

17.2 Fatigue behavior of uncracked components

Tests are carried out by cycling the material either in tension (compression) or in rotating bending (Figure 17.1). The stress, in general, varies sinusoidally with time, though modern servo-hydraulic testing machines allow complete control of the wave shape.

We define:

$$\Delta\sigma = \sigma_{max} - \sigma_{min}; \quad \sigma_m = \frac{\sigma_{max} + \sigma_{min}}{2}; \quad \sigma_a = \frac{\sigma_{max} - \sigma_{min}}{2}$$

where N = number of fatigue cycles and N_f = number of cycles to failure. We will consider fatigue under zero mean stress ($\sigma_m = 0$) first, and later generalize the results to non-zero mean stress.

For *high-cycle fatigue of uncracked components*, where neither σ_{max} nor $|\sigma_{min}|$ are above the yield stress, it is found empirically that the experimental data can be fitted to an equation of form

$$\Delta\sigma N_f^a = C_1 \qquad (17.1)$$

This relationship is called *Basquin's Law*. Here, a is a constant (between $\frac{1}{8}$ and $\frac{1}{15}$ for most materials) and C_1 is a constant also.

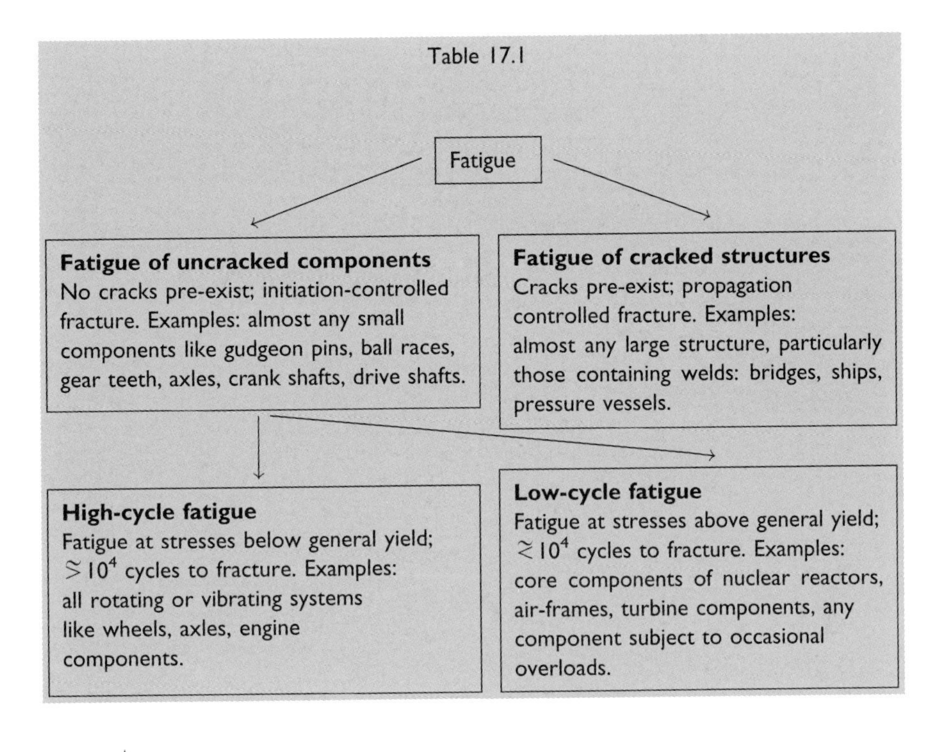

Table 17.1

Fatigue

Fatigue of uncracked components
No cracks pre-exist; initiation-controlled fracture. Examples: almost any small components like gudgeon pins, ball races, gear teeth, axles, crank shafts, drive shafts.

Fatigue of cracked structures
Cracks pre-exist; propagation controlled fracture. Examples: almost any large structure, particularly those containing welds: bridges, ships, pressure vessels.

High-cycle fatigue
Fatigue at stresses below general yield; $\gtrsim 10^4$ cycles to fracture. Examples: all rotating or vibrating systems like wheels, axles, engine components.

Low-cycle fatigue
Fatigue at stresses above general yield; $\lesssim 10^4$ cycles to fracture. Examples: core components of nuclear reactors, air-frames, turbine components, any component subject to occasional overloads.

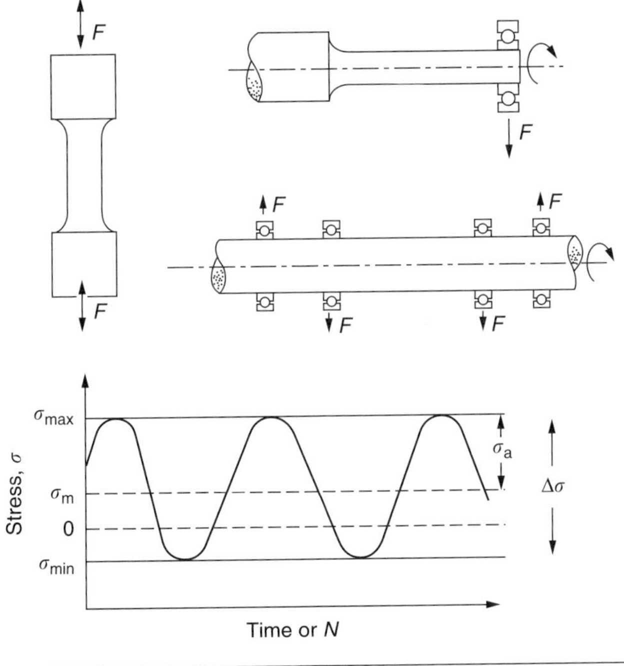

Figure 17.1 Fatigue testing.

For *low-cycle fatigue of uncracked components* where σ_{max} or $|\sigma_{min}|$ are above σ_y, Basquin's Law no longer holds, as Figure 17.2 shows. But a linear plot is obtained if the plastic strain range $\Delta\epsilon^{pl}$, defined in Figure 17.3, is plotted, on logarithmic scales, against the cycles to failure, N_f (Figure 17.4). This result is known as the *Coffin–Manson Law:*

$$\Delta\epsilon^{pl} N_f^b = C_2 \tag{17.2}$$

where b (0.5–0.6) and C_2 are constants.

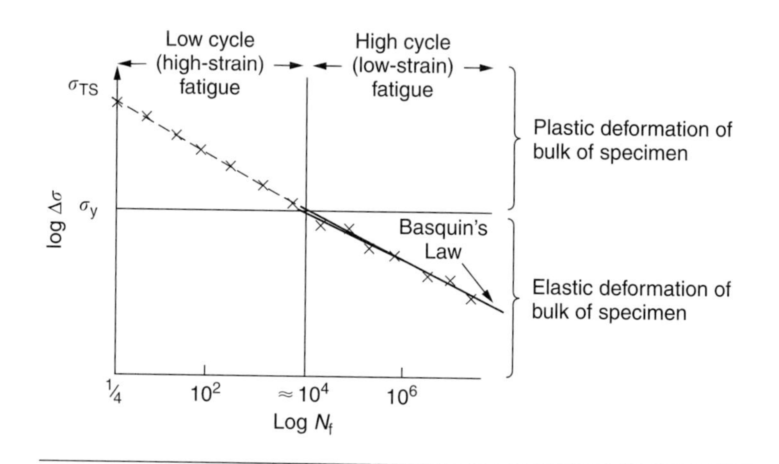

Figure 17.2 Initiation-controlled high-cycle fatigue — Basquin's Law.

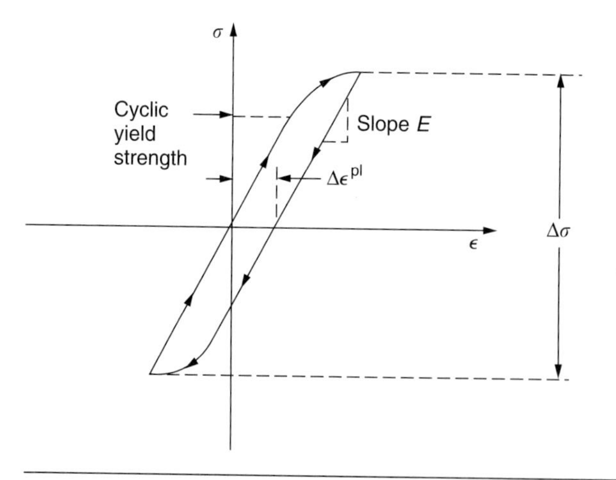

Figure 17.3 The plastic strain range, $\Delta\epsilon^{pl}$, in low-cycle fatigue.

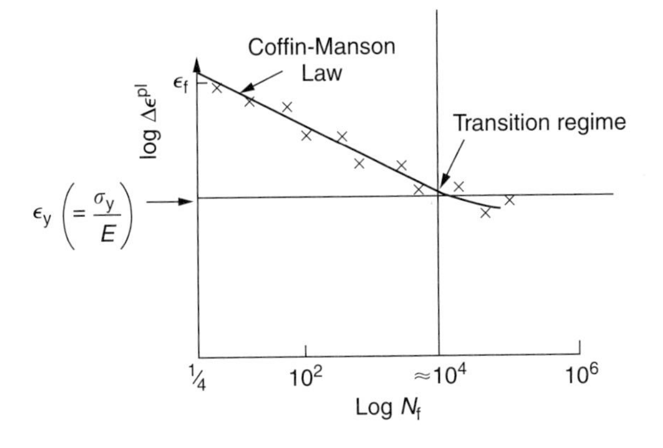

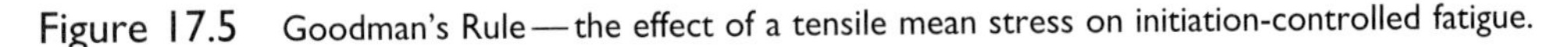

Figure 17.4 Initiation-controlled low-cycle fatigue — the Coffin–Manson Law.

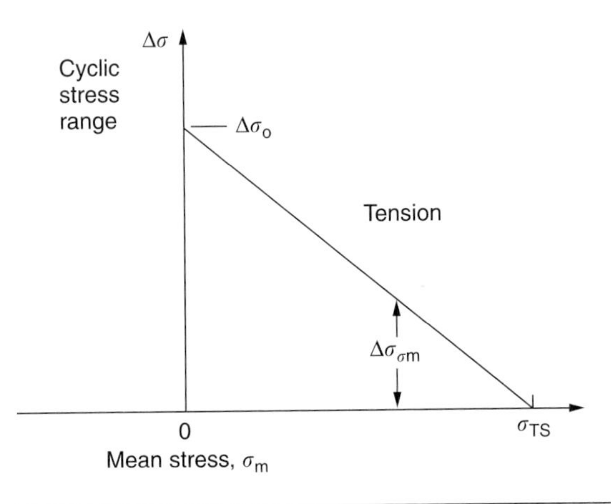

Figure 17.5 Goodman's Rule — the effect of a tensile mean stress on initiation-controlled fatigue.

These two laws (given data for a, b, C_1, and C_2) adequately describe the fatigue failure of unnotched components, cycled at constant amplitude about a mean stress of zero. What do we do when $\Delta\sigma$, and σ_m, vary?

When material is subjected to a mean tensile stress (i.e. $\sigma_m > 0$) the stress range must be decreased to preserve the same N_f according to *Goodman's Rule* (Figure 17.5)

$$\Delta\sigma_{\sigma m} = \Delta\sigma_0\left(1 - \frac{\sigma_m}{\sigma_{TS}}\right) \tag{17.3}$$

(Here $\Delta\sigma_0$ is the cyclic stress range for failure in N_f cycles under zero mean stress, and $\Delta\sigma_{\sigma m}$ is the same thing for a mean stress of σ_m.) Goodman's Rule is empirical, and does not always work—then tests simulating service conditions must be carried out, and the results used for the final design. But preliminary designs are usually based on this rule.

When, in addition, $\Delta\sigma$ varies during the lifetime of a component, the approach adopted is to sum the damage according to *Miner's Rule of cumulative damage* (Figure 17.6):

$$\sum_i \frac{N_i}{N_{fi}} = 1 \tag{17.4}$$

Here N_{fi} is the number of cycles to fracture under the stress cycle in region i, and N_i/N_{fi} is the fraction of the lifetime used up after N_i cycles in that region. Failure occurs when the sum of the fractions is unity (equation (17.4)). This rule, too, is an empirical one. It is widely used in design against fatigue failure; but if the component is a critical one, Miner's Rule should be checked by tests simulating service conditions.

17.3 Fatigue behavior of cracked components

Large structures—particularly welded structures like bridges, ships, oil rigs, nuclear pressure vessels—always contain cracks. All we can be sure of is that the initial length of these cracks is less than a given length—the length we can reasonably detect when we check or examine the structure. To assess the safe life of the structure we need to know how long (for how many cycles) the structure can last before one of these cracks grows to a length at which it propagates catastrophically.

Data on fatigue crack propagation are gathered by cyclically loading specimens containing a sharp crack like that shown in Figure 17.7. We define

$$\Delta K = K_{max} - K_{min} = \Delta\sigma\sqrt{\pi a}$$

The cyclic stress intensity ΔK increases with time (at constant load) because the crack grows in tension. It is found that the crack growth per cycle, da/dN, increases with ΔK in the way shown in Figure 17.8.

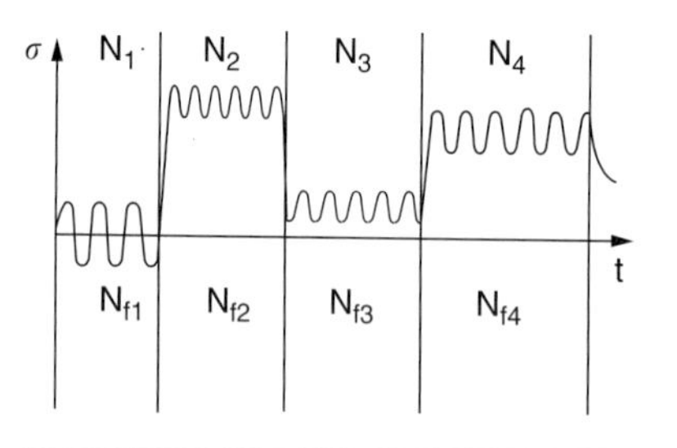

Figure 17.6 Summing damage due to initiation-controlled fatigue.

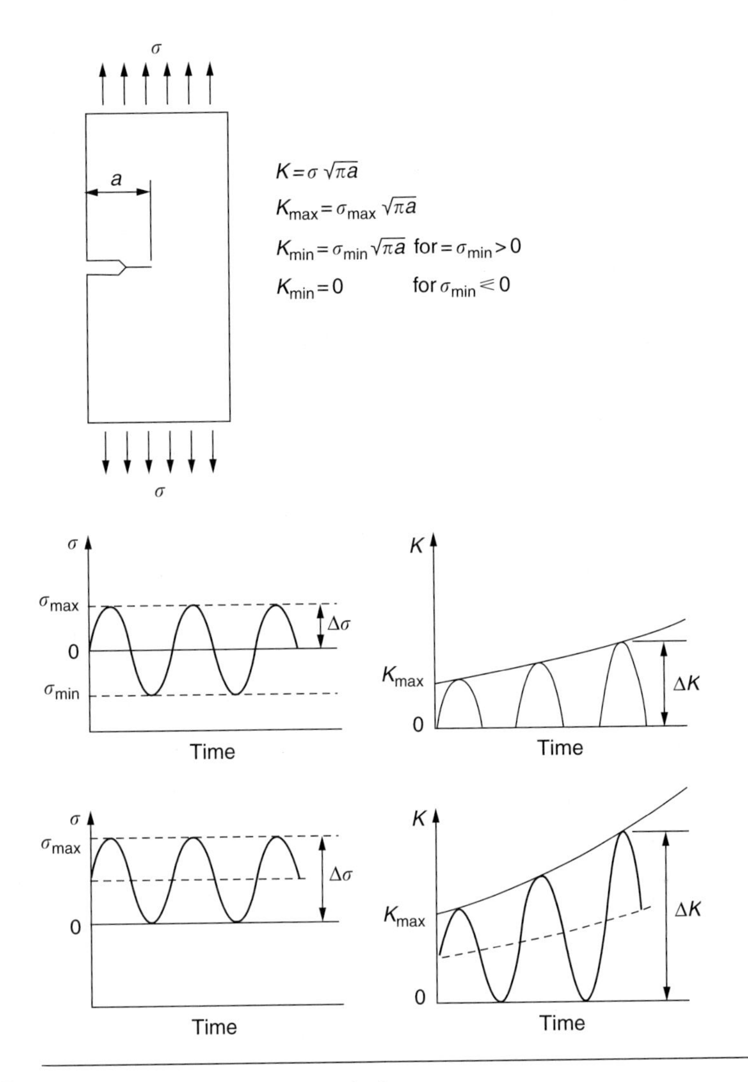

Figure 17.7 Fatigue-crack growth in pre-cracked components.

In the steady-state régime, the crack growth rate is described by

$$\frac{da}{dN} = A\Delta K^m \tag{17.5}$$

where A and m are material constants. Obviously, if a_0 (the initial crack length) is given, and the final crack length (a_f) at which the crack becomes unstable and runs rapidly is known or can be calculated, then the safe number

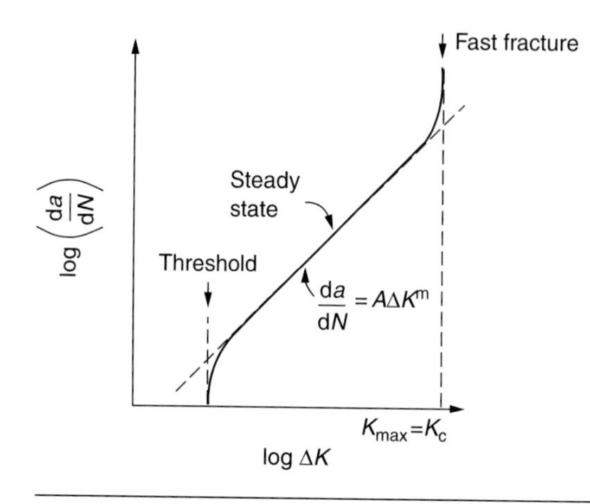

Figure 17.8 Fatigue crack-growth rates for pre-cracked material.

of cycles can be estimated by integrating the equation

$$N_f = \int_0^{N_f} dN = \int_{a_0}^{a_f} \frac{da}{A(\Delta K)^m} \tag{17.6}$$

remembering that $\Delta K = \Delta\sigma\sqrt{\pi a}$. Case Study 3 of Chapter 19 gives a worked example of this method of estimating fatigue life.

17.4 Fatigue mechanisms

Cracks grow in the way shown in Figure 17.9. In a pure metal or polymer (left-hand diagram), the tensile stress produces a plastic zone (Chapter 14) which makes the crack tip stretch open by the amount δ, creating new surface there. As the stress is removed the crack closes and the new surface folds forward extending the crack (roughly, by δ). On the next cycle the same thing happens again, and the crack inches forward, roughly at $da/dN \approx \delta$. Note that the crack cannot grow when the stress is compressive because the crack faces come into contact and carry the load (crack closure).

We mentioned in Chapter 14 that real engineering alloys always have little inclusions in them. Then (right-hand diagram of Figure 17.9), within the plastic zone, holes form and link with each other, and with the crack tip. The crack now advances a little faster than before, aided by the holes.

In *pre-cracked structures* these processes determine the fatigue life. In uncracked components subject to *low-cycle fatigue*, the general plasticity quickly roughens the surface, and a crack forms there, propagating first along a slip path ("Stage 1" crack) and then, by the mechanism we have described, normal to the tensile axis (Figure 17.10).

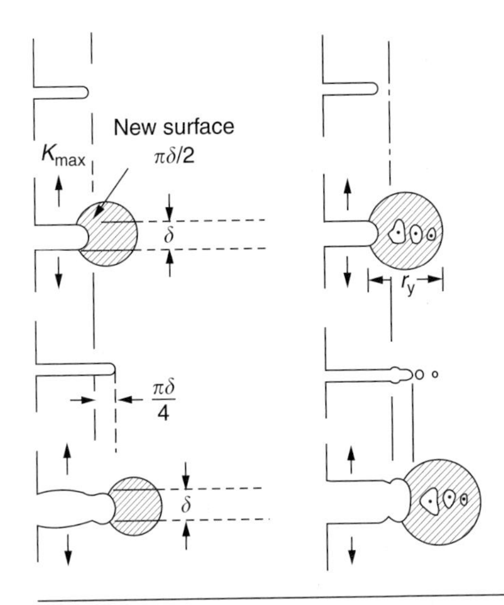

Figure 17.9 How fatigue cracks grow.

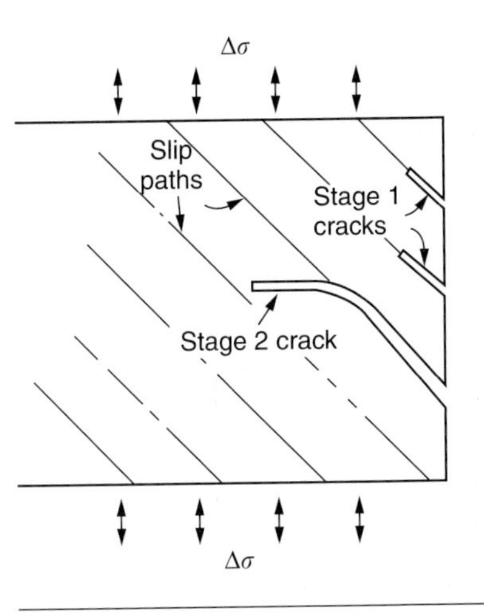

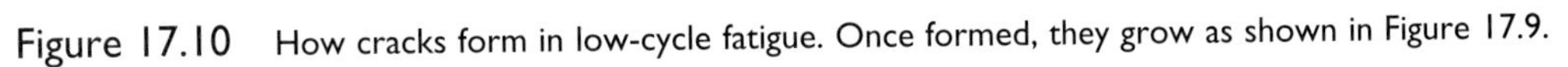

Figure 17.10 How cracks form in low-cycle fatigue. Once formed, they grow as shown in Figure 17.9.

High-cycle fatigue is different. When the stress is below general yield, almost all of the life is taken up in initiating a crack. Although there is no *general* plasticity, there is *local* plasticity wherever a notch or scratch or change of section concentrates stress. A crack ultimately initiates in the zone of one of these stress concentrations (Figure 17.11) and propagates, slowly at first, and then faster, until the component fails. For this reason, sudden changes of section or scratches are very dangerous in high-cycle fatigue, often reducing the fatigue life by orders of magnitude.

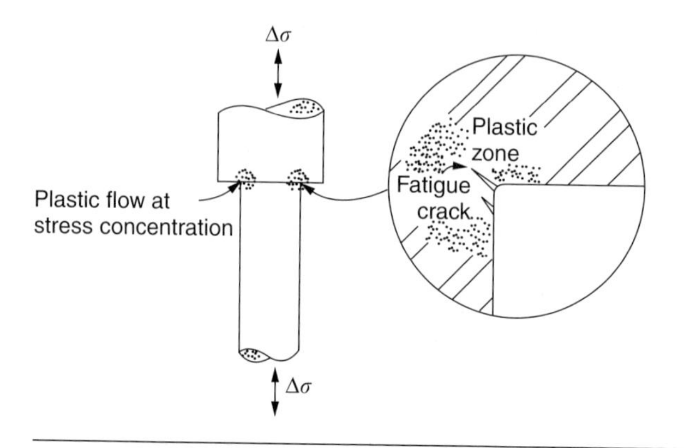

Figure 17.11 How cracks form in high-cycle fatigue.

Examples

17.1 A component is made of a steel for which $K_c = 54\,\text{MN m}^{-3/2}$. Non-destructive testing by ultrasonic methods shows that the component contains cracks of up to $2a = 0.2$ mm in length. Laboratory tests show that the crack-growth rate under cyclic loading is given by

$$\frac{da}{dN} = A(\Delta K)^4$$

where $A = 4 \times 10^{-13}\,(\text{MN m}^{-2})^{-4}\,\text{m}^{-1}$. The component is subjected to an alternating stress of range

$$\Delta\sigma = 180\,\text{MN m}^{-2}$$

about a mean tensile stress of $\Delta\sigma/2$. Given that $\Delta K = \Delta\sigma\sqrt{\pi a}$, calculate the number of cycles to failure.

Answer

2.4×10^6 cycles.

17.2 An aluminum alloy for an airframe component was tested in the laboratory under an applied stress which varied sinusoidally with time about a mean stress of zero. The alloy failed under a stress range, $\Delta\sigma$, of $280\,\text{MN m}^{-2}$ after 10^5 cycles; under a range of $200\,\text{MN m}^{-2}$, the alloy failed after 10^7 cycles. Assuming that the fatigue behaviour of the alloy can be represented by

$$\Delta\sigma(N_f)^a = C$$

where a and C are material constants, find the number of cycles to failure, N_f, for a component subjected to a stress range of $150\,\text{MN m}^{-2}$.

Answer

(a) 5.2×10^8 cycles.

17.3 When a fast-breeder reactor is shut down quickly, the temperature of the surface of a number of components drops from 600°C to 400°C in less than a second. These components are made of a stainless steel, and have a thick section, the bulk of which remains at the higher temperature for several seconds. The low-cycle fatigue life of the steel is described by

$$N_f^{1/2} \Delta\epsilon^{pl} = 0.2$$

where N_f is the number of cycles to failure and $\Delta\epsilon^{pl}$ is the plastic strain range. Estimate the number of fast shut-downs the reactor can sustain before serious cracking or failure will occur. (The thermal expansion coefficient of stainless steel is $1.2 \times 10^{-5}\,K^{-1}$; the yield strain at 400°C is 0.4×10^{-3}.)

Answer

10^4 shut-downs.

17.4 (a) A cylindrical steel pressure vessel of 7.5 m diameter and 40 mm wall thickness is to operate at a working pressure of $5.1\,MN\,m^{-2}$. The design assumes that small thumb-nail shaped flaws in the inside wall will gradually extend through the wall by fatigue.

If the fracture toughness of the steel is $200\,MN\,m^{-3/2}$ would you expect the vessel to fail in service by leaking (when the crack penetrates the thickness of the wall) or by fast fracture? Assume $K = \sigma\sqrt{\pi a}$, where a is the length of the edge-crack and σ is the hoop stress in the vessel.

(b) During service the growth of a flaw by fatigue is given by

$$\frac{da}{dN} = A(\Delta K)^4$$

where $A = 2.44 \times 10^{-14}\,(MN\,m^{-2})^{-4}\,m^{-1}$. Find the minimum pressure to which the vessel must be subjected in a proof test to guarantee against failure in service in less than 3000 loading cycles from zero to full load and back.

Answers

(a) vessel leaks; (b) $9.5\,MN\,m^{-2}$.

17.5 An electric iron for smoothing clothes was being used when there was a loud bang and a flash of flame from the electric flex next to the iron. An inspection showed that the failure had occurred at the point where the flex entered a polymer tube which projected about 70 mm from the body of the iron. There was a break in the live wire and the ends of the wire showed signs of fusion. The fuse in the electric plug was intact. Explain the failure. Relevant data are given as follows:

Rating of appliance	1.2 kW
Power supply	250 V, 50 Hz AC
Fuse rating of plug	13 A
Flex	3 conductors (live, neutral and earth) each rated 13 A
Individual conductor	23 strands of copper wire in a polymer sheath; each strand 0.18 mm in diameter
Age of iron	14 years
Estimated number of movements of iron	10^6

17.6 The photograph shows the fracture surfaces of two broken tools from a pneumatic drill. The circular fracture surface is 35 mm in diameter, and the "rectangular" fracture surface measures 24 mm × 39 mm. The shape of the fatigue crack at any given time is indicated by the "clam-shell marks", which are clearly visible on the fatigue part of the fracture surface. Indicate the following features on each fracture surface:

(a) the point where the fatigue crack initiated,
(b) the position of the crack just before the final fast fracture event.

The fatigue crack traverses much more of the cross section in the circular tool than in the rectangular tool. What does this tell you about the maximum stress in the fatigue cycle?

17.7 The photograph shows a small part of a fatigue fracture surface taken in the scanning electron microscope (SEM). The material is aluminum alloy. The direction of crack growth was from bottom to top. The position of the crack at the end of each stress cycle is indicated by parallel lines, or "striations", which are spaced a few micrometers apart. By taking measurements from the photograph, estimate the striation spacing. [Note: This fracture surface comes from the fatigue region of the broken roller arm in Case Study 1 of Chapter 19.]

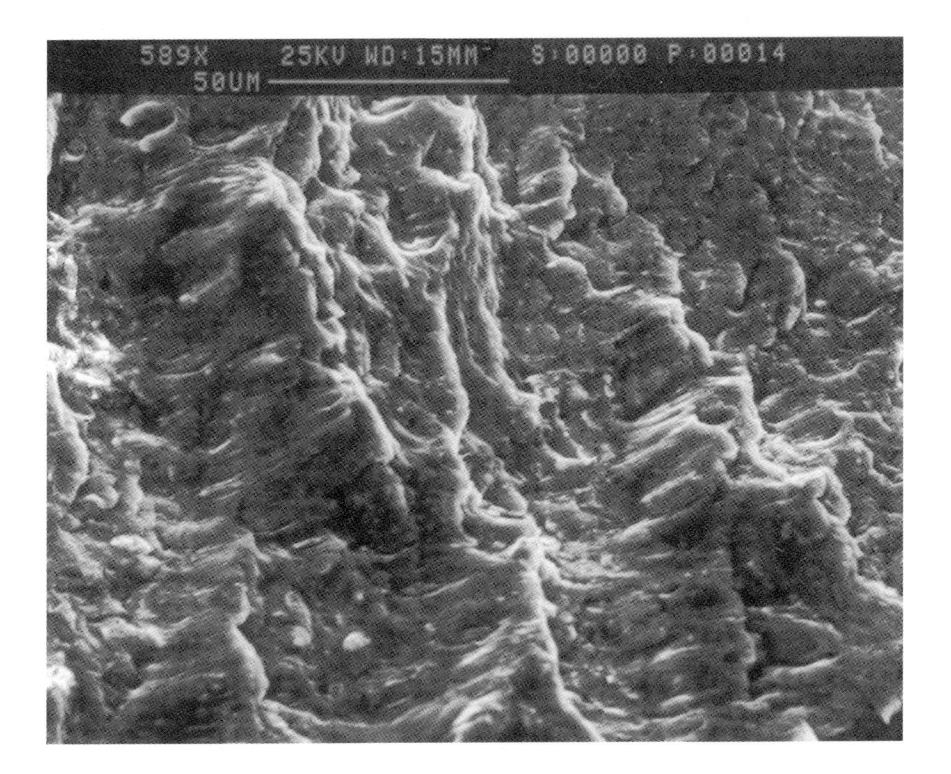

Chapter 18

Fatigue design

Chapter contents

18.1 Introduction

In this chapter we look at a number of aspects of fatigue which are relevant to designing structures or components against fatigue failure in service. We give data for the fatigue strengths of metals and alloys (useful for designing mechanical components) and for welded joints (important in large structures like bridges and oil rigs). We look at the problems of stress concentrations produced by abrupt changes in cross section (such as shoulders or holes). We see how fatigue strength can be improved by better surface finish, better component geometry and compressive residual surface stress. We look at how the preloading of bolts is essential in bolted connections such as car engine big-end bearings. Finally, we see how pressure vessels can be checked for fatigue cracking by hydrostatic testing.

18.2 Fatigue data for uncracked components

Table 18.1 gives high-cycle fatigue data for uncracked specimens tested about zero mean stress. The data are for specimens with an excellent surface finish tested in clean dry air. Fatigue strengths can be considerably less than these if the surface finish is poor, or if the environment is corrosive.

Obviously, if we have a real component with an excellent surface finish in clean dry air, then if it is to survive 10^8 cycles of constant-amplitude fatigue loading about zero mean stress, the stress amplitude $\Delta\sigma/2$ in service must be less than that given in Table 18.1 by a suitable safety factor. If the mean stress is not zero, then Goodman's rule must be used to calculate the fatigue strength under conditions of non-zero mean stress (equation (17.3)). In the absence of

Table 18.1 Approximate fatigue strengths of metals and alloys

Metal or alloy	Stress amplitude $\Delta\sigma_0/2$ for failure after 10^8 cycles (zero mean stress) ($\mathrm{MN\,m^{-2}}$)
Aluminum	35–60
Aluminum alloys	50–170
Copper	60–120
Copper alloys	100–300
Magnesium alloys	50–100
Nickel	230–340
Nickel alloys	230–620
Steels	170–500
Titanium	180–250
Titanium alloys	250–600

specific data, it is useful to know that $\Delta\sigma_0/2 = C\sigma_{TS}$. The value of the constant C is typically 0.3–0.5 depending on the material.

18.3 Stress concentrations

Any abrupt change in the cross section of a loaded component causes the local stress to increase above that of the background stress. The ratio of the maximum local stress to the background stress is called the stress concentration factor, or SCF for short. Figure 18.1 gives details of the SCF for two common changes in section—a hole in an axially loaded plate, and a shouldered shaft in bending.

The hole gives a SCF of 3. The SCF of the shaft is critically dependent on the ratio of the fillet radius r to the minor shaft diameter d—to minimize the

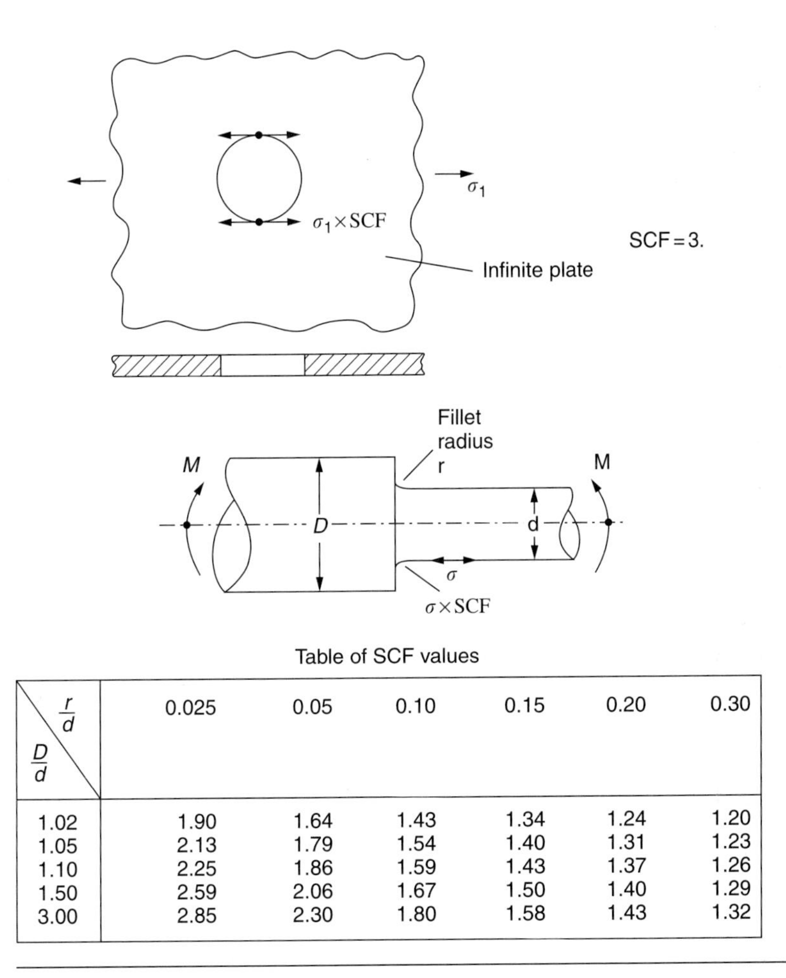

D/d \ r/d	0.025	0.05	0.10	0.15	0.20	0.30
1.02	1.90	1.64	1.43	1.34	1.24	1.20
1.05	2.13	1.79	1.54	1.40	1.31	1.23
1.10	2.25	1.86	1.59	1.43	1.37	1.26
1.50	2.59	2.06	1.67	1.50	1.40	1.29
3.00	2.85	2.30	1.80	1.58	1.43	1.32

Figure 18.1 Typical stress concentration factors (SCFs).

SCF, r/d should be maximized. Obviously, fatigue failure will occur preferentially at sites of local stress concentration. The Comet aircraft disasters were caused by fatigue cracks in the fuselage skin which initiated at rivet holes next to the windows (essentially a stress concentration within a stress concentration). If a component has a SCF then it is the maximum local stress which must be kept below the material fatigue strength, and not the background stress.

18.4 The notch sensitivity factor

Taking the shouldered shaft as an example, we can see that as the r/d ratio decreases towards zero (a sharp corner) the SCF should increase towards infinity. This implies that any component with a sharp corner, or notch, will always fail by fatigue no matter how low the background stress! Clearly, this is not correct, because there are many components with sharp corners which are used successfully in fatigue loading (although this is very bad practice).

In fatigue terminology, we define an effective stress concentration factor, SCF_{eff} such that $SCF_{eff} < SCF$. The two are related by the equation

$$SCF_{eff} = S(SCF - 1) + 1 \tag{18.1}$$

where S, the notch sensitivity factor, lies between 0 and 1. If the material is fully notch sensitive, $S = 1$ and $SCF_{eff} = SCF$. If the material is not notch sensitive, $S = 0$ and $SCF_{eff} = 1$.

Figure 18.2 shows that S increases with increasing σ_{TS} and fillet radius r. We would expect S to increase with σ_{TS}. As we saw in Chapter 14 for sharp cracks, material at the fillet radius can yield in response to the local stress, and this will limit the maximum local stress to the yield stress. In general, increasing σ_{TS} increases σ_y. In turn, this increases the maximum local stress which can be sustained before yielding limits the stress, and helps keep $SCF_{eff} \approx SCF$.

The decrease in S with decreasing r has a different origin. As Figure 18.2 shows, as r tends towards zero, S also tends towards zero for all values of σ_{TS}. This is because a sharp notch produces a small process zone (the zone in which the fatigue crack initiates) and this makes it harder for a fatigue crack to grow. We saw in Chapter 16 that the tensile strength of a brittle component increases as the volume decreases. This size effect also applies to the formation of fatigue cracks—the smaller the process zone, the larger the fatigue strength of the component.

The notch sensitivity curves in Figure 18.2 have an interesting implication for designing components with small fillet radii. One would think that increasing the tensile strength (and hence the fatigue strength) of the material would increase the fatigue strength of the component. However, this is largely offset by the increase in notch sensitivity, which increases the value of the effective SCF by equation (18.1). Fortunately, as we shall see later, there are other ways of increasing the fatigue strength of notched components.

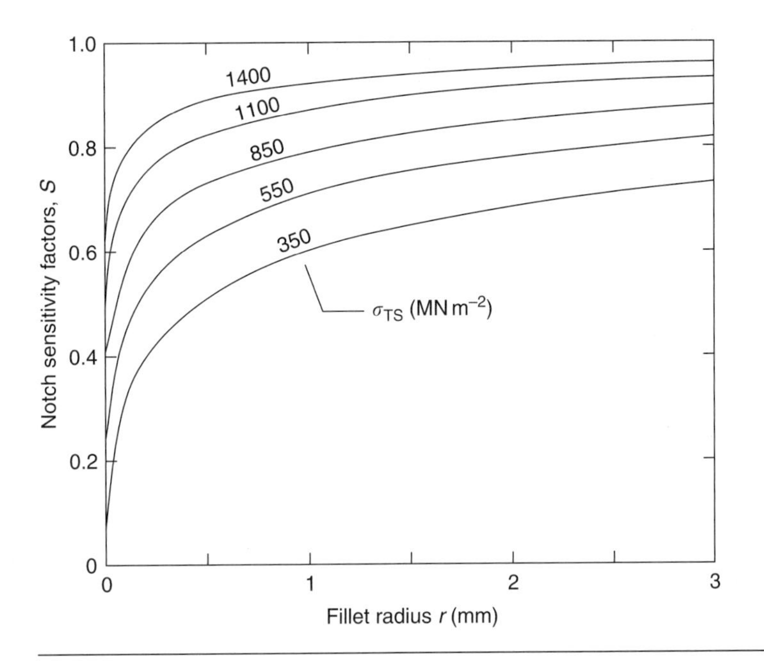

Figure 18.2 Effect of tensile strength and fillet radius on notch sensitivity factor.

18.5 Fatigue data for welded joints

Welding is the preferred method for joining structural steels and aluminum alloys in many applications. The world is awash with welds subjected to fatigue loading—bridges, oil rigs, ships, boats, chemical plant, and so on. Because welded joints are so important (and because they have some special features) there is a large amount of data in constructional standards for weld fatigue strength.

Figure 18.3 shows how the various types of welded joints can be categorized into standard weld classes. Figure 18.4 gives the fatigue strengths of the classes for structural steel. The 97.7% survival lines are used for design purposes, and the 50% lines for analyzing welds which actually failed. It is important to note that the vertical axis of the fatigue lines in Figure 18.4 is the full stress range $\Delta\sigma$, and not the $\Delta\sigma/2$ conventionally used for high-cycle fatigue data (see Table 18.1). From Table 18.1, we can see that $\Delta\sigma_0$ for steel (10^8 cycles) is at least $2 \times 170 = 340\,\text{MN}\,\text{m}^{-2}$. The $\Delta\sigma$ for a class G weld (10^8 cycles) is only $20\,\text{MN}\,\text{m}^{-2}$. This huge difference is due mainly to three special features of the weld—the large SCF, the rough surface finish and the presence of small crack-like defects produced by the welding process.

It is important to note that the fatigue strength of welds does not depend on the value of the mean stress in the fatigue cycle. Goodman's Rule (equation (17.3))

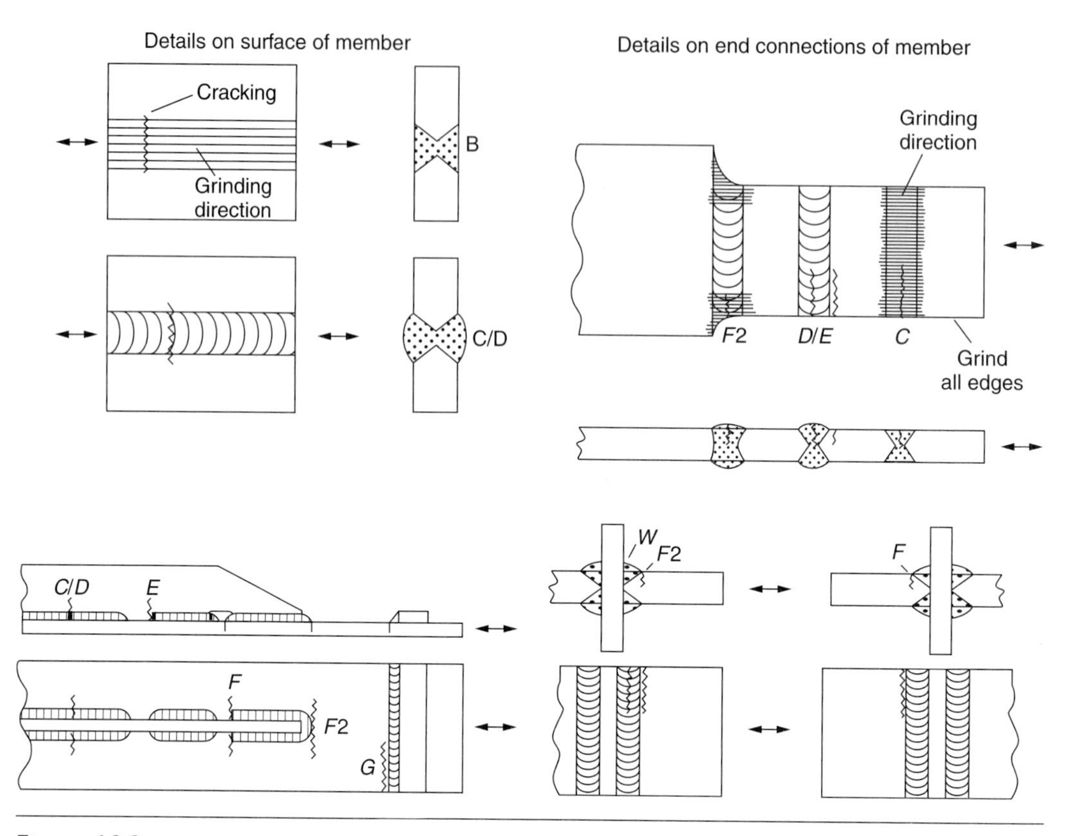

Figure 18.3 Standard weld classes.

should not be used for welds. This makes life much easier for the designer—the data in Figure 18.4 work for any mean stress, and the input required is simply the stress range. This major difference from conventional fatigue data is again due to a special feature of welded joints. Welds contain tensile residual stresses which are usually equal to the yield stress (these residual stresses are produced when the weld cools and contracts after the weld bead has been deposited). Whatever the applied stress cycle, the actual stress cycle in the weld itself always has a maximum stress of σ_y and a minimum stress of $\sigma_y - \Delta\sigma$.

18.6 Fatigue improvement techniques

We have already seen that the fatigue strength of a component can be increased by minimizing stress concentration factors, and having a good surface finish (a rough surface is, after all, just a collection of small stress concentrations). However, it is not always possible to remove SCFs completely. A good example

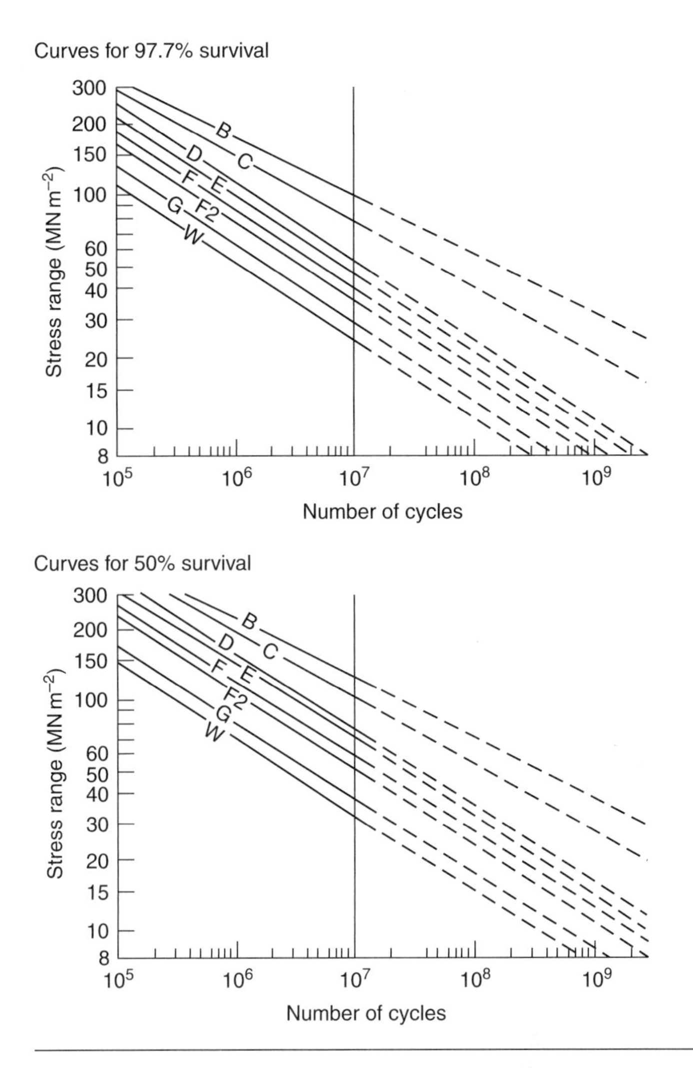

Figure 18.4 Fatigue strengths of the standard weld classes for structural steel.

is a screw thread, or the junction between the shank and the head of a bolt, which cannot be removed without destroying the functionality of the component. The answer here is to introduce a residual compressive stress into the region of potential crack initiation. This can be done using thread rolling (for screw threads), roller peening (for fillet radii on bolts or shafts), hole expansion (for pre-drilled holes) and shot peening (for relatively flat surfaces). The compressive stress makes it harder for fatigue cracks to grow away from the initiation sites in the surface.

Figure 18.5 shows how the fatigue strengths of welds can be improved. The first step is to improve the class of weld, if this is possible. By having a full

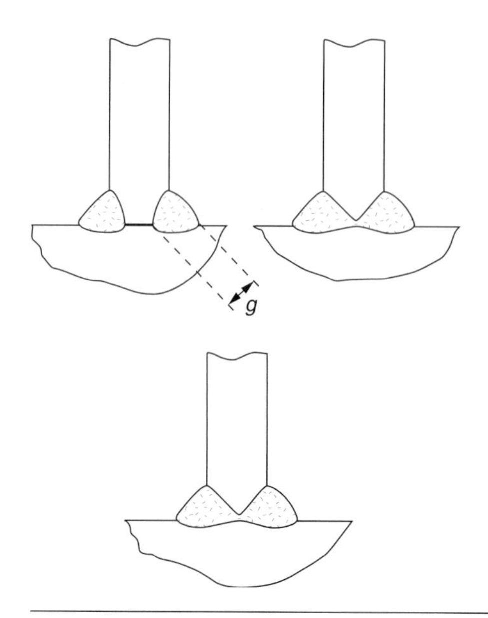

Figure 18.5 Improving the fatigue strength of a typical welded connection.

penetration weld, the very poor class W weld is eliminated, and the class of the connection is raised to class F. Further improvements are possible by grinding the weld bead to improve surface finish, reduce SCFs and remove welding defects. Finally, shot peening can be used to put the surface into residual compression.

18.7 Designing-out fatigue cycles

In some applications, the fatigue strength of the component cannot easily be made large enough to avoid failure under the applied loading. But there may be design-based solutions, which involve reducing or even eliminating the stress range which the loading cycle produces in the component. A good example is the design of bolted connections in the bearing housings of automotive crankshafts and con-rod big ends. Looking at Figure 18.6, it is easy to see that if the bolts are left slightly slack on assembly, the whole of the applied loading is taken by the two bolts (there is nothing else to take a tensile load). The load in each bolt therefore cycles from 0 to P to 0 with each cycle of applied loading.

The situation is quite different if the bolts are torqued-up to produce a large tension (or preload) in the bolts at assembly. The situation can be modelled very clearly as shown in Figure 18.7. Here, the bolt is represented

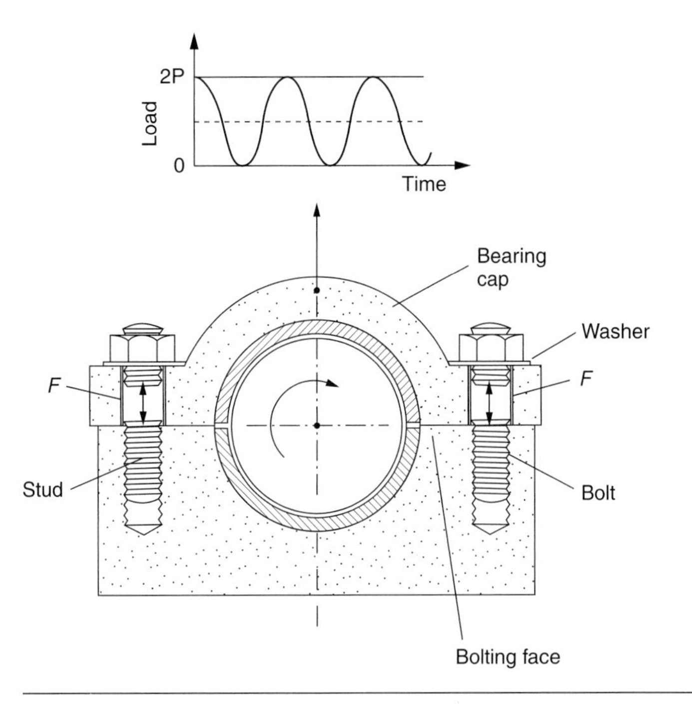

Figure 18.6 Typical bearing housing, with studs or bolts used to secure the bearing cap.

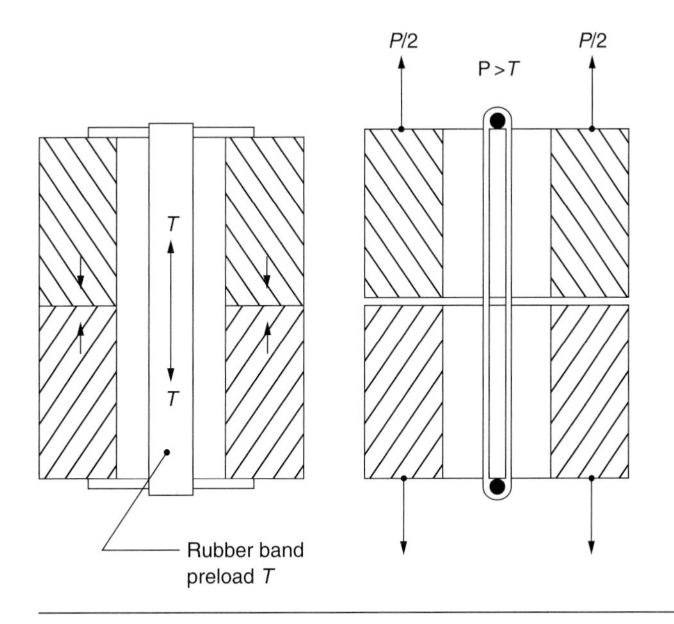

Figure 18.7 Model of a bolted connection, with assembly preload T.

by a rubber band, assembled with a tension T. There is an equal and opposite compressive force at the interface between the two halves of the housing. As the connection is loaded, nothing obvious happens, until the applied load P reaches T. Then, the two halves of the housing begin to separate. Provided $P < T$, the bolt sees no variation in stress at all. The variations in the applied load are provided by variations in the compressive force at the interface between the two halves of the housing. Of course, in a real connection, the bolt does not behave in such a springy way as a rubber band. But provided the bolts are long and have a small diameter, it is possible to make them springy enough that they are shielded from most of the variations in the applied loading cycle. Incidentally, this is one reason why bolts for fatigue critical applications are made from high tensile steel—they must not break in tension under the high preload even if they have a small diameter.

18.8 Checking pressure vessels for fatigue cracking

We saw in example 17.4 that cracks in pressure vessels can grow by fatigue under repeated cycles of pressurization, leading ultimately to failure—non-catastrophic in the case of leak before fast fracture, catastrophic in the case of fast fracture before leak. Cracks can also initiate and grow as the result of thermal fatigue (example 17.3), stress corrosion cracking (Chapter 15, Case Study 1) and creep (Chapters 20–23). Corrosion leads to fissuring, pitting or general loss of thickness (Chapters 26 and 27). It is worrying that a vessel which is safe when it enters service may become unsafe because of progressive crack growth or loss of thickness. Obviously, if a vessel starts to leak from a crack or perforation before it can suffer fast fracture, then the defect should be detected before disaster strikes. But leak before fast fracture cannot be relied on—it depends critically on the geometry of the crack. A thumbnail shaped crack, as in example 17.4, may well give leak before fast fracture (although this depends on the critical crack size and the wall thickness). But a crack which forms along the length of a weld (like the stress corrosion crack in Chapter 15, Case Study 1) can never give leak before fast fracture—unless the vessel is tested for cracks this type of defect will ultimately result in catastrophic failure. In fact the standard procedure is to test pressure vessels regularly by filling them with water, and pressurising them to typically 1.5–2 times the working pressure. Steam boilers (Figure 18.8) are tested this way annually. If failure does not occur at twice working pressure, then there is a margin of safety of 2 against crack growth or loss of thickness over the following year. If failure does occur under hydraulic testing, nobody will get hurt because the stored energy in compressed water is small.

Figure 18.8 A pressure vessel in action — the boiler of the articulated steam locomotive *Merddin Emrys*, built in 1879 and still hauling passengers on the Ffestiniog narrow-gauge railway in North Wales.

Examples

18.1 An aircraft using the airframe components of example 17.2 has encountered an estimated 4×10^8 cycles at a stress range of $150\,\mathrm{MN\,m^{-2}}$. It is desired to extend the airframe life by another 4×10^8 cycles by reducing the performance of the aircraft. Find the decrease in stress range necessary to achieve the additional life. You may assume a simple cumulative-damage law of the form

$$\sum_i \frac{N_i}{N_{fi}} = 1$$

for this calculation.

Answer

$13\,\mathrm{MNm^{-2}}$.

18.2 Indicate briefly how the following affect fatigue life:

(a) a good surface finish;
(b) the presence of a rivet hole;

(c) a significant mean tensile stress;
(d) a corrosive atmosphere.

18.3 The diagram shows the crank-pin end of the connection rod on a large-scale miniature steam locomotive. The locomotive weighs about 900 kg and is

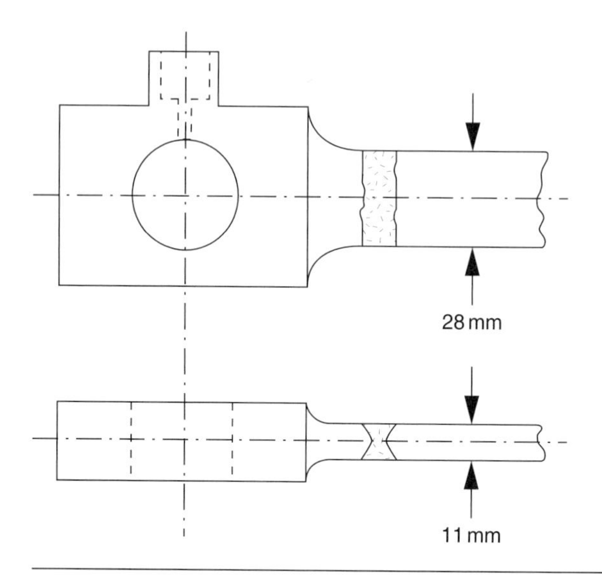

designed for hauling passengers around a country park. In the full-size prototype the connecting rod and big-end were forged from a single billet of steel. However, to save cost in building the miniature version, it is intended to weld the two parts together with a full-penetration double-sided weld, grinding the surface flush to hide the joint. Do you think that this design solution will have the required fatigue properties? Design and operational data are given below:

Diameter of double-acting cylinder	90 mm
Diameter of driving wheel	235 mm
Steam pressure inside cylinder at point of admission	7 bar gauge
Estimated annual distance travelled	6000 km
Design life	20 years minimum

18.4 Vibrating screens are widely used in the mining industry for sizing, feeding and washing crushed mineral particles. A typical screen consists of a box fabricated from structural steel plate which contains a mesh screen. During operation the box is shaken backwards and forwards at a frequency of up to 20 Hz. Although the major parts of the box are often fixed together using bolts or rivets, individual sub-assemblies such as a side of the box frequently have welded joints, particularly where frame stiffeners or gussets are added. Owing to the inertial forces generated by the rapid shaking the stresses in the unit can be significant and the fatigue design of the welded connections has to be considered carefully.

The side of one box consists of a relatively thin plate which is stiffened with triangular gussets as shown in the diagram. Strain gauges attached to the plate during operation show that the maximum principal stress range in the plate near end of the gusset is $8\,\mathrm{MN\,m^{-2}}$. Given that the screen is

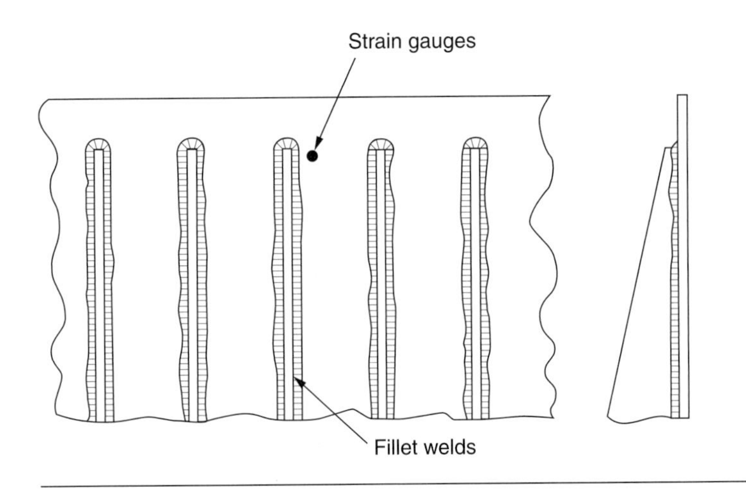

expected to work for 12 h per day and 6 days per week estimate the time that it will take for there to be a 50% chance of a crack forming in the plate at the end of each gusset. What would be the time for a 2.3% chance of cracking?

Answers

11 years; 4 years.

Chapter 19

Case studies in fatigue failure

Chapter contents

19.1 Introduction

In this chapter we look at three real situations in which failure occurred, or could have occurred, by fatigue cracking. The first two case studies describe fatigue failures in which the component was a round bar, which had been subjected to a cyclically varying bending moment. In both cases, fatigue had initiated at a machined step in the diameter of the bar. In the first incident, a number of failures occurred in the mechanism of a recently rebuilt pipe organ. In the second, an undersea submersible vehicle and its deployment platform were lost overboard because a pulley block broke away from the hoisting crane. The third case study is a structural integrity assessment of a large reciprocating engine which might contain a crack.

19.2 Case study 1: high-cycle fatigue of an uncracked component—failure of a pipe organ mechanism

Background

The first case study describes a rather unusual example of fatigue, which led to failures in the mechanism of a pipe-organ only two years after it had been rebuilt. In order to put the failures into context we first give a brief summary of the construction and operation of the instrument.

Figure 19.1 shows a general view of the organ, Figure 19.2 is a close-up of the area where the player sits, and Figure 19.3 is a schematic cross-section of the instrument showing its main parts. The organ has three musical departments: the "Great", the "Choir" and the "Pedal". The Great provides the basic musical tone quality of the organ. The Choir is used for added colour or musical contrast. The Pedal supplies the base notes for the whole instrument. The organist plays notes on the Great by pressing down keys on a keyboard (or "manual") using the fingers. The Choir has its own separate manual which is mounted just above the Great manual. Mounted below the bench is a "pedalboard" which is effectively a large keyboard designed so it can be played with the feet; as one would expect, the pedals are used to play the notes on the Pedal organ.

Each of the departments is based on a wind chest which is fabricated from boards of wood. The wind chest is supplied with low-pressure air from an electric fan-blower. On top of each wind chest are rows (or "ranks") of organ pipes made either from wood or metal. Each rank has one pipe for every semitone in the musical scale. All the pipes in a particular rank have a unified acoustic quality, which is characteristic of that particular rank. Although Figure 19.3 only shows a few ranks in each department there is in fact a total of 23 ranks on the whole instrument,

Figure 19.1 General view of the organ, showing the wooden case which contains the pipes, and below, the organ loft where the player sits.

Figure 19.2 Close-up of the organ loft, showing the two keyboards, or "manuals". The wooden knobs on either side of the manuals are the stops. The pedalboard is out of sight below the bench.

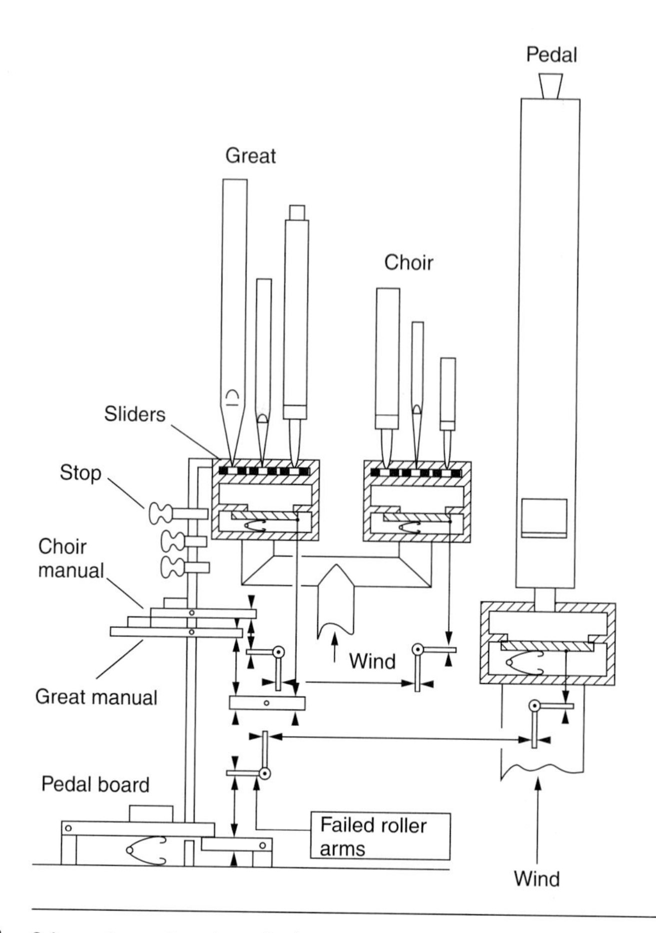

Figure 19.3 Schematic section through the organ (not to scale). Note the mechanical action which links each pallet to either a key or a pedal.

with 9 on the Great, 7 on the Choir and 7 on the Pedal. It is possible to play a tune on the Great, for example, using only one rank (which would be rather quiet), or every rank (which would be very loud) or any combination of ranks in between. The same can be done on the Choir, and on the Pedal. Since all three departments can be played independently the organist can produce a wide variety of sounds and therefore has to use considerable judgment when deciding which combination is best suited to the piece of music being performed.

Each wind chest is divided in half vertically. The bottom half is always under air pressure. The top half is partitioned along the length of the wind chest into a large number of separate boxes, one for each note in the musical scale. Each box is separated from the air in the lower half by a wooden flap valve called a pallet. The pallet is normally kept closed by a spring. The air pressure also helps to press the pallet shut against its seating. Each pallet is connected to either a key on one of the manuals or to a pedal on the pedalboard by a mechanical linkage. The arrangement is shown schematically in Figure 19.3. We can see from the drawing that if a note or a pedal is pressed down the linkage will pull the pallet away from its seating and let air into the box above. This means that air is now available to make the pipes above the box "speak".

In order to play notes on a particular rank the organist pulls out a small wooden knob called a "stop". Because there are 23 ranks on the organ there are also 23 stops. These are positioned on either side of the manuals as shown in Figure 19.2. Each stop is connected by a mechanical linkage to a long, narrow strip of wood or "slider" which slides underneath the rank of pipes. The slider has a row of holes bored through it, one for each pipe in the rank. When the rank is shut off the holes in the slider are exactly out of phase with the holes at the feet of the pipes: air cannot get to the pipes even if the pallets are opened. To activate the rank the stop must be pulled out. This moves the slider a small distance along the length of the chest so that every hole in it lines up with the hole at the foot of a pipe.

The failures all occurred in the linkages between the pedalboard and the pallets in the Pedal wind chest. A total of about five linkages failed, affecting five separate musical notes. An unexpected failure during an organ recital can be very frustrating both for the organist and the listeners and it was important to establish the cause of the breakages.

Operational history

The organ was installed in 1705 by the famous English organ builder Bernard Smith. The original carved organ case survives, as do some of the pipes. Soon after 1785 the organ became derelict and some of the pipes were stored away in boxes behind the empty organ case. The organ was reconstructed by Hill and Son in 1865, and Choir and Pedal departments were added. In 1909 the organ was rebuilt again by Norman and Beard, who added a third manual and greatly increased the number of pipes. By 1983 the organ had become unreliable and the opportunity was taken of rebuilding it along the lines of the more economical Hill version of 1865. Of the original 1705 instrument only the case and some pipes survived; all other components, including the mechanical linkages, were renewed. "Modern" constructional materials (such as aluminum alloy and polymers) were used instead of wood for much of the mechanism. In the event, the failures were traced to the inappropriate use of an aluminum–copper alloy in a small part of the pedal action.

Details of the failures

The general location of the failures is marked in Figure 19.3. Figure 19.4 shows the details of the mechanical action in this area. The pedal is connected to the end of the roller arm by a vertical rod of wood called a tracker. When the pedal is pushed down the tracker pulls the end of the roller arm down and this makes the roller rotate. The arm at the other end of the roller rotates as well, actuating the horizontal tracker. Both the arms and the roller were made from aluminum alloy.

Figure 19.5 is a cross-section of the roller-arm assembly which shows the position of the fractures. The arm was made from round rod 4.75 mm in diameter. The end of the rod had been turned down to a diameter of 4 mm to fit the hole in the roller and had been riveted over to hold it in place.

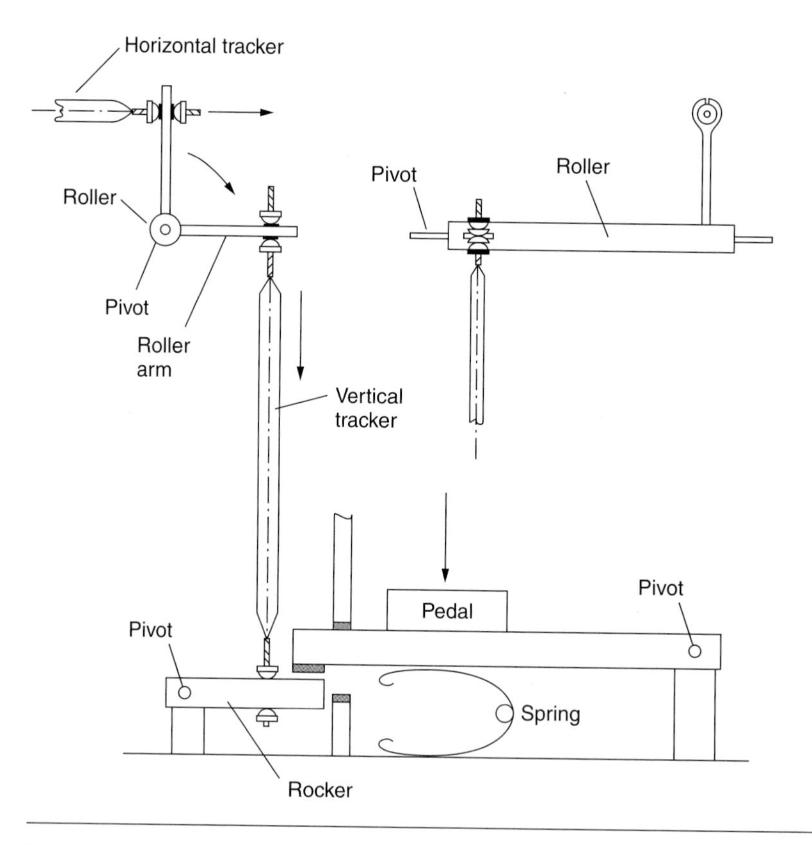

Figure 19.4 Schematic of the mechanical action in the region of the failed components (not to scale). The vertical tracker and the rocker are both made from light wood, and their weight can be neglected in the loading calculations.

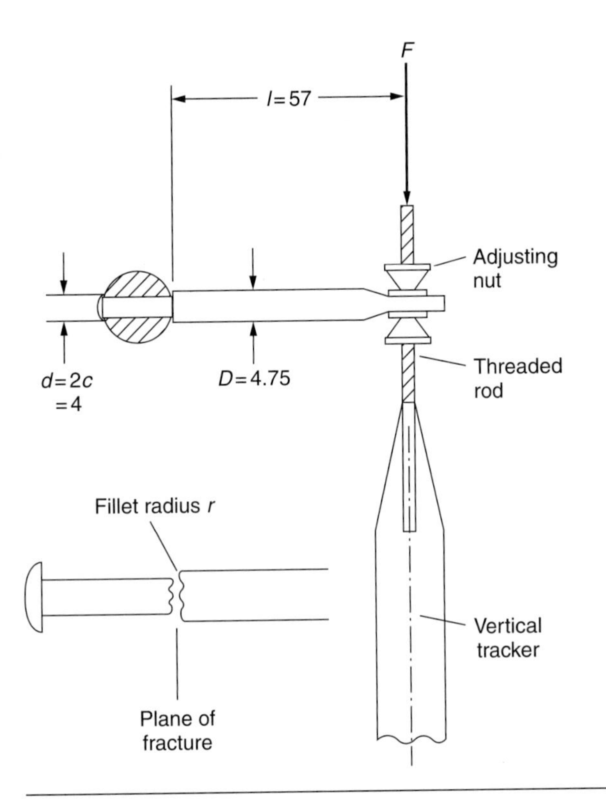

Figure 19.5 Fatigue fractures occurred at a sharp change of section in the roller arms. To scale—dimensions in mm.

The turning operation had been performed with a sharp-ended lathe tool, which resulted in a negligible fillet radius where the 4 mm section met the main body of the rod. The fatigue crack had initiated at this sharp change of section.

Figure 19.6 shows a general view of the fracture surface taken in the scanning electron microscope (SEM). The top of the arm is at the top of the photograph. This is the region which was put into tension when the tracker was pulled down. The photograph shows that the fatigue crack initiated in this region of maximum tensile stress and then propagated downwards across the 4 mm section. The progress of the fatigue crack is clearly marked by "clam-shell" marks. SEM photographs taken of the fatigue surfaces at high magnification showed the characteristic features of fatigue striations, burnished areas, and secondary cracks. SEM photographs taken of the final fracture at high magnification showed the characteristic features of ductile fracture by microvoid coalescence.

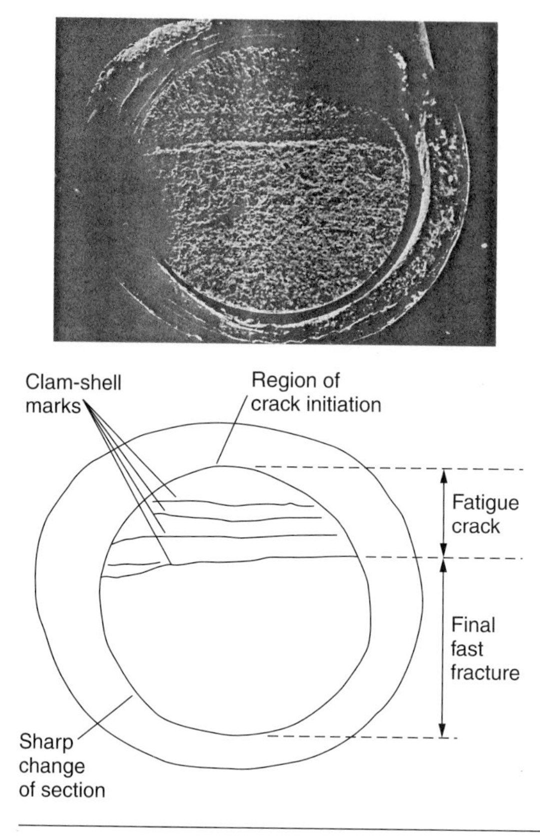

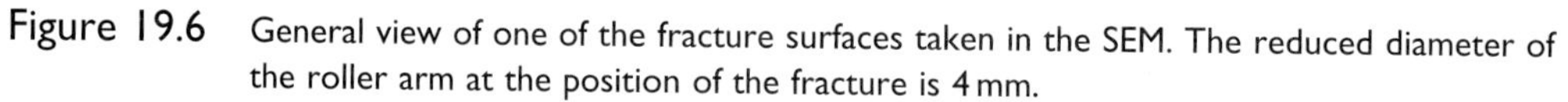

Figure 19.6 General view of one of the fracture surfaces taken in the SEM. The reduced diameter of the roller arm at the position of the fracture is 4 mm.

Materials data

The Vickers hardness of the arm was 69. The endurance limit for an aluminum alloy of this strength is approximately $\pm 70\,\text{MN m}^{-2}$ about zero mean stress for a life of 5×10^8 cycles. The yield and tensile strengths are approximately 100 and $200\,\text{MN m}^{-2}$.

Service loadings

Tests were carried out with the organ running to find the force which was needed to depress the end of the roller arm. At the bottom end of the pedal-board the force was about 1 kgf, and at the top end it was about 0.5 kgf. The difference was caused by a systematic decrease in the size of the pallets along the length of the wind chest. Because the pipes at the bottom end of the wind chest are larger than those at the top they need more wind to make them sound properly. To ensure that there is an adequate supply of air to all the pipes the pallets at the bottom end of the wind chest are made larger than those at the top end. Naturally, the force which is needed to open a pallet against the pressure of the air decreases with decreasing size of pallet.

It was not easy to obtain an accurate figure for the number of loading cycles experienced by the roller arms. An estimate of 3.6×10^5 was used, based on reasonable assumptions about the pattern of use.

Stress calculations

The maximum bending stress in the reduced section of the roller arm can be found from the standard equation

$$\sigma = \frac{4F\ell}{\pi c^3}$$

Setting $F = 9.81$ N (for 1 kgf), $\ell = 57$ mm and $c = 2$ mm gives a tensile stress of $90\,\mathrm{MN\,m^{-2}}$ at the point of fatigue crack initiation. The fatigue cycle was therefore $\pm 45\,\mathrm{MN\,m^{-2}}$ about a tensile mean stress of $45\,\mathrm{MN\,m^{-2}}$. The effect of the mean stress was corrected for by using the Goodman equation

$$\Delta\sigma_{\sigma m} = \Delta\sigma_0\left(1 - \frac{\sigma_m}{\sigma_{TS}}\right)$$

Setting $\Delta\sigma_{\sigma m} = 90\,\mathrm{MN\,m^{-2}}$, $\sigma_m = 45\,\mathrm{MN\,m^{-2}}$ and $\sigma_{TS} = 200\,\mathrm{MN\,m^{-2}}$, we obtain $\Delta\sigma_0 = 120\,\mathrm{MN\,m^{-2}}$. The endurance limit required to avoid failure about zero mean stress is therefore $\pm 60\,\mathrm{MN\,m^{-2}}$. This is 86 percent of the endurance limit of $\pm 70\,\mathrm{MN\,m^{-2}}$ about zero mean stress for the alloy in the un-notched condition. In reality, the sharp notch at the change of section would have reduced the endurance limit of the alloy, which explains why the arms failed after only 3.6×10^5 cycles.

Design modifications

The problem can be solved by eliminating the notch, and using a material with high enough fatigue strength. The notch is most easily disposed of by making replacement arms from rod with a uniform diameter of 4 mm, and fixing them into the rollers with anaerobic adhesive (this has the counter-intuitive consequence that removing material actually increases strength). If the design life of an organ is taken to be 100 years ($\approx 2 \times 10^7$ cycles) then mild steel (endurance limit $\approx \pm 200\,\mathrm{MN\,m^{-2}}$) should give an ample safety margin.

19.3 Case study 2: low-cycle fatigue of an uncracked component—failure of a submersible lifting eye

Background

Figure 19.7 shows a pulley block which was used for lowering and hoisting an underwater platform (used for deploying a submersible vehicle). The mass of the platform was around 1800 kg. The block was supported from the boom of a deck-mounted hydraulic crane. The platform was lowered and hoisted by a wire rope, which was taken from a power winch mounted alongside the crane. The pull of the crane boom was transmitted to the head of the block through a lifting eye. During the hoisting operation the crane exerted a straight pull of 2760 kgf along the axis of the eye. However, when the platform had been winched up to the level of the deck, it was secured directly to the side of the block as shown schematically in Figure 19.8. The winch was slackened off, the

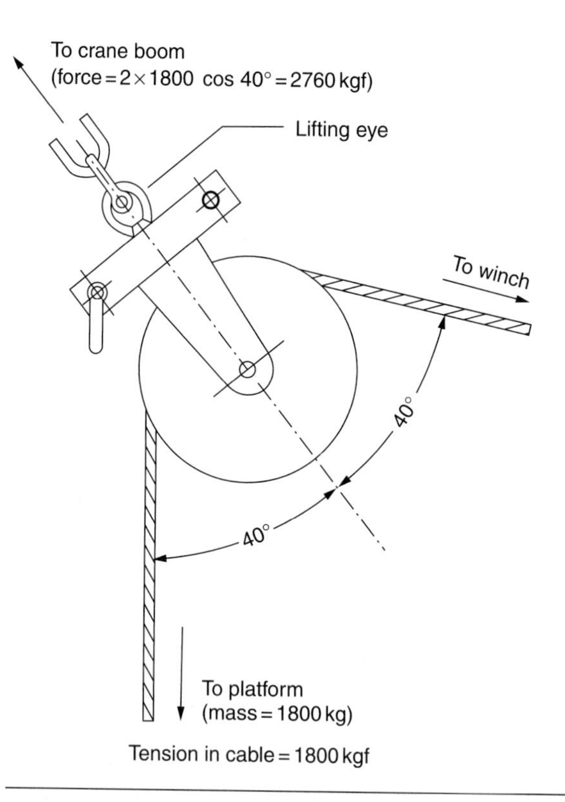

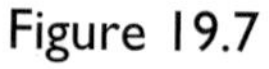
Figure 19.7 Schematic of pulley block during a hoisting operation. The crane exerts a straight pull along the axis of the lifting eye. Not to scale—the sheave is 0.75 m in diameter.

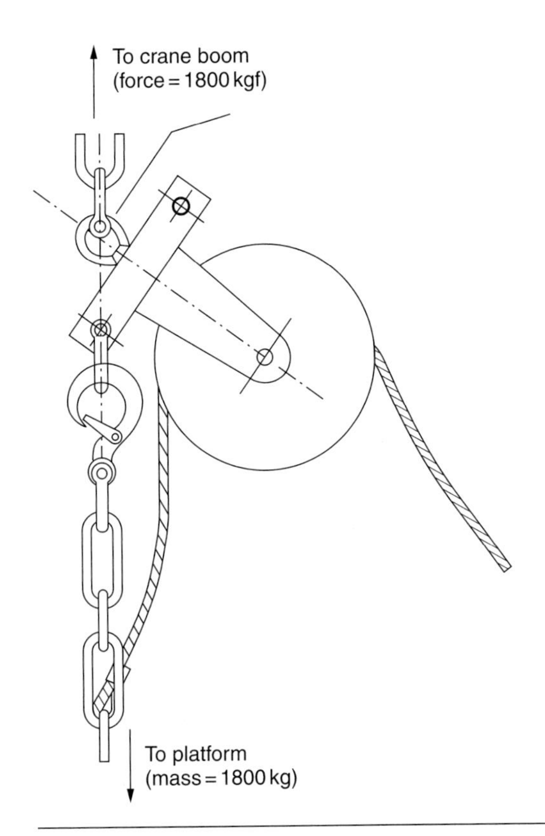

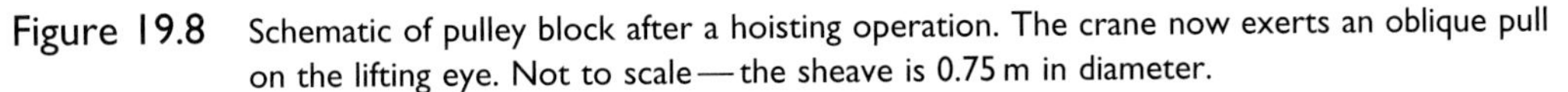

Figure 19.8 Schematic of pulley block after a hoisting operation. The crane now exerts an oblique pull on the lifting eye. Not to scale — the sheave is 0.75 m in diameter.

crane was slewed around and the platform was lowered onto the deck. During this operation the crane boom exerted an oblique pull of 1800 kgf on the eye.

Figure 19.9 shows the detailed construction of the lifting eye. The eye itself was a steel forging, supplied as a standard item. It was fixed to the pulley block by a steel bolt which had been manufactured for the purpose. The threaded end of the bolt was screwed into the tapped hole in the eye until the shoulder on the bolt was in contact with the flat surface at the bottom of the eye. No specifications were given for the assembly torque but the bolt was secured against unscrewing by a pin. The shank of the bolt was a running fit in the block.

After about 200 hoisting operations the eye broke away at the top of the block and the platform and its submersible were lost overboard. The details of the fracture are shown in Figure 19.10. The failure had occurred in the bolt at the position where the thread met the shank. In order to allow the screw-cutting tool to run out at the end of the thread a sharp groove had been machined next to the shoulder. The failure had initiated at this sharp change of

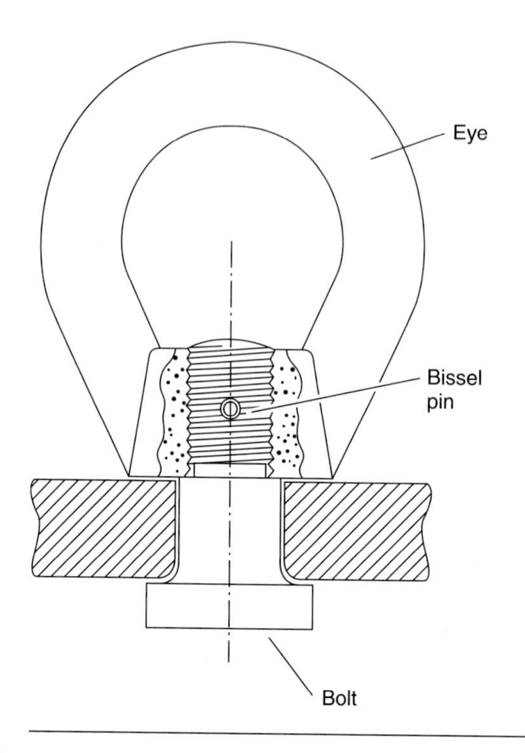

Figure 19.9 Details of lifting eye. Note the sharp change of section in the bolt between shank and thread. To scale—the shank is 35 mm in diameter.

section as a fatigue crack. The crack had obviously propagated in from the surface of the groove until the remaining cross section became unable to support the load and failed by fast fracture. The fatigue surface was flat and had circumferential "clam-shell" markings. When it was examined in the SEM it had a burnished appearance. The fast-fracture surface looked bright and examination in the SEM confirmed that it consisted of cleavage facets.

Material properties

The steel used for the bolt had a yield strength of $540\,\text{MN}\,\text{m}^{-2}$ and a tensile strength of $700\,\text{MN}\,\text{m}^{-2}$.

Stresses in the bolt

It is easy to show that the straight pull of 2760 kgf simply generates a uniform tensile stress of $55\,\text{MN}\,\text{m}^{-2}$ in the bolt at the position of the groove. In contrast, the oblique pull of 1800 kgf generates three stress states (see

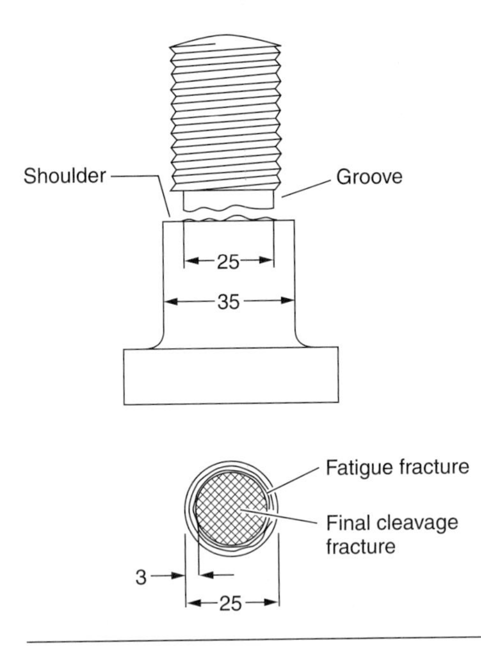

Figure 19.10 Fatigue failure occurred at the sharp change of section in the bolt. To scale—dimensions in mm.

Figure 19.11). The vertical component of 1160 kgf generates a uniform tensile stress at the groove of 23 $MN\,m^{-2}$. The horizontal component of 1380 kgf generates a uniform shear stress of 28 $MN\,m^{-2}$. The overall force of 1800 kgf also applies a bending moment of 1.06×10^6 N mm to the groove.

The moment M_{el} at which the cross-section starts to yield can be found from the standard equation

$$M_{el} = \frac{\pi \sigma_y c^3}{4}.$$

Setting $\sigma_y = 540\,MN\,m^{-2}$ and $c = 12.5$ mm, we find that $M_{el} = 0.83 \times 10^6$ N mm. This is 22 percent less than the actual bending moment, so the bolt should have experienced plastic deformation in service. The uniform tensile and shear stresses are negligible compared to the maximum bending stress and the bending moment is the dominant term.

Failure mechanism

The failure is a classic example of low-cycle fatigue. The plastic strain range at the corner of the groove was presumably large enough for a crack to form and grow to the critical depth for fast fracture after only 200 cycles. Certainly

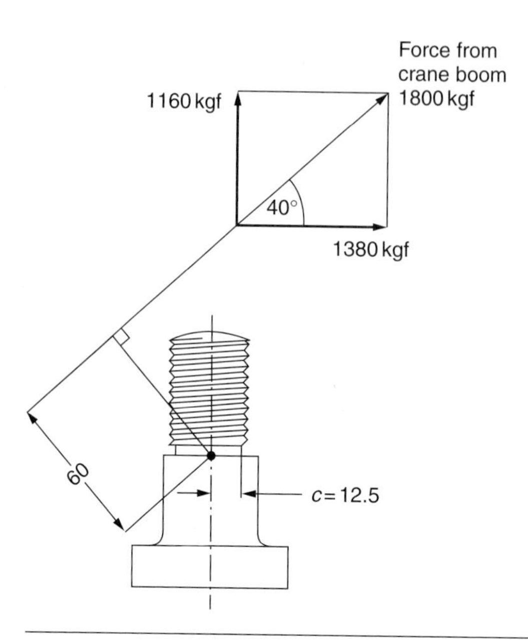

Figure 19.11 Forces acting on the bolt after hoisting. The oblique pull from the crane generates a large bending moment in the reduced section of the bolt. To scale — dimensions in mm.

the burnished look of the fatigue surface suggests that the fatigue crack was subjected to a large compressive stress during part of each stress cycle. And the small depth of the critical crack gives extra confirmation that the bolt failed at a high stress.

Design modifications

Even in the absence of the groove the roots of the threads provide an unavoidable source of stress concentration. Given the need for a large margin of safety in this application the only satisfactory solution is to design-out the bending load, and this was done by the manufacturers.

19.4 Case study 3: fatigue of a cracked component — the safety of the Stretham engine

The Stretham steam pumping engine (Figure 19.12) was built in 1831 as part of an extensive project to drain the Fens for agricultural use. In its day it was one of the largest beam engines in the Fens, having a maximum power of

Figure 19.12 Part of the Stretham pumping engine. In the foreground are the crank and the lower end of the connecting rod. Also visible are the flywheel (with separate spokes and rim segments, all pegged together), the eccentric drive to the valve-gear and, in the background, an early treadle-driven lathe for on-the-spot repairs.

105 horsepower at 15 rpm (it could lift 30 tons of water per revolution, or 450 tons per minute); it is now the sole surviving steam pump of its type in East Anglia.*

The engine could still be run for demonstration purposes. Suppose that you are called in to assess its safety. We will suppose that a crack 2 cm

* Until a couple of centuries ago much of the eastern part of England which is now called East Anglia was a vast area of desolate marshes, or fens, which stretched from the North Sea as far inland as Cambridge.

deep has been found in the connecting rod—a cast-iron rod, 21 feet long, with a section of $0.04\,\mathrm{m}^2$. Will the crack grow under the cyclic loads to which the connecting rod is subjected? And what is the likely life of the structure?

Mechanics

The stress in the crank shaft is calculated approximately from the power and speed as follows. Bear in mind that approximate calculations of this sort may be in error by up to a factor of 2—but this makes no difference to the conclusions reached below. Referring to Figure 19.13:

$$\begin{aligned}\text{Power} &= 105\ \text{horsepower}\\ &= 7.8\times10^4\,\mathrm{J\,s^{-1}}\\ \text{Speed} &= 15\ \text{rpm} = 0.25\ \text{rev}\,\mathrm{s^{-1}}\\ \text{Stroke} &= 8\ \text{feet} = 2.44\,\mathrm{m}\end{aligned}$$

Force × 2 × stroke × speed ≈ power

$$\therefore\ \text{Force} \approx \frac{7.8\times10^4}{2\times2.44\times0.25} \approx 6.4\times10^4\,\mathrm{N}$$

Nominal stress in the connecting rod $= F/A = 6.4\times10^4/0.04 = 1.6\,\mathrm{MN\,m^{-2}}$ approximately.

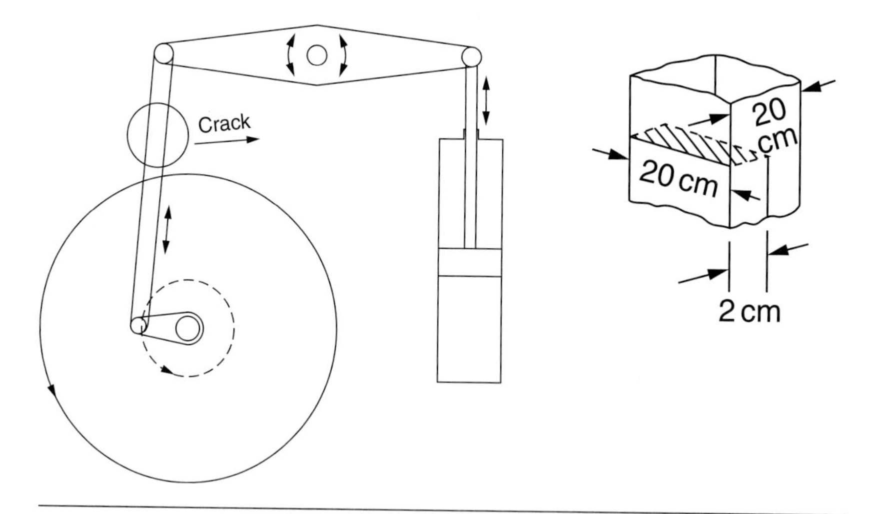

Figure 19.13 Schematic of the Stretham engine.

Failure by fast fracture

For cast iron, $K_c = 18\ \mathrm{MN\,m^{-3/2}}$.

First, could the rod fail by fast fracture? The stress intensity is:

$$K = \sigma\sqrt{\pi a} = 1.6\sqrt{\pi.0.02}\ \mathrm{MN\,m^{-3/2}} = 0.40\ \mathrm{MN\,m^{-3/2}}$$

It is so much less than K_c that there is no risk of fast fracture, even at peak load.

Failure by fatigue

The growth of a fatigue crack is described by

$$\frac{da}{dN} = A(\Delta K)^m \tag{19.1}$$

For cast iron,

$$A = 4.3 \times 10^{-8}\mathrm{m(MN\,m^{-3/2})^{-4}}$$
$$m = 4$$

We have that

$$\Delta K = \Delta\sigma\sqrt{\pi a}$$

where $\Delta\sigma$ is the range of the tensile stress (Figure 19.14). Although $\Delta\sigma$ is constant (at constant power and speed), ΔK increases as the crack grows. Substituting in equation (19.1) gives

$$\frac{da}{dN} = A\Delta\sigma^4\pi^2a^2$$

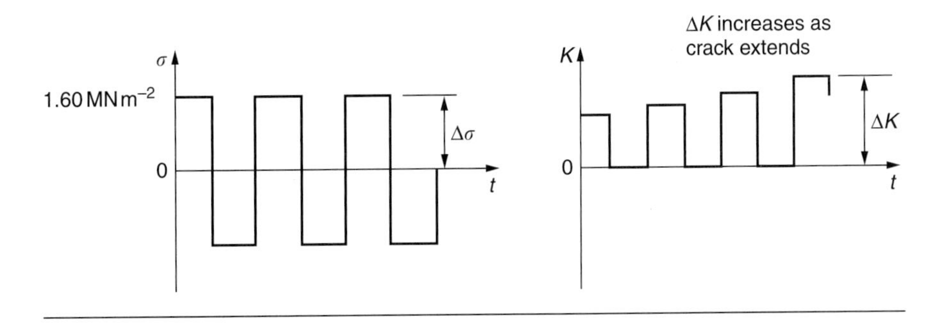

Figure 19.14 Crack growth by fatigue in the Stretham engine.

and

$$dN = \frac{1}{(A\Delta\sigma^4\pi^2)}\frac{da}{a^2}$$

Integration gives the number of cycles to grow the crack from a_1 to a_2:

$$N = \frac{1}{(A\Delta\sigma^4\,\pi^2)}\left\{\frac{1}{a_1} - \frac{1}{a_2}\right\}$$

for a range of a small enough that the crack geometry does not change appreciably. Let us work out how long it would take our crack to grow from 2 to 3 cm. Then

$$N = \frac{1}{4.3 \times 10^{-8}(1.6)^4\,\pi^2}\left\{\frac{1}{0.02} - \frac{1}{0.03}\right\}$$

$$= 5.9 \times 10^6 \text{cycles}$$

This is sufficient for the engine to run for 8 h on each of 832 open days for demonstration purposes, i.e. to give 8 hours of demonstration each weekend for 16 years. A crack of 3 cm length is still far too small to go critical, and thus the engine will be perfectly safe after the 5.9×10^6 cycles. Under demonstration the power delivered will be far less than the full 105 horsepower, and because of the $\Delta\sigma^4$ dependence of N, the number of cycles required to make the crack grow to 3 cm might be as much as 30 times the one we have calculated.

The estimation of the total lifetime of the structure is more complex—substantial crack growth will make the crack geometry change significantly; this will have to be allowed for in the calculations by incorporating a correction factor, Y.

Examples

19.1 The photographs show a bicycle crank. The end of the crank has suffered a fatigue failure where the pedal was screwed into the crank (the pedal is missing). In the close-up photograph of the fracture, the bottom of the photograph corresponds to the outside (pedal side) of the crank arm. Answer the following questions:

(a) Why do you think the failure was caused by fatigue?
(b) Identify the fatigue fracture surface(s). Explain your choice.
(c) Identify the final fast fracture surface(s). Explain your choice.
(d) Where do you think the fatigue crack(s) initiated?
(e) In your opinion, was the level of applied stress at final failure high, moderate or low?

19.2 The diagrams show the general arrangement of a swing door and a detailed drawing of the top pivot assembly. The full weight of the door was taken by the bottom pivot. The function of the top pivot was simply to keep the door upright by applying a horizontal force to the top of the door. When the door swung back and forth, it rotated about the top and bottom pivot pins. After a year in service, the top pivot pin failed by fatigue. As a result, the top door housing separated from the top frame housing, and the door fell on a person passing by.

(a) Where in the top pivot pin do you think the fatigue fracture occurred?
(b) Why do you think the fatigue fracture occurred where it did?
(c) What was the origin of the fatigue loading cycle at the failure location?
(d) How did the door housing manage to become separated from the frame housing?

[In answering these questions, please note the following information: the pivot pin was a very loose fit in both the door housing and the frame housing; the

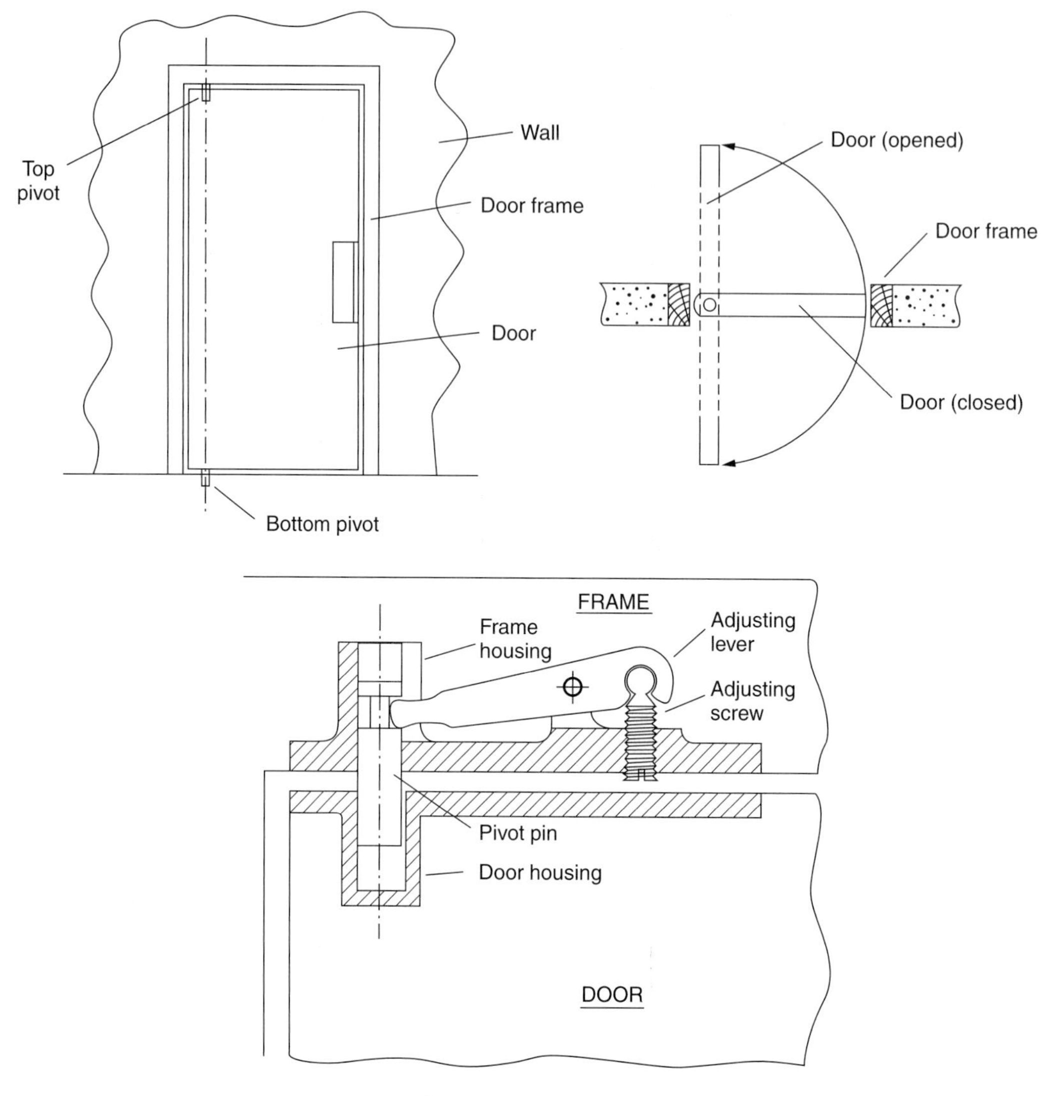

Elevation of pivot assembly

pivot pin tended not to rotate in the frame housing — instead, the door housing tended to rotate about the pivot pin.]

19.3 With reference to Case Study 2, how would you redesign the pulley block to design-out the bending load on the lifting eye?

19.4 With reference to Case Study 3, why do we take into account only the *tensile* part of the stress cycle when calculating ΔK?

Part F

Creep deformation and fracture

Chapter 20

Creep and creep fracture

Chapter contents

20.1 Introduction

So far we have concentrated on mechanical properties at room temperature. Many structures—particularly those associated with energy conversion, like turbines, reactors, steam, and chemical plant—operate at much higher temperatures.

At room temperature, most metals and ceramics deform in a way which depends on stress but which, for practical purposes, is independent of time:

$$\epsilon = f(\sigma) \text{ elastic/plastic solid}$$

As the temperature is raised, loads which give no permanent deformation at room temperature cause materials to *creep*. Creep is slow, continuous deformation with time: the strain, instead of depending only on the stress, now depends on temperature and time as well:

$$\epsilon = f(\sigma, t, T) \text{ creeping solid}$$

It is common to refer to the former behavior as "low-temperature" behavior, and the latter as "high temperature". But what is a "low" temperature and what is a "high" temperature? Tungsten, used for lamp filaments, has a very high melting point—well over 3000°C. Room temperature, for tungsten, is a very low temperature. If made hot enough, however, tungsten will creep—that is the reason that lamps ultimately burn out. Tungsten lamps run at about 2000°C—this, for tungsten, is a high temperature. If you examine a lamp filament which has failed, you will see that it has sagged under its own weight until the turns of the coil have touched—that is, it has deformed by creep.

Figure 20.1 and Table 20.1 give melting points for metals and ceramics and softening temperatures for polymers. Most metals and ceramics have high melting points and, because of this, they start to creep only at temperatures well above room temperature—this is why creep is a less familiar phenomenon than elastic or plastic deformation. But the metal *lead*, for instance, has a melting point of 600 K; room temperature, 300 K, is exactly half its absolute melting point. Room temperature for lead is a high temperature, and it creeps—as Figure 20.2 shows. And the ceramic *ice* melts at 0°C. "Temperate" glaciers (those close to 0°C) are at a temperature at which ice creeps rapidly—that is why glaciers move. Even the thickness of the Antarctic ice cap, which controls the levels of the earth's oceans, is determined by the creep spreading of the ice at about −30°C.

The point, then, is that the temperature at which materials start to creep depends on their melting point. As a general rule, it is found that creep starts when

$T > 0.3$ to $0.4 T_M$ for metals,
$T > 0.4$ to $0.5 T_M$ for ceramics,

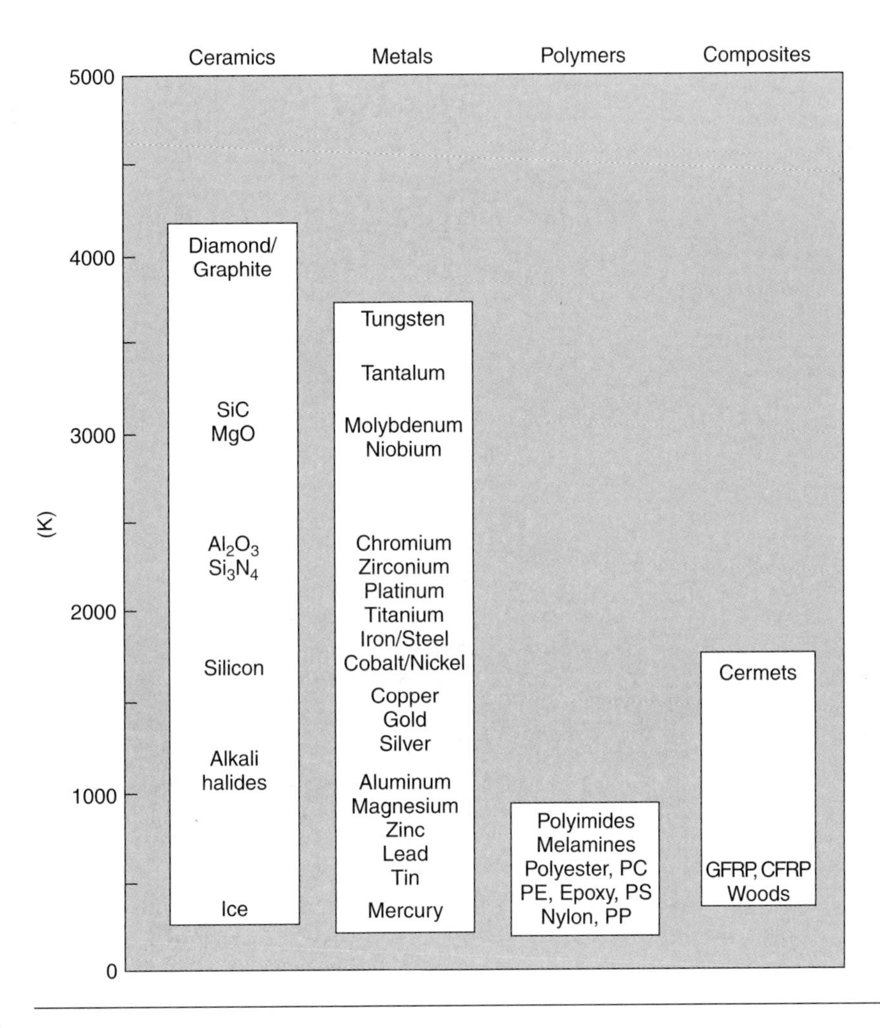

Figure 20.1 Melting or softening temperature.

where T_M is the melting temperature in kelvin. However, special alloying procedures can raise the temperature at which creep becomes a problem.

Polymers, too, creep — many of them do so at room temperature. As we said in Chapter 5, most common polymers are not crystalline, and have no well-defined melting point. For them, the important temperature is the glass temperature, T_G, at which the Van der Waals bonds solidify. Above this temperature, the polymer is in a leathery or rubbery state, and creeps rapidly under load. Below, it becomes hard (and sometimes brittle) and, for practical purposes, no longer creeps. T_G is near room temperature for most polymers, so creep is a problem.

Table 20.1 Melting or softening[(S)] temperature

Material	T(K)	*Material*	T(K)
Diamond, graphite	4000	Gold	1336
Tungsten alloys	3500–3683	Silver	1234
Tantalum alloys	2950–3269	Silica glass	1100[(S)]
Silicon carbide, SiC	3110	Aluminum alloys	750–933
Magnesia, MgO	3073	Magnesium alloys	730–923
Molybdenum alloys	2750–2890	Soda glass	700–900[(S)]
Niobium alloys	2650–2741	Zinc alloys	620–692
Beryllia, BeO	2700	Polyimides	580–630[(S)]
Iridium	2682–2684	Lead alloys	450–601
Alumina, Al_2O_3	2323	Tin alloys	400–504
Silicon nitride, Si_3N_4	2173	Melamines	400–480[(S)]
Chromium	2148	Polyesters	450–480[(S)]
Zirconium alloys	2050–2125	Polycarbonates	400[(S)]
Platinum	2042	Polyethylene, high-density	300[(S)]
Titanium alloys	1770–1935	Polyethylene, low-density	360[(S)]
Iron	1809	Foamed plastics, rigid	300–380[(S)]
Carbon steels	1570–1800	Epoxy, general purpose	340–380[(S)]
Cobalt alloys	1650–1768	Polystyrenes	370–380[(S)]
Nickel alloys	1550–1726	Nylons	340–380[(S)]
Cermets	1700	Polyurethane	365[(S)]
Stainless steels	1660–1690	Acrylic	350[(S)]
Silicon	1683	GFRP	340[(S)]
Alkali halides	800–1600	CFRP	340[(S)]
Beryllium alloys	1540–1551	Polypropylene	330[(S)]
Uranium	1405	Ice	273
Copper alloys	1120–1356	Mercury	235

In design against creep, we seek the material and the shape which will carry the design loads, without failure, for the design life at the design temperature. The meaning of "failure" depends on the application. We distinguish four types of failure, illustrated in Figure 20.3.

(a) Displacement-limited applications, in which precise dimensions or small clearances must be maintained (as in the discs and blades of turbines).
(b) Rupture-limited applications, in which dimensional tolerance is relatively unimportant, but fracture must be avoided (as in pressure-piping).
(c) Stress relaxation-limited applications in which an initial tension relaxes with time (as in the pretensioning of cables or bolts).
(d) Buckling-limited applications, in which slender columns or panels carry compressive loads (as in the upper wing skin of an aircraft, or an externally pressurized tube).

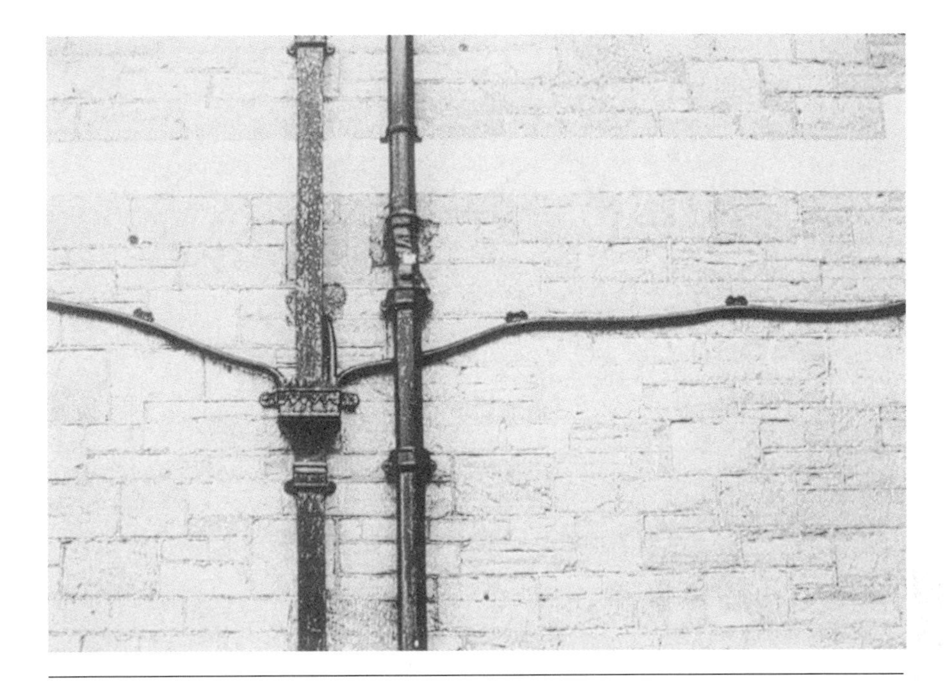

Figure 20.2 Lead pipes often creep noticeably over the years.

To tackle any of these we need *constitutive equations* which relate the strain-rate $\dot{\epsilon}$ or time-to-failure t_f for the material to the stress σ and temperature T to which it is exposed. These come next.

20.2 Creep testing and creep curves

Creep tests require careful temperature control. Typically, a specimen is loaded in tension or compression, usually at constant load, inside a furnace which is maintained at a constant temperature, T. The extension is measured as a function of time. Figure 20.4 shows a typical set of results from such a test. Metals, polymers, and ceramics all show creep curves of this general shape.

Although the *initial elastic* and the *primary creep* strain cannot be neglected, they occur quickly, and they can be treated in much the way that elastic deflection is allowed for in a structure. But thereafter, the material enters *steady state,* or *secondary* creep, and the strain increases steadily with time. In designing against creep, it is usually this steady accumulation of strain with time that concerns us most.

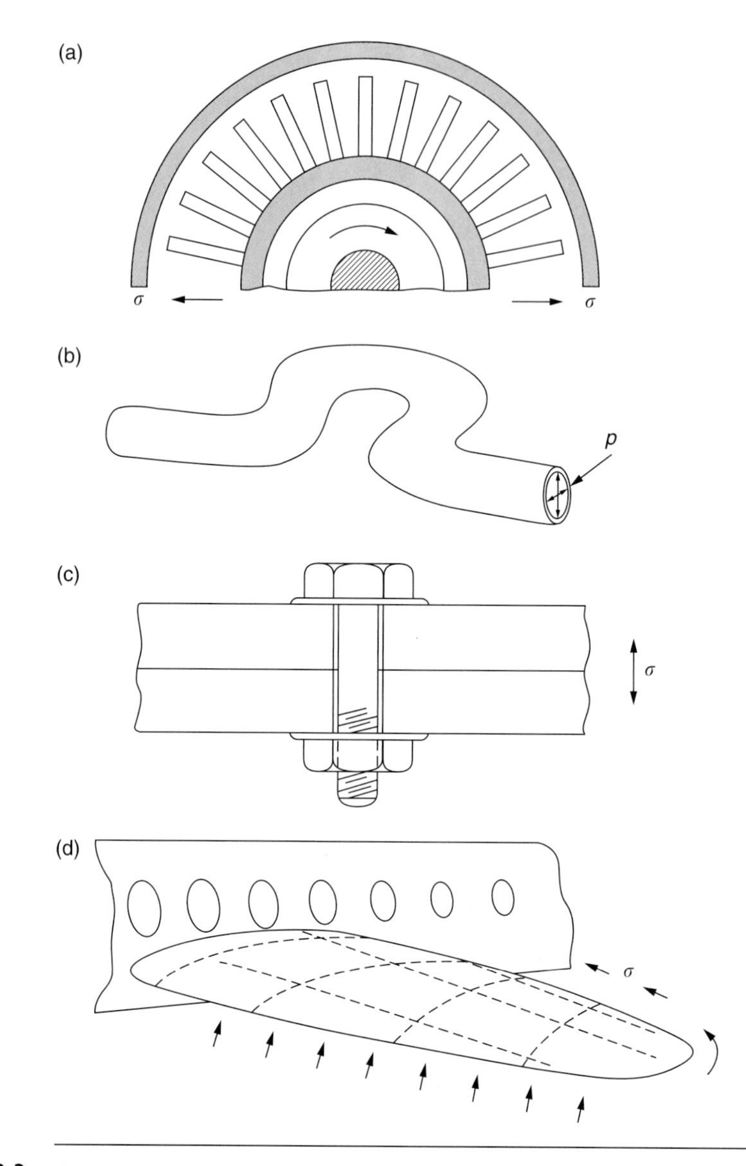

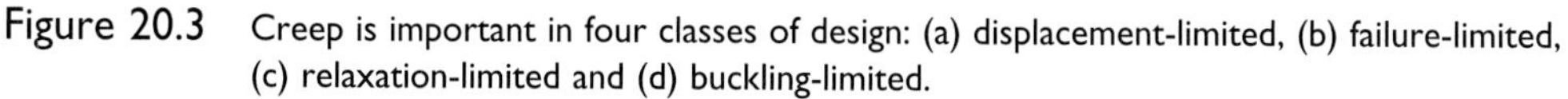

Figure 20.3 Creep is important in four classes of design: (a) displacement-limited, (b) failure-limited, (c) relaxation-limited and (d) buckling-limited.

By plotting the log of the steady creep rate, $\dot{\epsilon}_{ss}$, against log (stress, σ), at constant T, as shown in Figure 20.5 we can establish that

$$\dot{\epsilon}_{ss} = B\sigma^n \tag{20.1}$$

where n, the *creep exponent,* usually lies between 3 and 8. This sort of creep is called "*power-law*" *creep*. (At low σ, a different régime is entered

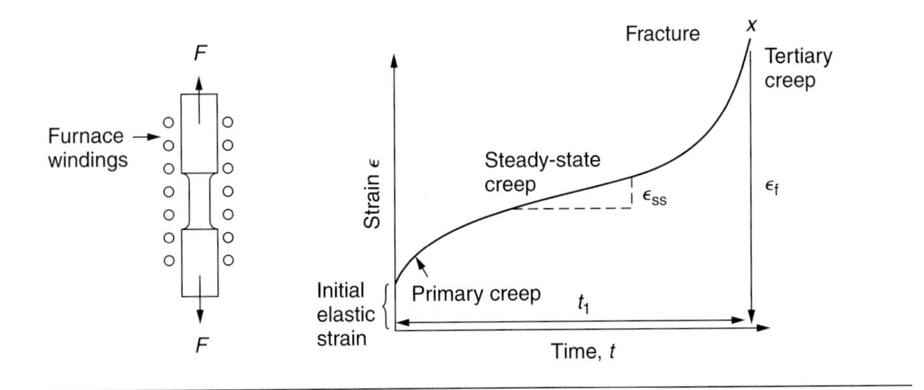

Figure 20.4 Creep testing and creep curves.

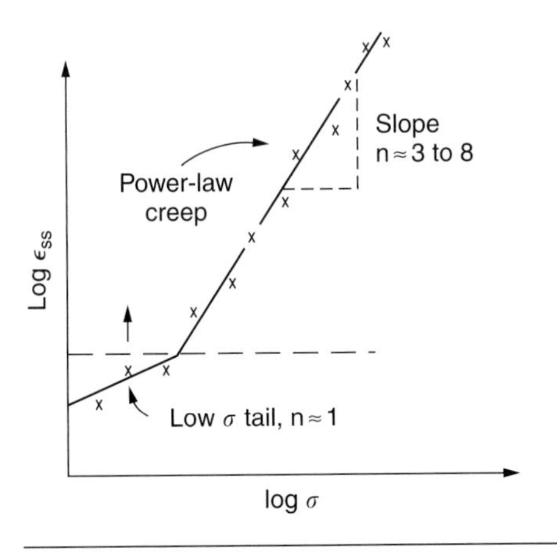

Figure 20.5 Variation of creep rate with stress.

where $n \approx 1$; we shall discuss this low-stress deviation from power-law creep in Chapter 22, but for the moment we shall not comment further on it.)

By plotting the *natural* logarithm (ln) of $\dot{\epsilon}_{ss}$ against the reciprocal of the *absolute* temperature ($1/T$) at constant stress, as shown in Figure 20.6, we find that:

$$\dot{\epsilon}_{ss} = Ce^{-(Q/\overline{R}T)} \tag{20.2}$$

Here $\overline{R}$ is the Universal Gas Constant ($8.31\,\mathrm{J\,mol^{-1}K^{-1}}$) and Q is called the *Activation Energy for Creep*—it has units of $\mathrm{J\,mol^{-1}}$. Note that the creep rate increases exponentially with temperature (Figure 20.6, inset). An increase in temperature of 20°C can *double* the creep rate.

Combining these two dependences of $\dot{\epsilon}_{ss}$ gives, finally,

$$\dot{\epsilon}_{ss} = A\sigma^n e^{-(Q/\overline{R}T)} \tag{20.3}$$

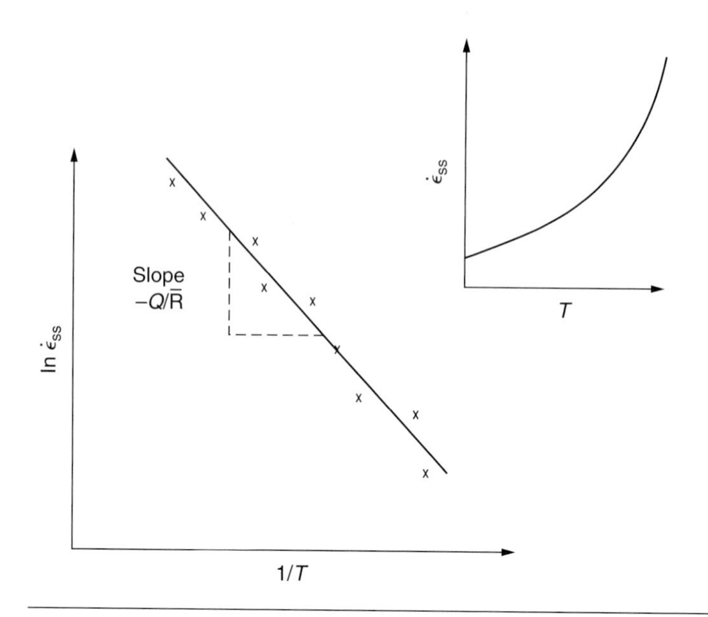

Figure 20.6 Variation of creep rate with temperature.

where A is the creep constant. The values of the three constants A, n and Q charactize the creep of a material; if you know these, you can calculate the strain rate at any temperature and stress by using the last equation. They vary from material to material, and have to be found experimentally.

20.3 Creep relaxation

At constant displacement, creep causes stresses to relax with time. Bolts in hot turbine casings must be regularly tightened. Plastic paper clips are not, in the long-term, as good as steel ones because, even at room temperature, they slowly lose their grip.

The relaxation time (arbitrarily defined as the time taken for the stress to relax to half its original value) can be calculated from the power-law creep data as follows. Consider a bolt which is tightened onto a rigid component so that the initial stress in its shank is σ_i. In this geometry (Figure 20.3(*c*)) the length of the shank must remain constant—that is, the *total* strain in the shank ϵ^{tot} must remain constant. But creep strain ϵ^{cr} can *replace* elastic strain ϵ^{el}, causing the stress to relax. At any time t

$$\epsilon^{tot} = \epsilon^{el} + \epsilon^{cr} \tag{20.4}$$

But

$$\epsilon^{el} = \sigma/E \tag{20.5}$$

and (at constant temperature)

$$\dot{\epsilon}^{cr} = B\sigma^n$$

Since ε^{tot} is constant, we can differentiate equation (20.4) with respect to time and substitute the other two equations into it to give

$$\frac{1}{E}\frac{d\sigma}{dt} = -B\sigma^n \tag{20.6}$$

Integrating from $\sigma = \sigma_i$ at $t = 0$ to $\sigma = \sigma$ at $t = t$ gives

$$\frac{1}{\sigma^{n-1}} - \frac{1}{\sigma_i^{n-1}} = (n-1)BEt \tag{20.7}$$

Figure 20.7 shows how the initial elastic strain σ_i/E is slowly replaced by creep strain, and the stress in the bolt relaxes. If, as an example, it is a casing bolt in a large turbogenerator, it will have to be retightened at intervals to prevent steam leaking from the turbine. The time interval between retightening, t_r, can be calculated by evaluating the time it takes for σ to fall to (say) one-half of its initial value. Setting $\sigma = \sigma_i/2$ and rearranging gives

$$t_r = \frac{(2^{n-1} - 1)}{(n-1)BE\sigma_i^{n-1}} \tag{20.8}$$

Experimental values for n, A, and Q for the material of the bolt thus enable us to decide how often the bolt will need retightening. Note that overtightening the bolt does not help because t_r decreases rapidly as σ_i increases.

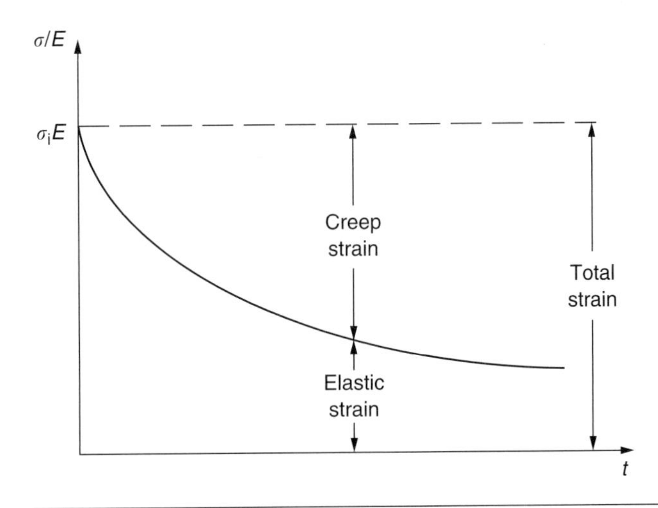

Figure 20.7 Replacement of elastic strain by creep strain with time at high temperature.

20.4 Creep damage and creep fracture

During creep, damage, in the form of internal cavities, accumulates. The damage first appears at the start of the tertiary stage of the creep curve and grows at an increasing rate thereafter. The shape of the tertiary stage of the creep curve (Figure 20.4) reflects this: as the cavities grow, the section of the sample decreases, and (at constant load) the stress goes up. Since $\dot{\epsilon} \propto \sigma^n$, the creep rate goes up even faster than the stress does (Figure 20.8).

It is not surprising—since creep causes creep fracture—that the time-to-failure, t_f is described by a constitutive equation which looks very like that for creep itself:

$$t_f = A'\sigma^{-m}e^{+(Q/\bar{R}T)}$$

Here A', m and Q are the creep-failure constants, determined in the same way as those for creep (the exponents have the opposite sign because t_f is a time whereas $\dot{\epsilon}_{ss}$ is a rate).

In many high-strength alloys this creep damage appears early in life and leads to failure after small creep strains (as little as 1 percent). In high-temperature design it is important to make sure:

(a) that the *creep strain* ϵ^{cr} during the design life is acceptable;
(b) that the *creep ductility* ϵ_f^{cr} (strain to failure) is adequate to cope with the acceptable creep strain;
(c) that the *time-to-failure*, t_f, at the design loads and temperatures is longer (by a suitable safety factor) than the design life.

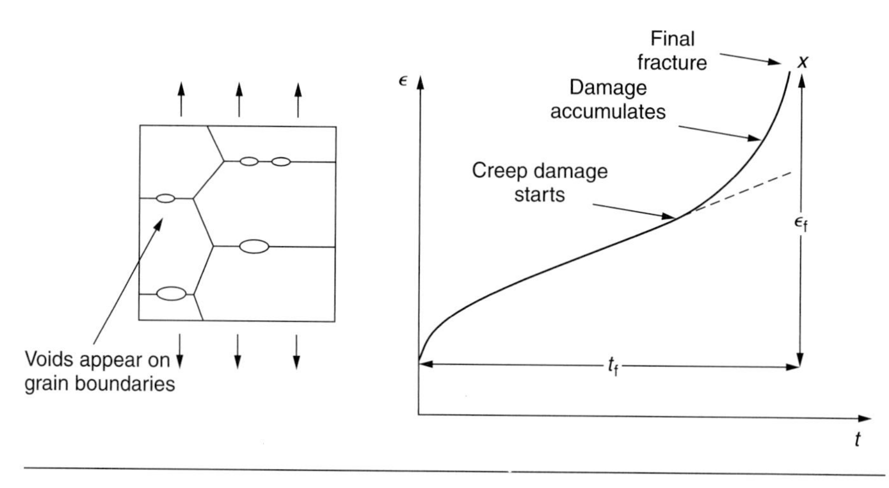

Figure 20.8 Creep damage.

Times-to-failure are normally presented as *creep-rupture* diagrams (Figure 20.9). Their application is obvious: if you know the stress and temperature you can read off the life; if you wish to design for a certain life at a certain temperature, you can read off the design stress.

20.5 Creep-resistant materials

From what we have said so far it should be obvious that the first requirement that we should look for in choosing materials that are resistant to creep is that they should have high melting (or softening) temperatures. If the material can then be used at less than 0.3 of its melting temperature creep will not be a problem. If it has to be used above this temperature, various alloying procedures can be used to increase creep resistance. To understand these, we need to know more about the mechanisms of creep—the subject of the next two chapters.

Examples

20.1 A cylindrical tube in a chemical plant is subjected to an excess internal pressure of $6\,\mathrm{MN\,m^{-2}}$, which leads to a circumferential stress in the tube wall. The tube wall is required to withstand this stress at a temperature of 510°C for 9 years. A designer has specified tubes of 40 mm bore and 2 mm wall thickness made from a stainless alloy of iron with 15 percent by weight of chromium. The manufacturer's specification for this alloy gives the following information:

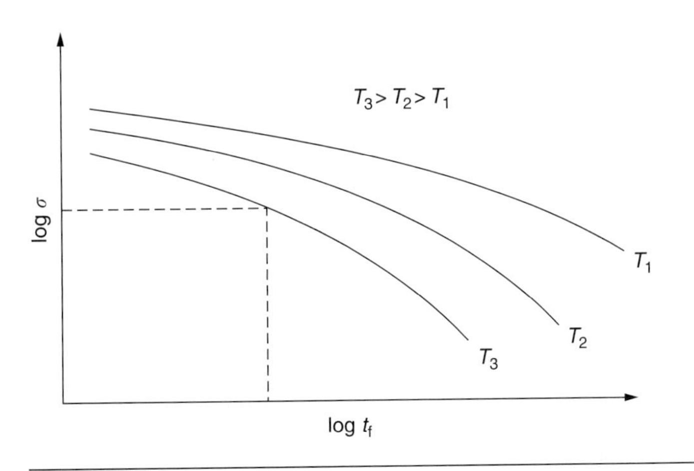

Figure 20.9 Creep-rupture diagram.

Temperature (°C)	618	640	660	683	707
Steady state creep rate $\dot{\epsilon}(s^{-1})$, for an applied tensile stress σ of 200 MNm^{-2}	1.0×10^{-7}	1.7×10^{-7}	4.3×10^{-7}	7.7×10^{-7}	2.0×10^{-6}

Over the present ranges of stress and temperature the alloy can be considered to creep according to the equation

$$\dot{\epsilon} = A\sigma^5 e^{-Q/\bar{R}T}$$

where A and Q are constants, $\bar{R}$ is the universal gas constant, and T is the absolute temperature. Given that failure is imminent at a creep strain of 0.01 for the present alloy, comment on the safety of the design.

Answer

Strain over 9 years = 0.00057; design safe.

20.2 An alloy tie bar in a chemical plant has been designed to withstand a stress, σ, of 25 $MN\,m^{-2}$ at 620°C. Creep tests carried out on specimens of the alloy under these conditions indicated a steady-state creep rate, $\dot{\epsilon}$, of $3.1 \times 10^{-12}\,s^{-1}$. In service it was found that, for 30 percent of the running time, the stress and temperature increased to 30 $MN\,m^{-2}$ and 650°C. Calculate the average creep rate under service conditions. It may be assumed that the alloy creeps according to the equation

$$\dot{\epsilon} = A\sigma^5 e^{-Q/\bar{R}T}$$

where A and Q are constants, $\bar{R}$ is the universal gas constant and T is the absolute temperature. Q has a value of 160 $kJ\,mol^{-1}$.

Answer

$6.82 \times 10^{-12}\,s^{-1}$

20.3 The window glass in old buildings often has an uneven surface, with features which look like flow marks. The common explanation is that the glass has suffered creep deformation over the years under its own weight. Explain why this scenario is complete rubbish (why do you think the glass does appear to have "flow marks"?).

20.4 Why do bolted joints in pressure vessels operating at high temperature have to be retorqued at regular intervals?

20.5 Why are creep–rupture figures useful when specifying materials for high-temperature service?

20.6 Structural steelwork in high-rise buildings is generally protected against fire by a thick coating of refractory material even though a fire would not melt the steel. Explain.

Chapter 21

Kinetic theory of diffusion

Chapter contents

21.1 Introduction

We saw in the last chapter that the rate of steady-state creep, $\dot{\epsilon}_{ss}$, varies with temperature as

$$\dot{\epsilon}_{ss} = Ce^{-(Q/\overline{R}T)}. \tag{21.1}$$

Here Q is the activation energy for creep (J mol^{-1} or, more usually, kJ mol^{-1}), $\overline{R}$ is the universal gas constant ($8.31\ \text{J mol}^{-1}\text{K}^{-1}$) and T is the absolute temperature (K). This is an example of *Arrhenius's law*. Like all good laws, it has great generality. Thus it applies not only to the rate of creep, but also to the rate of oxidation (Chapter 24), the rate of corrosion (Chapter 26), the rate of diffusion (this chapter), even to the rate at which bacteria multiply and milk turns sour. It states that the *rate of the process* (creep, corrosion, diffusion, etc.) *increases exponentially with temperature* (or, equivalently, that the time for a given amount of creep, or of oxidation, *decreases* exponentially with temperature) in the way shown in Figure 21.1. It follows that, if the rate of a process which follows Arrhenius's law is plotted on a $\log_e$ scale against $1/T$, a straight line with a slope of $-Q/\overline{R}$ is obtained (Figure 21.2). The value of Q characterizes the process—a feature we have already used in Chapter 20.

In this chapter, we discuss the origin of Arrhenius's law and its application to *diffusion*. In the next, we examine how it is that the rate of diffusion determines that of creep.

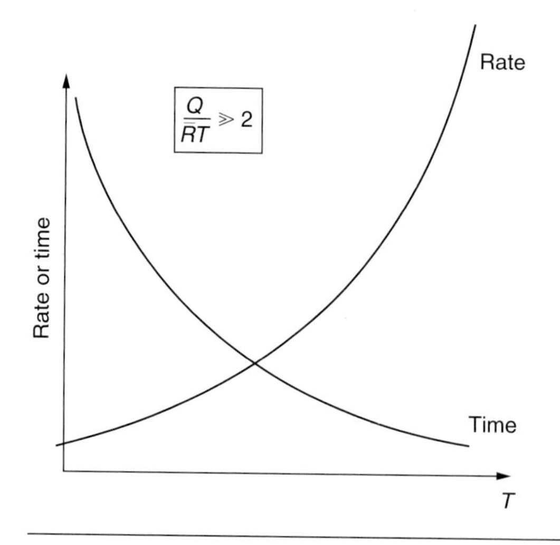

Figure 21.1 Consequences of Arrhenius's law.

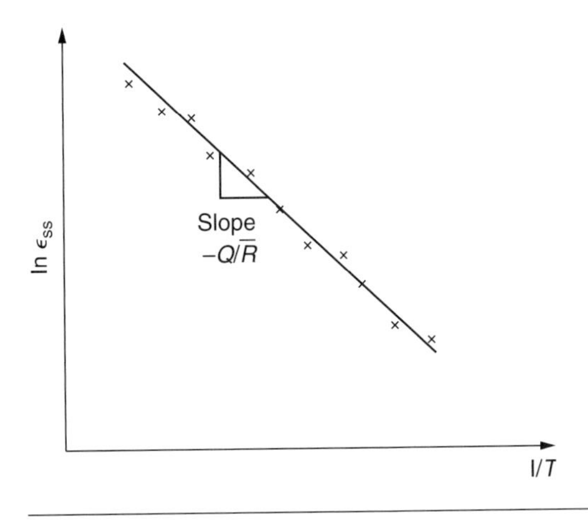

Figure 21.2 Creep rates follow Arrhenius's law.

21.2 Diffusion and Fick's law

First, what do we mean by *diffusion*? If we take a dish of water and drop a blob of ink very gently into the middle of it, the ink will spread sideways into the water. Measure the distance from the initial blob to the edge of the dish by the coordinate x. Provided the water is stagnant, the spreading of the ink is due to the movement of ink molecules by random exchanges with the water molecules. Because of this, the ink molecules move from regions where they are concentrated to regions where they are less concentrated; put another way: the ink diffuses down the ink *concentration gradient*. This behavior is described by *Fick's first law of diffusion*:

$$J = -D\frac{dc}{dx} \qquad (21.2)$$

Here J is the number of ink molecules diffusing down the concentration gradient *per second per unit area*; it is called the *flux* of molecules (Figure 21.3). The quantity c is the concentration of ink molecules in the water, defined as the *number* of ink molecules per *unit volume* of the ink—water solution; and D is the *diffusion coefficient* for ink in water—it has units of $m^2\,s^{-1}$.

This diffusive behavior is not just limited to ink in water—it occurs in all liquids, and more remarkably, in all solids as well. As an example, in the alloy *brass*—a mixture of zinc in copper—zinc atoms diffuse through the solid copper in just the way that ink diffuses through water. Because the materials of engineering are mostly solids, we shall now confine ourselves to talking about diffusion in the solid state.

Physically, diffusion occurs because atoms, even in a solid, are able to move—to jump from one atomic site to another. Figure 21.4 shows a solid in

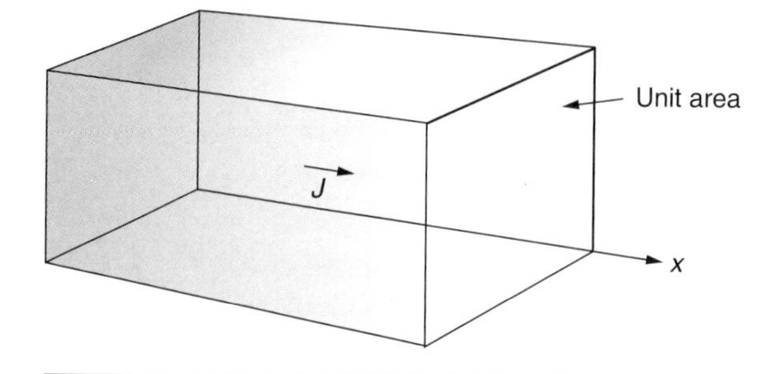

Figure 21.3 Diffusion down a concentration gradient.

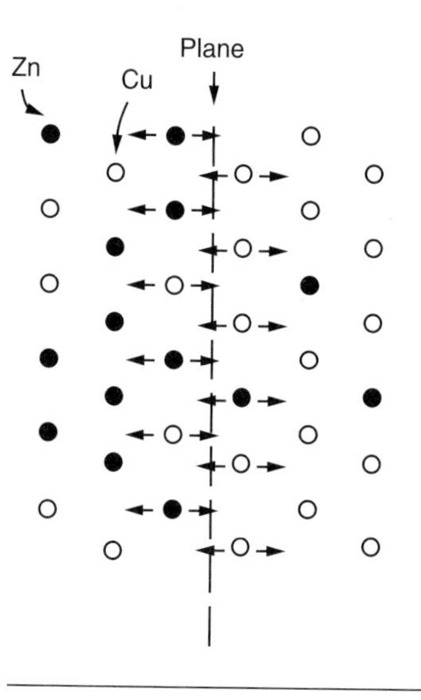

Figure 21.4 Atom jumps across a plane.

which there is a concentration gradient of black atoms: there are more to the left of the broken line than there are to the right. If atoms jump across the broken line at random, then there will be a *net flux* of black atoms to the right (simply because there are more on the left to jump), and, of course, a net flux of white atoms to the left. Fick's law describes this. It is derived in the following way.

The atoms in a solid vibrate, or oscillate, about their mean positions, with a frequency ν (typically about $10^{13}\,s^{-1}$). The crystal lattice defines these mean positions. At a temperature T, the average energy (kinetic plus potential) of a vibrating atom is $3kT$ where k is Boltzmann's constant (1.38×10^{-23} J $atom^{-1}\,K^{-1}$). But this is only the average energy. As atoms (or molecules) vibrate, they collide, and energy is continually transferred from one to another. Although the *average* energy is $3kT$, at any instant, there is a certain probability that an atom has more or less than this. A very small fraction of the atoms have, at a given instant, much more — enough, in fact, to jump to a neighboring atom site. It can be shown from statistical mechanical theory that the probability, p, that an atom will have, at any instant, an energy $\geqslant q$ is

$$p = e^{-q/kT} \tag{21.3}$$

Why is this relevant to the diffusion of zinc in copper? Imagine two adjacent lattice planes in the brass with two slightly different zinc concentrations, as shown in exaggerated form in Figure 21.5. Let us denote these two planes as A and B. Now for a zinc atom to diffuse from A to B, down the concentration gradient, it has to "squeeze" between the copper atoms (a simplified statement — but we shall elaborate on it in a moment). This is another way of saying: the zinc atom has to overcome an *energy barrier* of height q, as shown in Figure 21.5. Now, the number of zinc atoms in layer A is n_A, so that the number of zinc atoms that have enough energy to climb over the barrier from A to B at any instant is

$$n_A p = n_A e^{-q/kT} \tag{21.4}$$

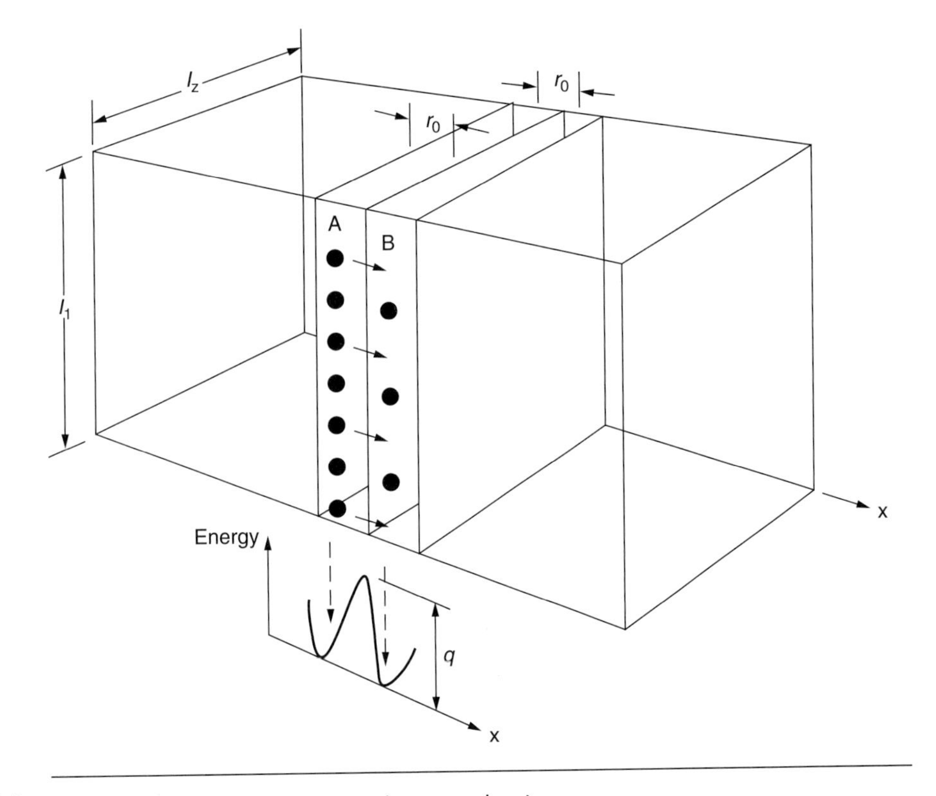

Figure 21.5 Diffusion requires atoms to cross the energy barrier q.

In order for these atoms to *actually* climb over the barrier from A to B, they must of course be moving in the right direction. The number of times each zinc atom oscillates towards B is $\approx \nu/6\ \mathrm{s}^{-1}$ (there are six possible directions in which the zinc atoms can move in three dimensions, only one of which is from A to B). Thus the *number* of atoms that *actually* jump from A to B per second is

$$\frac{\nu}{6} n_A e^{-q/kT} \tag{21.5}$$

But, meanwhile, some zinc atoms jump back. If the number of zinc atoms in layer B is n_B, the number of zinc atoms that can climb over the barrier from B to A per second is

$$\frac{\nu}{6} n_B e^{-q/kT} \tag{21.6}$$

Thus the net number of zinc atoms climbing over the barrier per second is

$$\frac{\nu}{6}(n_A - n_B) e^{-q/kT} \tag{21.7}$$

The area of the surface across which they jump is $l_1\, l_2$, so the net *flux* of atoms, using the definition given earlier, is:

$$J = \nu \frac{(n_A - n_B)}{6 l_1 l_2} e^{-q/kT} \tag{21.8}$$

Concentrations, c, are related to numbers, n, by

$$c_A = \frac{n_A}{l_1 l_2 r_0}, \quad c_B = \frac{n_B}{l_1 l_2 r_0} \tag{21.9}$$

where c_A and c_B are the zinc concentrations at A and B and r_0 is the atom size. Substituting for n_A and n_B in equation (21.8) gives

$$J = \frac{\nu}{6}\, r_0 (c_A - c_B) e^{-q/kT} \tag{21.10}$$

But $-(c_A - c_B)/r_0$ is the concentration gradient, $\mathrm{d}c/\mathrm{d}x$. The quantity q is inconveniently small and it is better to use the larger quantities $Q = N_A q$ and $\bar{R} = N_A k$ where N_A is Avogadro's number. The quantity $\nu r_0^2/6$ is usually written as D_0. Making these changes gives:

$$J = -D_0 e^{-Q/\bar{R}T} \left(\frac{\mathrm{d}c}{\mathrm{d}x}\right) \tag{21.11}$$

and this is just Fick's law (equation 21.2) with

$$D = D_0 e^{-Q/\bar{R}T} \tag{21.12}$$

This method of writing D emphasizes its exponential dependence on temperature, and gives a conveniently sized activation energy (expressed per mole of diffusing atoms rather than per atom). Thinking again of creep, the thing about equation (21.12) is that the exponential dependence of D on temperature has exactly the same form as the dependence of $\dot{\epsilon}_{ss}$ on temperature that we seek to explain.

21.3 Data for diffusion coefficients

Diffusion coefficients are best measured in the following way. A thin layer of a *radioactive isotope* of the diffusing atoms or molecules is plated onto the bulk material (e.g. radioactive zinc onto copper). The temperature is raised to the diffusion temperature for a measured time, during which the isotope diffuses into the bulk. The sample is cooled and sectioned, and the concentration of isotope measured as a function of depth by measuring the radiation it emits. D_0 and Q are calculated from the diffusion profile.

Materials handbooks list data for D_0 and Q for various atoms diffusing in metals and ceramics; Table 21.1 gives some of the most useful values. Diffusion occurs in polymers, composites, and glasses, too, but the data are less reliable.

Table 21.1 Data for bulk diffusion

Material	D_0(m^{2}s^{-1})	Q(kJmol^{-1})	$Q/\bar{R}T_M$
BCC metals			
Tungsten	5.0×10^{-4}	585	19.1
Molybdenum	5.0×10^{-5}	405	14.9
Tantalum	1.2×10^{-5}	413	16.9
Alpha-iron	2.0×10^{-4}	251	16.6
CPH metals			
Zinc	1.3×10^{-5}	91.7	15.9
Magnesium	1.0×10^{-4}	135	17.5
Titanium	8.6×10^{-10}	150	9.3
FCC metals			
Copper	2.0×10^{-5}	197	17.5
Aluminum	1.7×10^{-4}	142	18.3
Lead	1.4×10^{-4}	109	21.8
Gamma-iron	1.8×10^{-5}	270	17.9
Oxides			
MgO	1.4×10^{-6}	460	17.7
Al_2O_3	3.0×10^{-2}	556	28.0
FeO	1.0×10^{-2}	326	23.9
Interstitial diffusion in iron			
C in α Fe	2.0×10^{-6}	84	5.6
C in γ Fe	2.3×10^{-5}	147	9.7
N in α Fe	3.0×10^{-7}	76	5.1
N in γ Fe	9.1×10^{-5}	168	11.6
H in α Fe	1.0×10^{-7}	14	1.0

Table 21.2 Average values of D_0 and $Q/\bar{R}T_M$ for material classes

Material class	$D_0(m^2s^{-1})$	$Q/\bar{R}T_M$
BCC metals (W, Mo, Fe below 911°C, etc.)	1.6×10^{-4}	17.8
CPH metals (Zn, Mg. Ti, etc.)	5×10^{-5}	17.3
FCC metals (Cu, Al, Ni, Fe above 911°C, etc.)	5×10^{-5}	18.4
Alkali halides (NaCl, LiF, etc.)	2.5×10^{-3}	22.5
Oxides (MgO, FeO, Al_2O_3, etc.)	3.8×10^{-4}	23.4

The last column of Table 21.1 shows the *normalized activation energy*, $Q/\bar{R}T_M$, where T_M is the melting point (in kelvin). It is found that, for a given *class of material* (e.g. f.c.c. metals, or refractory oxides) the diffusion parameter D_0 for mass transport—and this is the one that is important in creep—is roughly constant; and that the activation energy is proportional to the melting temperature T_M (K) so that $Q/\bar{R}T_M$, too, is a constant (which is why creep is related to the melting point). This means that *many* diffusion problems can be solved approximately using the data given in Table 21.2, which shows the average values of D_0 and $Q/\bar{R}T_M$, for material classes.

21.4 Mechanisms of diffusion

In our discussion so far we have begged the question of just how the atoms in a solid move around when they diffuse. There are several ways in which this can happen. For simplicity, we shall talk only about crystalline solids, although diffusion occurs in amorphous solids as well, and in similar ways.

Bulk diffusion: interstitial and vacancy diffusion

Diffusion in the bulk of a crystal can occur by two mechanisms. The first is *interstitial* diffusion. Atoms in all crystals have spaces, or *interstices*, between them, and *small* atoms dissolved in the crystal can diffuse around by squeezing between atoms, jumping—when they have enough energy—from one interstice to another (Figure 21.6). Carbon, a small atom, diffuses through steel in this way; in fact C, O, N, B, and H diffuse interstitially in most crystals. These small atoms diffuse very quickly. This is reflected in their exceptionally small values of $Q/\bar{R}T_M$, seen in the last column of Table 21.1.

The second mechanism is that of *vacancy* diffusion. When zinc diffuses in brass, for example, the zinc atom (comparable in size to the copper atom) cannot fit into the interstices—the zinc atom has to wait until a *vacancy*, or

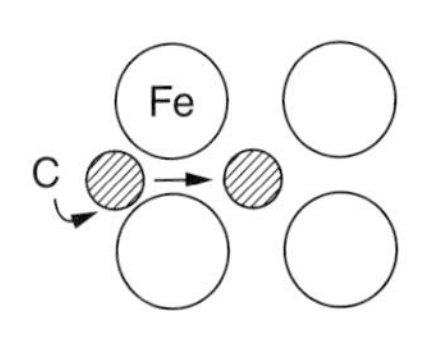

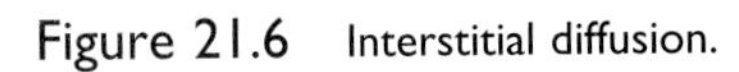

Figure 21.6 Interstitial diffusion.

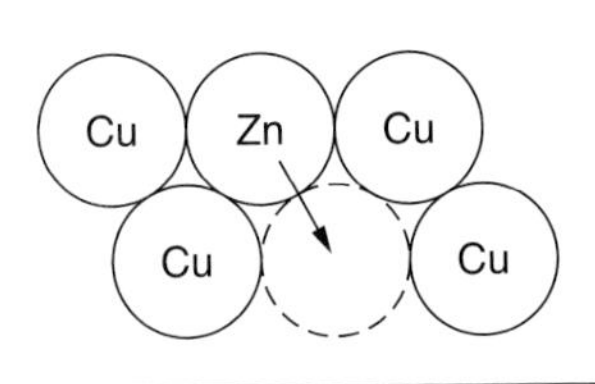

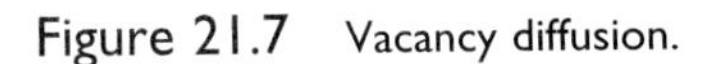

Figure 21.7 Vacancy diffusion.

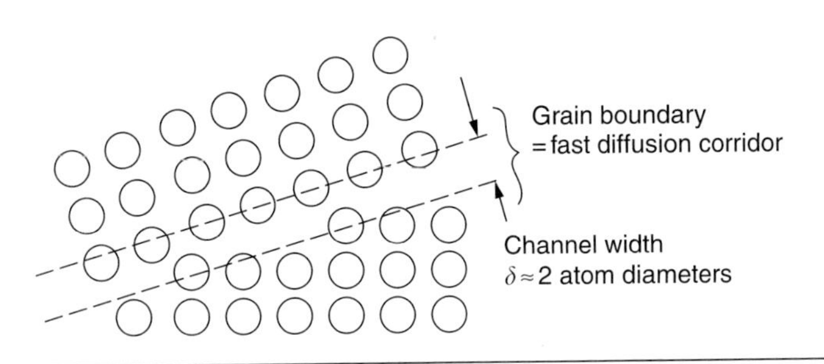

Figure 21.8 Grain-boundary diffusion.

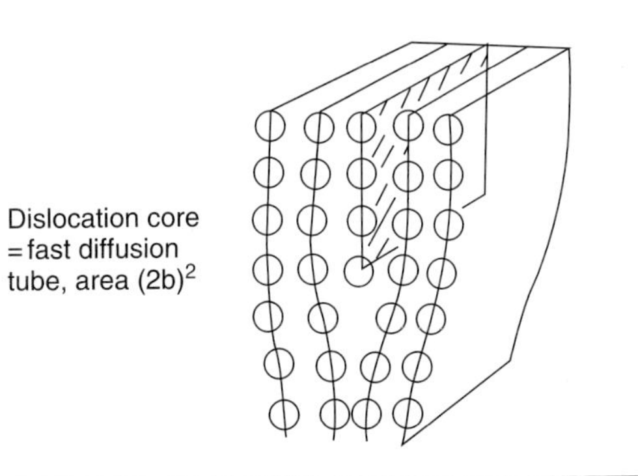

Figure 21.9 Dislocation-core diffusion.

missing atom, appears next to it before it can move. This is the mechanism by which most diffusion in crystals takes place (Figures 21.7 and 10.4).

Fast diffusion paths: grain boundary and dislocation core diffusion

Diffusion in the bulk crystals may sometimes be *short circuited* by diffusion down grain boundaries or dislocation cores. The boundary acts as a planar channel, about two atoms wide, with a local diffusion rate which can be as much as 10^6 times greater than in the bulk (Figures 21.8 and 10.4). The dislocation core, too, can act as a high conductivity "wire" of cross-section about $(2b)^2$, where b is the atom size (Figure 21.9). Of course, their contribution to the total diffusive flux depends also on how many grain boundaries or dislocations there are: when grains are small or dislocations numerous, their contribution becomes important.

A useful approximation

Because the diffusion coefficient D has units of $\mathrm{m^2\,s^{-1}}$, $\sqrt{D \times \text{time (in seconds)}}$ has units of meters. The length scale of $\sqrt{Dt}$ is in fact a good approximation to the distance that atoms diffuse over a period of time t. The proof of this result involves non-steady diffusion (Fick's second law of diffusion) and we will not go through it here. However, the approximation

$$x = \sqrt{Dt}$$

is a most useful result for estimating diffusion distances in a wide range of situations.

Examples

21.1 It is found that a force F will inject a given weight of a thermosetting polymer into an intricate mould in 30 s at 177°C and in 81.5 s at 157°C. If the viscosity of the polymer follows an *Arrhenius* law, with a rate of process proportional to $e^{-Q/RT}$, calculate how long the process will take at 227°C.

Answer

3.5 s.

21.2 Explain what is meant by *diffusion* in materials. Account for the variation of diffusion rates with (a) temperature, (b) concentration gradient and (c) grain size.

21.3 Use your knowledge of diffusion to account for the following observations:

(a) Carbon diffuses fairly rapidly through iron at 100 °C, whereas chromium does not.
(b) Diffusion is more rapid in polycrystalline silver with a small grain size than in coarse-grained silver.

21.4 Give an approximate expression for the time t required for significant diffusion to take place over a distance x in terms of x, and the diffusion coefficient, D. A component is made from an alloy of copper with 18% by weight of zinc. The concentration of zinc is found to vary significantly over distances of 10 μm. Estimate the time required for a substantial levelling out of the zinc concentration at 750°C. The diffusion coefficient for zinc in the alloy is given by

$$D = D_0 e^{-Q/\overline{R}T}$$

where $\overline{R}$ is the universal gas constant and T is the absolute temperature. The constants D_0 and Q have the values $9.5\,\text{mm}^2\,\text{s}^{-1}$ and $159\,\text{kJ}\,\text{mol}^{-1}$, respectively.

Answers

$t = x^2/D$; 23 min.

Chapter 22

Mechanisms of creep, and creep-resistant materials

Chapter contents

22.1 Introduction

In Chapter 20 we showed that, when a material is loaded at a high temperature, it creeps, that is, it deforms, continuously and permanently, at a stress that is less than the stress that would cause any permanent deformation at room temperature. In order to understand how we can make engineering materials more resistant to creep deformation and creep fracture, we must first look at how creep and creep-fracture take place on an atomic level, that is, we must identify and understand the *mechanisms* by which they take place.

There are two mechanisms of creep: *dislocation creep* (which gives power-law behavior) and *diffusional creep* (which gives linear-viscous creep). The rate of both is usually limited by diffusion, so both follow Arrhenius's law. Creep fracture, too, depends on diffusion. Diffusion becomes appreciable at about $0.3T_M$—that is why materials start to creep above this temperature.

22.2 Creep mechanisms: metals and ceramics

Dislocation creep (giving power-law creep)

As we saw in Chapter 10, the stress required to make a crystalline material deform plastically is that needed to make the dislocations in it move. Their movement is resisted by (a) the intrinsic lattice resistance and (b) the obstructing effect of obstacles (e.g. dissolved solute atoms, precipitates formed with undissolved solute atoms, or other dislocations). Diffusion of atoms can "unlock" dislocations from obstacles in their path, and the movement of these unlocked dislocations under the applied stress is what leads to dislocation creep.

How does this unlocking occur? Figure 22.1 shows a dislocation which cannot glide because a precipitate blocks its path. The glide force τb per unit length is balanced by the reaction f_0 from the precipitate. But unless the dislocation hits the precipitate at its mid-plane (an unlikely event) there is a component of force left over. It is the component $\tau b \tan \theta$, which tries to push the dislocation *out of its slip plane.*

The dislocation cannot *glide* upwards by the shearing of atom planes—the atomic geometry is wrong—but the dislocation *can* move upwards if atoms at the bottom of the half-plane are able to diffuse away (Figure 22.2). We have come across Fick's law in which diffusion is driven by differences in *concentration.* A *mechanical* force can do exactly the same thing, and this is what leads to the diffusion of atoms away from the "loaded" dislocation, eating away its extra half-plane of atoms until it can clear the precipitate. The process is called "climb", and since it requires diffusion, it can occur only when the temperature is above $0.3T_M$ or so. At the lower end of the creep regime

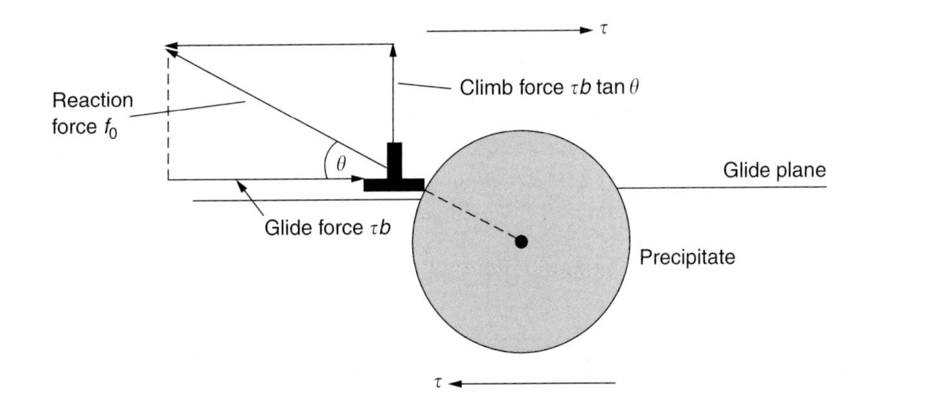

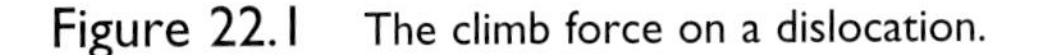

Figure 22.1 The climb force on a dislocation.

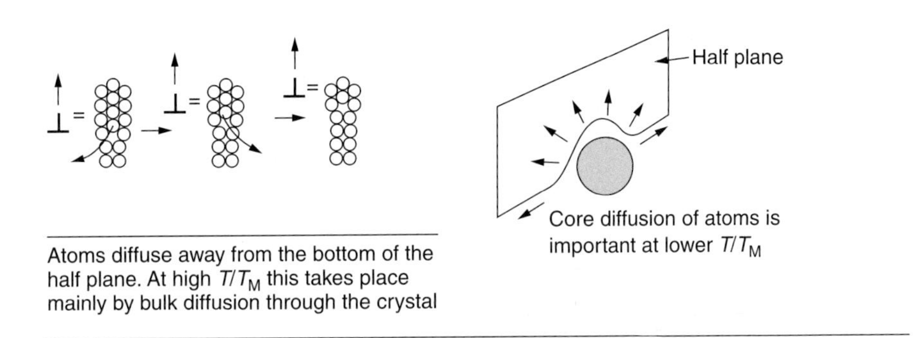

Half plane
Core diffusion of atoms is important at lower T/T_M
Atoms diffuse away from the bottom of the half plane. At high T/T_M this takes place mainly by bulk diffusion through the crystal

Figure 22.2 How diffusion leads to climb.

$(0.3–0.5T_M)$ core diffusion tends to be the dominant mechanism; at the higher end $(0.5T_M–0.99T_M)$ it is bulk diffusion (Figure 22.2).

Climb unlocks dislocations from the precipitates which pin them and further slip (or "glide") can then take place (Figure 22.3). Similar behavior takes place for pinning by solute, and by other dislocations. After a little glide, of course, the unlocked dislocations bump into the next obstacles, and the whole cycle repeats itself. This explains the *progressive, continuous,* nature of creep, and the role of diffusion, with diffusion coefficient

$$D = D_0 e^{-Q/\overline{R}T}$$

explains the dependence of creep rate on *temperature*, with

$$\dot{\epsilon}_{ss} = A\sigma^n e^{-Q/\overline{R}T} \tag{22.1}$$

The dependence of creep rate on applied *stress* σ is due to the climb force: the higher σ, the higher the climb force $\tau b \tan\theta$, the more dislocations become

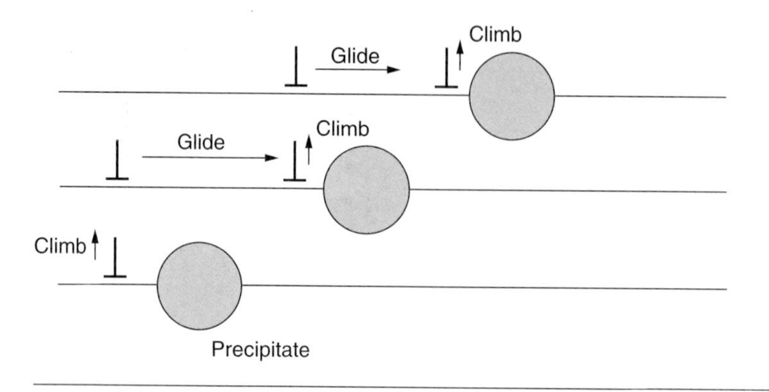

Figure 22.3 How the climb-glide sequence leads to creep.

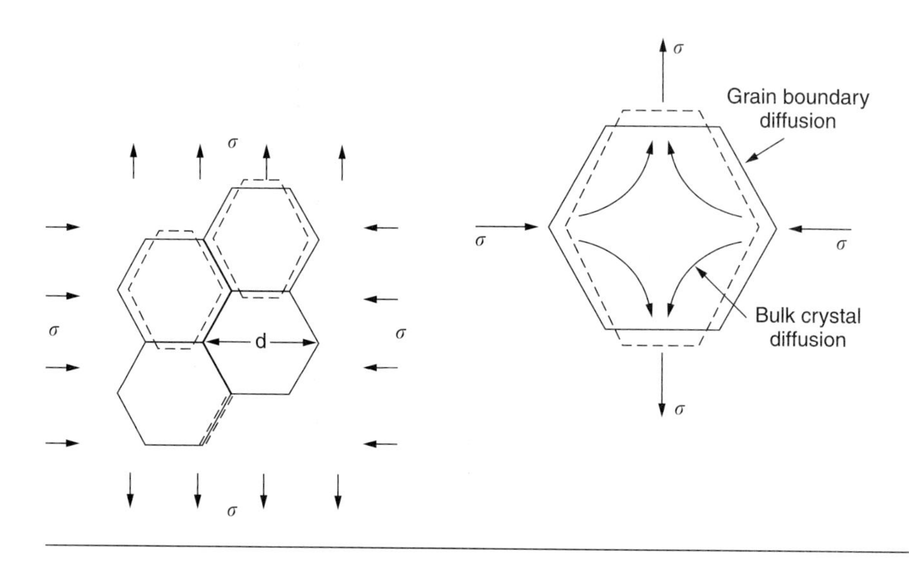

Figure 22.4 How creep takes place by diffusion.

unlocked per second, the more dislocations glide per second, and the higher is the strain rate.

Diffusion creep (giving linear-viscous creep)

As the stress is reduced, the rate of power-law creep (equation (22.1)) falls quickly (remember n is between 3 and 8). But creep does not stop; instead, an alternative mechanism takes over. As Figure 22.4 shows, a polycrystal can extend in response to the applied stress, σ, by grain elongation; here, σ acts again as a mechanical driving force, but this time atoms *diffuse* from one set of the grain faces to the other, and dislocations are not involved. At high T/T_M, this diffusion takes place through the crystal itself, that is, by bulk diffusion. The rate of creep is then obviously proportional to the *diffusion coefficient* D, (data in Table 21.1) and to the stress σ (because

Table 22.1 Temperature ranges and associated materials

Temperature range	*Principal materials**	*Applications*
Cryogenic	Copper alloys	Superconduction
−273 to −20°C	Austenitic (stainless) steels Aluminum alloys	Rocket casings, pipework, etc. Liquid O_2 or N_2 equipment
−20 to 150°C	Most polymers (max temp: 60 to 150°C) Magnesium alloys (up to 150°C) Aluminum alloys (up to 150°C) Monels and steels	Civil construction Household appliances Automotive Aerospace
150 to 400°C	PEEK, PEK, PI, PPD, PTFE and PES (up to 250°C) Fiber-reinforced polymers Copper alloys (up to 400°C) Nickel, monels and nickel-silvers	Food processing Automotive (engine)
400 to 575°C	Low alloy ferritic steels Titanium alloys (up to 450°C) Inconels and nimonics	Heat exchangers Steam turbines Gas turbine compressors
575 to 650°C	Iron-based super-alloys Ferritic stainless steels Austenitic stainless steels Inconels and nimonics	Steam turbines Superheaters Heat exchangers
650 to 1000°C	Austenitic stainless steels Nichromes, nimonics Nickel based super-alloys Cobalt based super-alloys	Gas turbines Chemical and petrochemical reactors Furnace components Nuclear construction
Above 1000°C	Refractory metals: Mo, W, Ta Alloys of Nb, Mo, W, Ta Ceramics: Oxides Al_2O_3, MgO, etc. Nitrides, carbides: Si_3N_4, SiC	Special furnaces Experimental turbines

* Copper alloys include brasses (Cu–Zn alloys), bronzes (Cu-Sn alloys), cupronickels (Cu–Ni alloys) and nickel-silvers (Cu–Sn–Ni–Pb alloys).
Titanium alloys generally mean those based on Ti–V–Al alloys.
Nickel alloys include monels (Ni–Cu alloys), nichromes (Ni–Cr alloys), nimonics and nickel-based super-alloys (Ni–Fe–Cr–Al–Co–Mo alloys).
Stainless steels include ferritic stainless (Fe–Cr–Ni alloys with $<6\%$ Ni) and austenitic stainless (Fe–Cr–Ni alloys with $>6.5\%$ Ni).
Low alloy ferritic steels contain up to 4% of Cr, Mo, and V.

σ drives diffusion in the same way that dc/dx does in Fick's law); and the creep rate varies as $1/d^2$ where d is the grain size (because when d gets larger, atoms have to diffuse further). Assembling these facts leads to the constitutive equation

$$\dot{\epsilon}_{ss} = C\frac{D\sigma}{d^2} = \frac{C'\sigma e^{-Q/\bar{R}T}}{d^2} \qquad (22.2)$$

where C and $C' = CD_0$ are constants. At lower T/T_M, when bulk diffusion is slow, grain-boundary diffusion takes over, but the creep rate is still proportional to σ. In order that holes do not open up between the grains, grain-boundary *sliding* is required as an accessory to this process.

Deformation mechanism diagrams

This competition between mechanisms is conveniently summarized on Deformation Mechanism Diagrams (Figures 22.5 and 22.6). They show the range of stress and temperature (Figure 22.5) or of strain-rate and stress (Figure 22.6) in which we expect to find each sort of creep (they also show where plastic yielding occurs, and where deformation is simply elastic). Diagrams like these are available for many metals and ceramics, and are a useful summary of creep behaviour, helpful in selecting a material for high-temperature applications.

Sometimes creep *is* desirable. Extrusion, hot rolling, hot pressing, and forging are carried out at temperatures at which power-law creep is the dominant mechanism of deformation. Then raising the temperature reduces the pressures required for the operation. The change in forming pressure for a given change in temperature can be calculated from equation (22.1).

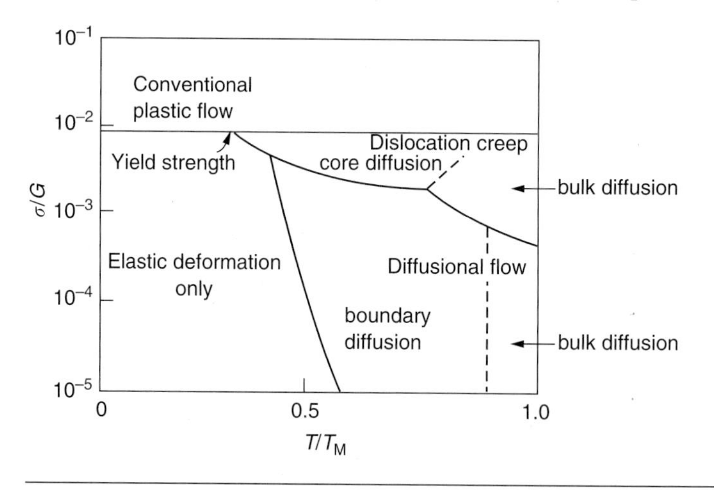

Figure 22.5 Deformation mechanisms at different stresses and temperatures.

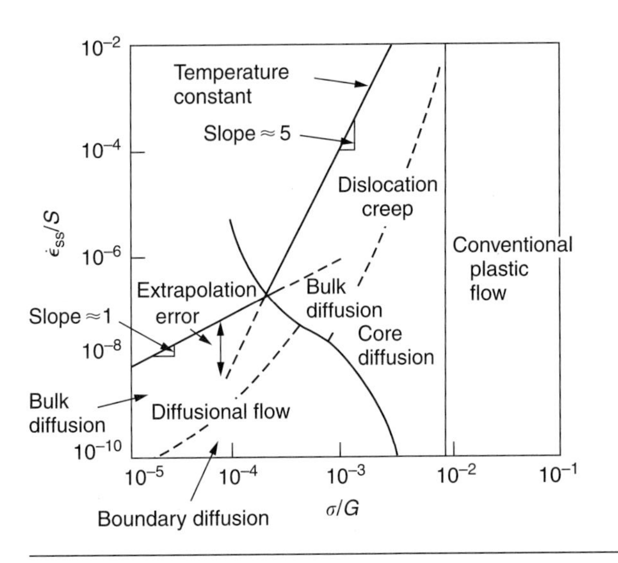

Figure 22.6 Deformation mechanisms at different strain-rates and stresses.

Creep fracture

Diffusion, we have seen, gives creep. It also gives creep fracture. If you stretch anything for long enough, it will break. You might think that a creeping material would—like toffee—stretch a long way before breaking in two, but, for crystalline materials, this is very rare. Indeed, creep fracture (in tension) can happen at unexpectedly small strains, often only 2–5%, by the mechanism shown in Figure 22.7. Voids appear on grain boundaries which lie normal to the tensile stress. These are the boundaries to which atoms diffuse to give diffusional creep, coming from the boundaries which lie parallel to the stress. But if the tensile boundaries have voids on them, they act as sources of atoms too, and in doing so, they grow. The voids cannot support load, so the *stress* rises on the remaining intact bits of boundary, the voids grow more and more quickly, until finally (bang) they link.

The life-time of a component—its time-to-failure, t_f—is related to the rate at which it creeps. As a general rule:

$$\dot{\epsilon}_{ss} t_f = C$$

where C is a constant, roughly 0.1. So, knowing the creep rate, the life can be estimated.

Many engineering components (e.g. tie bars in furnaces, super-heater tubes, high-temperature pressure vessels in chemical reaction plants) are expected to withstand moderate creep loads for long times (say 20 years) without failure. The safe loads or pressure they can safely carry are calculated by methods

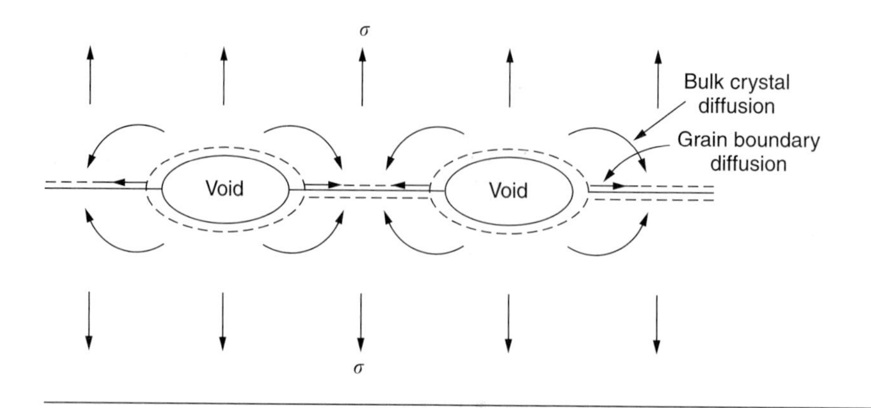

Figure 22.7 The growth of voids on grain boundaries by diffusion.

such as those we have just described. But there are dangers. One would like to be able to test new materials for these applications without having to run the tests for 20 years and more. It is thus tempting to speed up the tests by increasing the load to get observable creep in a short test time. Now, if this procedure takes us across the boundary between two different types of mechanism, we shall have problems about extrapolating our test data to the operating conditions. Extrapolation based on power-law creep will be on the dangerous side as shown in Figure 22.6. So: beware changes of mechanism in long extrapolations.

Designing metals and ceramics to resist power-law creep

If you are asked to select, or even to design, a material which will resist power-law creep, the criteria (all based on the ideas of this chapter and the last) are:

(a) choose a material with a high-melting point, since diffusion (and thus creep-rates) scale as T/T_M;
(b) maximize obstructions to dislocation motion by alloying to give a solid solution and precipitates—as much of both as possible; the precipitates must, of course, be stable at the service temperature;
(c) choose, if this is practical, a solid with a large lattice resistance: this means covalent bonding (as in many oxides, and in silicates, silicon carbide, silicon nitride, and related compounds).

Current creep-resistant materials are successful because they satisfy these criteria.

Designing metals and ceramics to resist diffusional flow

Diffusional flow is important when grains are small (as they often are in ceramics) and when the component is subject to high temperatures at low loads. To select a material which resists it, you should:

(a) choose a material with a high melting temperature;
(b) arrange that it has a large grain size, so that diffusion distances are long and grain boundaries do not help diffusion much — single crystals are best of all;
(c) arrange for precipitates at grain boundaries to impede grain-boundary sliding.

Metallic alloys are usually designed to resist power-law creep: diffusional flow is only rarely considered. One major exception is the range of directionally solidified ("DS") alloys described in the Case Study of Chapter 23: here special techniques are used to obtain very large grains.

Ceramics, on the other hand, often deform predominantly by diffusional flow (because their grains are small, and the high lattice resistance already suppresses power-law creep). Special heat treatments to increase the grain size can make them more creep-resistant.

22.3 Creep mechanisms: polymers

Creep of polymers is a major design problem. The glass temperature T_G, for a polymer, is a criterion of creep-resistance, in much the way that T_M is for a metal or a ceramic. For most polymers, T_G is close to room temperature. Well below T_G, the polymer is a glass (often containing crystalline regions — Chapter 5) and is a brittle, elastic solid — rubber, cooled in liquid nitrogen, is an example. Above T_G the Van der Waals bonds within the polymer melt, and it becomes a rubber (if the polymer chains are cross-linked) or a viscous liquid (if they are not). Thermoplastics, which can be molded when hot, are a simple example; well below T_G they are elastic; well above, they are viscous liquids, and flow like treacle.

Viscous flow is a sort of creep. Like diffusion creep, its rate increases linearly with stress and exponentially with temperature, with

$$\dot{\epsilon}_{ss} = C\sigma e^{-Q/\bar{R}T} \tag{22.3}$$

where Q is the activation energy for viscous flow.

The exponential term appears for the same reason as it does in diffusion; it describes the rate at which molecules can slide past each other, permitting flow. The molecules have a lumpy shape (see Figure 5.9) and the lumps key the molecules together. The activation energy, Q, is the energy it takes to push one lump of a molecule past that of a neighboring molecule. If we compare the last equation with that defining the *viscosity* (for the tensile deformation of a viscous material)

$$\eta = \frac{\sigma}{3\dot{\epsilon}} \tag{22.4}$$

we see that the viscosity is

$$\eta = \frac{1}{3C} e^{+Q/\bar{R}T} \tag{22.5}$$

(The factor 3 appears because the viscosity is defined for shear deformation— as is the shear modulus G. For tensile deformation we want the viscous equivalent of Young's modulus E. The answer is 3η, for much the same reason that $E \approx (8/3)G \approx 3G$—see Chapter 3.) Data giving C and Q for polymers are available from suppliers. Then equation (22.3) allows injection moulding or pressing temperatures and loads to be calculated.

The temperature range in which most polymers are used is that near T_G when they are neither simple elastic solids nor viscous liquids; they are *visco-elastic* solids. If we represent the elastic behavior by a spring and the viscous behavior by a dash-pot, then visco-elasticity (at its simplest) is described by a coupled spring and dash-pot (Figure 22.8). Applying a load causes creep, but at an ever-decreasing rate because the spring takes up the tension. Releasing the load allows slow reverse creep, caused by the extended spring.

Real polymers require more elaborate systems of springs and dash-pots to describe them. This approach of *polymer rheology* can be developed to provide criteria for design with structural polymers. At present, this is rarely done; instead, graphical data (showing the creep extension after time t at stress σ and temperature T) are used to provide an estimate of the likely deformation during the life of the structure.

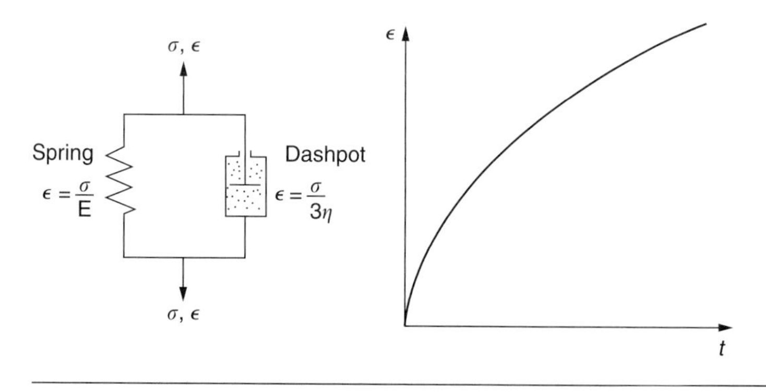

Figure 22.8 A model to describe creep in polymers.

Designing polymers to resist creep

The glass temperature of a polymer increases with the degree of cross-linking; heavily cross-linked polymers (e.g. epoxies) are therefore more creep-resistant at room temperature than those which are less cross-linked (like polyethylene). The viscosity of polymers above T_G increases with molecular weight, so that the rate of creep there is reduced by having a high molecular weight. Finally, crystalline or partly crystalline polymers (e.g. high-density polyethylene) are more creep-resistant than those which are entirely glassy (e.g. low-density polyethylene).

The creep-rate of polymers is reduced by filling them with glass or silica powders, roughly in proportion to the amount of filler added (PTFE on saucepans and polypropylene used for automobile components are both strengthened in this way). Much better creep resistance is obtained with composites containing continuous fibers (GFRP and CFRP) because much of the load is now carried by the fibers which, being very strong, do not creep at all.

22.4 Selecting materials to resist creep

Classes of industrial applications tend to be associated with certain characteristic temperature ranges. There is the cryogenic range, between −273°C and roughly room temperature, associated with the use of liquid gases like hydrogen, oxygen, or nitrogen. There is the regime at and near room temperature (−20 to +150°C) associated with conventional mechanical and civil engineering: household appliances, sporting goods, aircraft structures and housing are examples. Above this is the range 150–400°C, associated with automobile engines and with food and industrial processing. Higher still are the regimes of steam turbines and superheaters (typically 400–650°C) and of gas turbines and chemical reactors (650–1000°C). Special applications (lamp filaments, rocket nozzles) require materials which withstand even higher temperatures, extending as high as 2800°C.

Materials have evolved to fill the needs of each of these temperature ranges (Table 22.1). Certain polymers, and composites based on them, can be used in applications up to 250°C, and now complete with magnesium and aluminum alloys and with the much heavier cast irons and steels, traditionally used in those ranges. Temperatures above 400°C require special creep resistant alloys: ferritic steels, titanium alloys (lighter, but more expensive) and certain stainless steels. Stainless steels and ferrous superalloys really come into their own in the temperature range above this, where they are widely used in steam turbines and heat exchangers. Gas turbines require, in general, nickelbased or cobalt-based super-alloys. Above 1000°C, the refractory metals and ceramics become the only candidates. Materials used at high temperatures will, generally, perform perfectly well at lower temperatures too, but are not used there because of cost.

Examples

22.1 What is climb? Why does it require diffusion?

22.2 How do the processes of climb and glide produce creep deformation?

22.3 How does the rate of dislocation (power-law) creep depend on stress and temperature?

22.4 What is diffusion creep? How does the rate of diffusion creep depend on stress, temperature and grain size?

22.5 Why does core diffusion take over from bulk diffusion at lower temperatures in power-law creep?

22.6 Why does grain-boundary diffusion take over from bulk diffusion at lower temperatures in diffusion creep?

22.7 How would you design metals and ceramics to resist creep?

22.8 How do polymers creep? How would you design polymers to resist creep?

Chapter 23

The turbine blade — a case study in creep-limited design

Chapter contents

23.1 Introduction

In the last chapter we saw how a basic knowledge of the mechanisms of creep was an important aid to the development of materials with good creep properties. An impressive example is in the development of materials for the high-pressure stage of a modern aircraft gas turbine. Here we examine the properties such materials must have, the way in which the present generation of materials has evolved, and the likely direction of their future development.

As you may know, the *ideal* thermodynamic efficiency of a heat engine is given by

$$\frac{T_1 - T_2}{T_1} = 1 - \frac{T_2}{T_1} \tag{23.1}$$

where T_1 and T_2 are the absolute temperatures of the heat source and heat sink respectively. Obviously the greater T_1, the greater the maximum efficiency that can be derived from the engine. In practice the efficiency is a good deal less than ideal, but an increase in combustion temperature in a turbofan engine will, nevertheless, generate an increase in engine efficiency. Figure 23.1 shows the variation in efficiency of a turbofan engine plotted as a function of the turbine inlet temperature. In 1950 a typical aero engine operated at 700°C. The incentive then to increase the inlet temperature was strong, because of the steepness of the fuel-consumption curve at that temperature. By 1975 a typical engine (the RB211, for instance) operated at 1350°C, with a 50 percent saving in fuel per unit power output over the 1950 engines. But is it worth raising the temperature further? The shallowness of the consumption curve at 1400°C

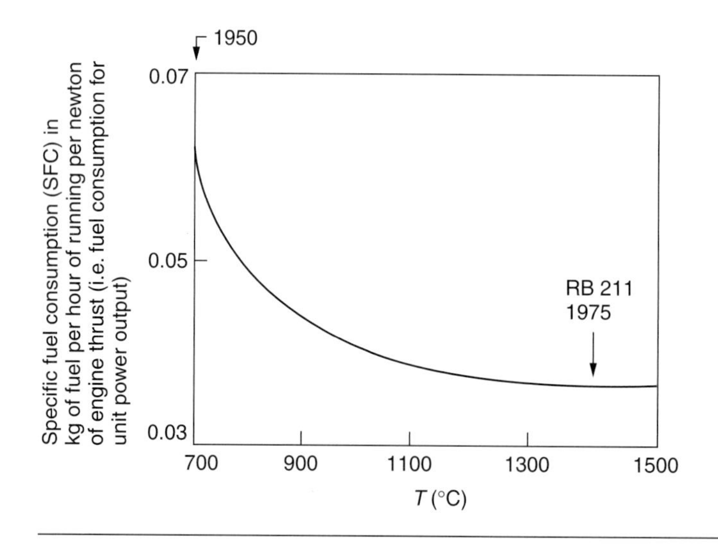

Figure 23.1 Turbofan efficiency at different inlet temperatures.

suggests that it might not be profitable; but there is a second factor: power-to-weight ratio Figure 23.2 shows a typical plot of the power output of a particular engine against turbine inlet temperature. This increases *linearly* with the temperature. If the turbine could both run at a higher temperature and be made of a lighter material there would be a double gain, with important financial benefits of increased payload.

23.2 Properties required of a turbine blade

Let us first examine the development of turbine-blade materials to meet the challenge of increasing engine temperatures. Although so far we have been stressing the need for excellent creep properties, a turbine-blade alloy must satisfy other criteria too. They are listed in Table 23.1.

The first—creep—is our interest here. The second—resistance to oxidation—is the subject of Chapter 24. Toughness and fatigue resistance (Chapters 13 and 17) are obviously important: blades must be tough enough to withstand the impact of birds and such like; and changes in the power level of

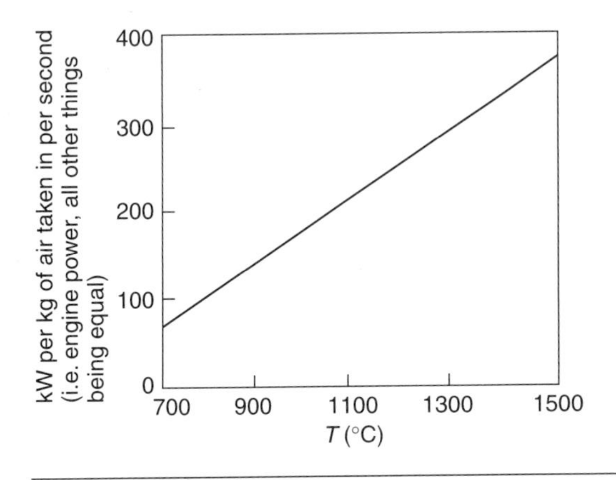

Figure 23.2 Turbofan power at different inlet temperatures.

Table 23.1 Alloy requirements

(a) Resistance to creep
(b) Resistance to high-temperature oxidation
(c) Toughness
(d) Thermal fatigue resistance
(e) Thermal stability
(f) Low density

the engine produce mechanical and thermal stresses which — if the blade material is wrongly chosen — will lead to *thermal fatigue*. The alloy composition and structure must remain *stable* at high temperature — precipitate particles can dissolve away if the alloy is overheated and the creep properties will then degenerate significantly. Finally, the *density* must be as low as possible — not so much because of blade weight but because of the need for stronger and hence heavier turbine discs to take the radial load.

These requirements severely limit our choice of creep-resistant materials. For example, ceramics, with their high softening temperatures and low densities, are ruled out for aero-engines because they are far too brittle (they are under evaluation for use in land-based turbines, where the risks and consequences of sudden failure are less severe — see below). Cermets offer no great advantage because their metallic matrices soften at much too low a temperature. The materials which best fill present needs are the *nickel-based super-alloys*.

23.3 Nickel-based super-alloys

The alloy used for turbine blades in the high pressure stage of an aircraft turbofan engine is a classic example of a material designed to be resistant to dislocation (power-law) creep at high stresses and temperatures. At take-off, the blade is subjected to stresses approaching 250 MN m^{-2}, and the design specification requires that this stress shall be supported for 30 h at 850°C without more than a 0.1 percent irreversible creep strain. In order to meet these stringent requirements, an alloy based on nickel has evolved with the rather mind-boggling specification given in Table 23.2.

No one tries to remember exact details of this or similar alloys. But the point of all these complicated additions of foreign atoms to the nickel is straightforward. It is: (a) to have as many atoms in solid solution as possible (the cobalt, the tungsten, and the chromium); (b) to form stable, hard precipitates of compounds like Ni_3Al, Ni_3Ti, MoC, TaC to obstruct the dislocations; and

Table 23.2 Composition of typical creep-resistant blade

Metals	*wt.%*	*Metals*	*wt.%*
Ni	59	Mo	0.25
Co	10	C	0.15
W	10	Si	0.1
Cr	9	Mn	0.1
Al	5.5	Cu	0.05
Ta	2.5	Zr	0.05
Ti	1.5	B	0.015
Hf	1.5	S	<0.008
Fe	0.25	Pb	<0.0005

(c) to form a protective surface oxide film of Cr_2O_3 to protect the blade itself from attack by oxygen (we shall discuss this in Chapter 25). Figure 23.3 (a and b) shows a piece of a nickel-based super-alloy cut open to reveal its complicated structure.

These super-alloys are remarkable materials. They resist creep so well that they can be used at 850°C—and since they melt at 1280°C, this is 0.72 of their (absolute) melting point. They are so hard that they cannot be machined easily by normal methods, and must be precision-cast to their final shape. This is done by *investment casting:* a precise wax model of the blade is embedded in an alumina paste which is then fired; the wax burns out leaving an accurate mould from which one blade can be made by pouring liquid super-alloy into it (Figure 23.4). Because the blades have to be made by this one-off method, they are expensive; the total cost of a rotor of 102 blades is UK£50,000 or US$90,000.

Cast in this way, the grain size of such a blade is small (Figure 23.4). The strengthening caused by alloying successfully suppresses power-law creep, but at $0.72T_M$, diffusional flow then becomes a problem (see the deformation-mechanism diagrams of Chapter 22). The way out is to increase

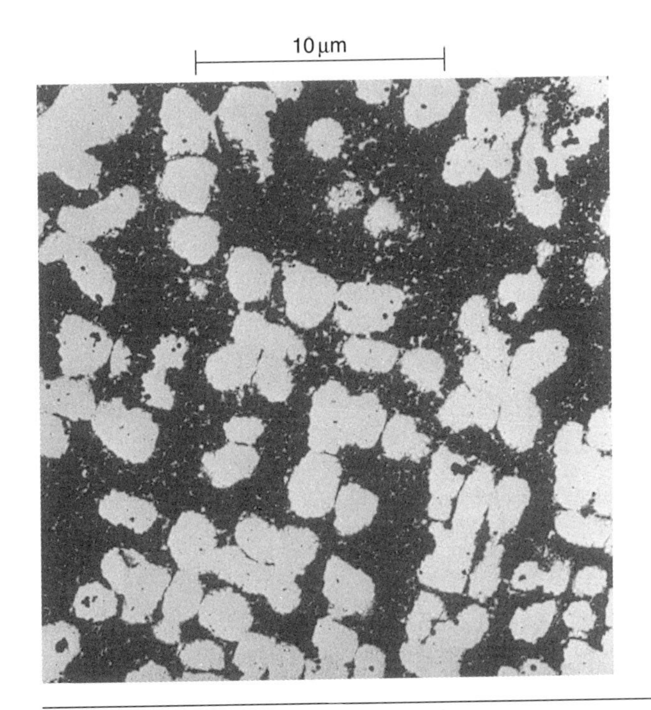

Figure 23.3(a) A piece of a nickel-based super-alloy cut open to show the structure: there are two sizes of precipitates in the alloy—the large white precipitates, and the much smaller black precipitates in between.

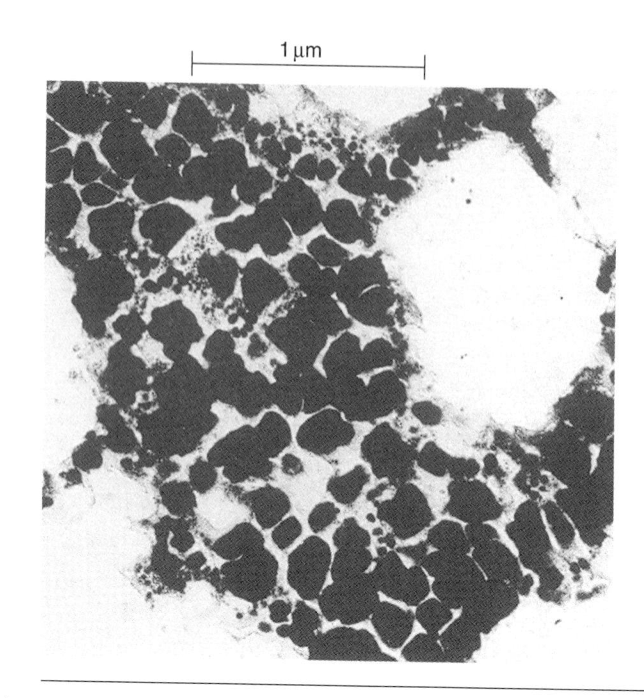

Figure 23.3(b) As Figure 23.3(a), but showing a much more magnified view of the structure, in which the small precipitates are more clearly identifiable.

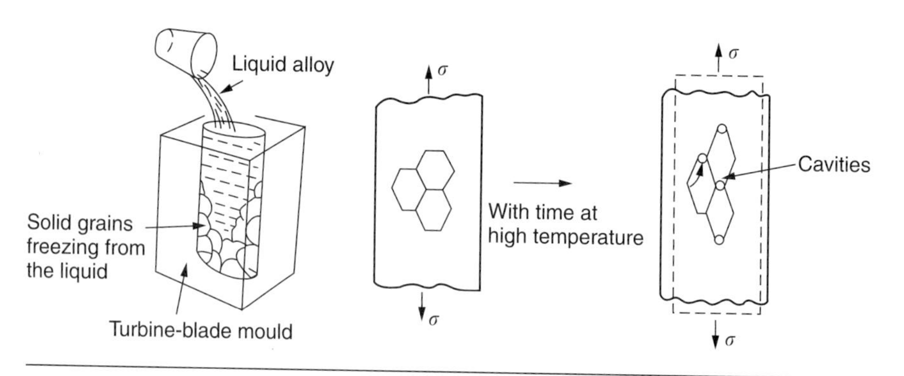

Figure 23.4 Investment casting of turbine blades. This produces a fine-grained material which may undergo a fair amount of diffusion creep, and which may fail rather soon by cavity formation.

the grain size, or even make blades with no grain boundaries at all. In addition, *creep damage* (Chapter 22) accumulates at grain boundaries; we can obviously stave off failure by eliminating grain boundaries, or aligning them parallel to the applied stress (see Figure 23.4). To do this, we *directionally solidify* the alloys (see Figure 23.5) to give long grains with grain boundaries parallel to the applied stress. The diffusional distances required for diffusional creep are

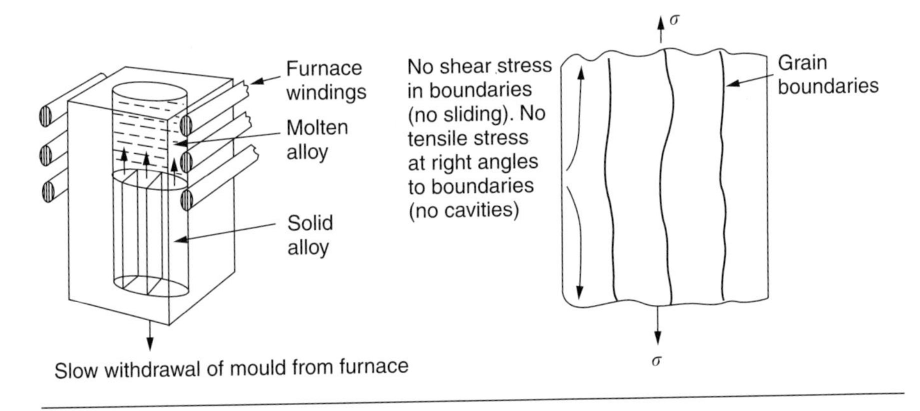

Figure 23.5 Directional solidification (DS) of turbine blades.

then very large (greatly cutting down the rate of diffusional creep); in addition, there is no driving force for grain boundary sliding or for cavitation at grain boundaries. Directionally solidified (DS) alloys are standard in high-performance engines and are now in use in civil aircraft also. The improved creep properties of the DS alloy will allow the engine to run at a flame temperature approximately 50°C higher than before, for a doubling in production cost.

How was this type of alloy discovered in the first place? Well, the fundamental principles of creep-resistant materials design that we talked about helps us to select the more promising alloy recipes and discard the less promising ones fairly easily. Thereafter, the approach is an empirical one. Large numbers of alloys having different recipes are made up in the laboratory, and tested for creep, oxidation, toughness, thermal fatigue, and stability. The choice eventually narrows down to a few alloys and these are subjected to more stringent testing, coupled with judicious tinkering with the alloy recipe. All this is done using a semi-intuitive approach based on previous experience, knowledge of the basic principles of materials design and a certain degree of hunch and luck! Small improvements are continually made in alloy composition and in the manufacture of the finished blades, which evolve by a sort of creepy Darwinism, the fittest (in the sense of Table 23.1) surviving.

Figure 23.6 shows how this evolutionary process has resulted in a continual improvement of creep properties of nickel alloys over time, and shows how the amounts of the major foreign elements have been juggled to obtain these improvements—keeping a watchful eye on the remaining necessary properties. The figure also shows how improvements in alloy manufacture—in this case the use of directional solidification—have helped to increase the operating temperature. Nevertheless, it is clear from the graph that improvements in nickel alloys are now nearing the point of diminishing returns.

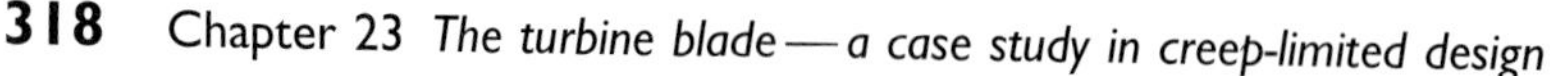

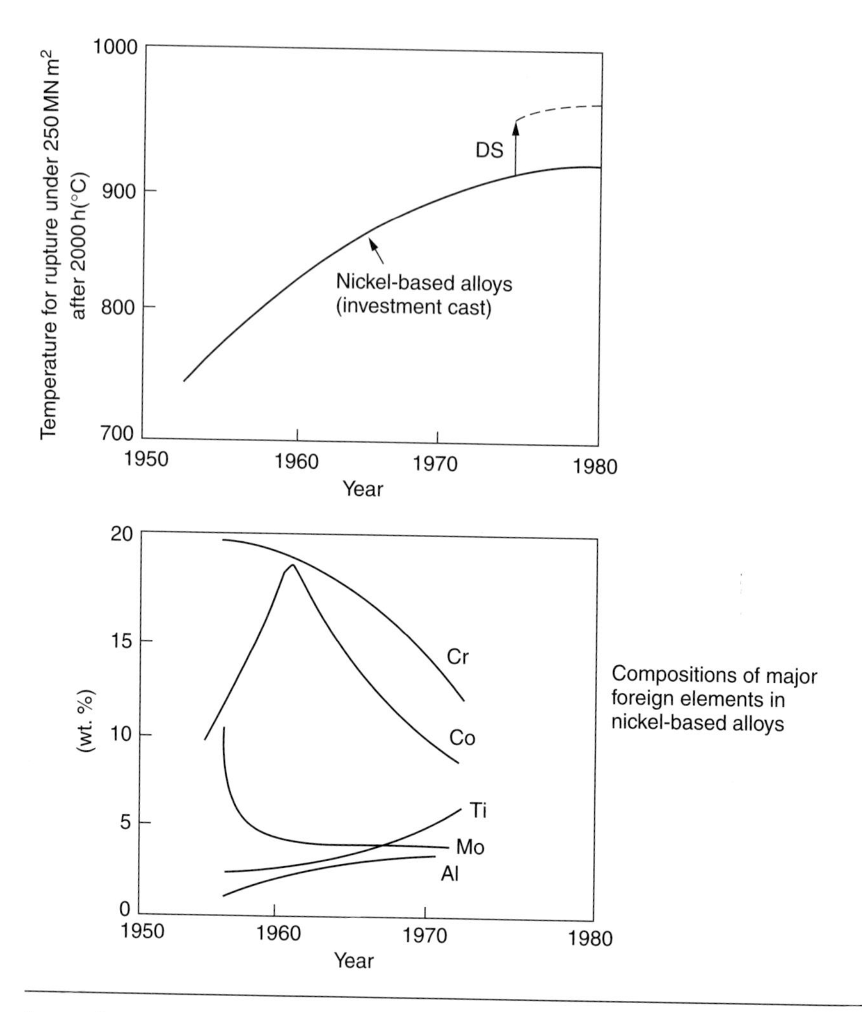

Figure 23.6 Super-alloy developments.

23.4 Engineering developments—blade cooling

Figure 23.7 shows that up to 1960 turbine inlet temperatures were virtually the same as the metal temperatures. After 1960 there was a sharp divergence, with inlet temperatures substantially above the temperatures of the blade metal itself—indeed, the gas temperature is greater than the *melting point* of the blades. Impossible? Not at all. It is done by air-cooling the blades, and in a cunning way. In the earliest form of cooled blade, cooling air from the compressor stage of the engine was fed through ports passing along the full length of the blade, and was ejected into the gas stream at the blade end

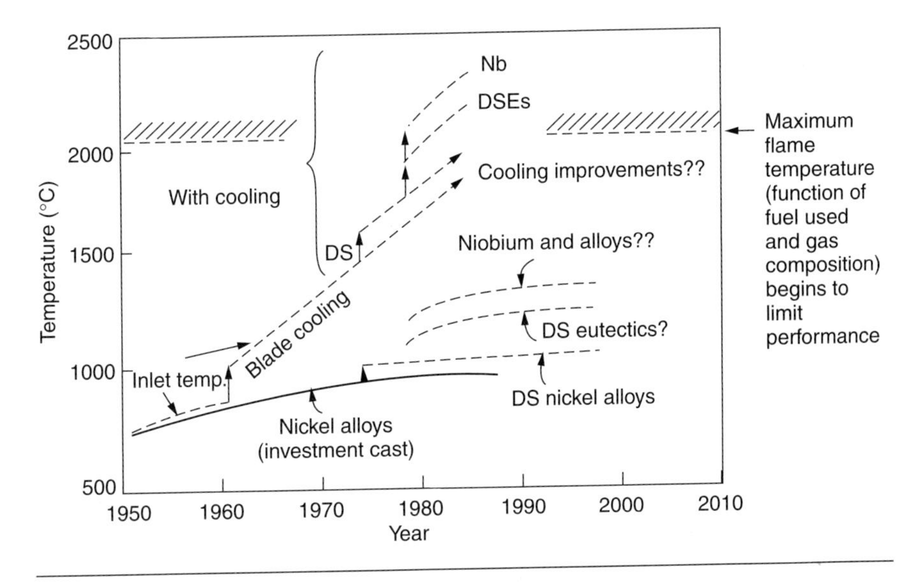

Figure 23.7 Temperature evolution and future materials trends in turbine blades.

(see Figure 23.8). This *internal cooling* of the blade enabled the inlet temperature to be increased immediately by 100°C with no change in alloy composition. A later improvement was *film cooling*, in which the air was ejected over the surface of the blade, through little holes connected to the central channel, giving a cool boundary layer between the blade and the hot gases. A program of continuous improvement in the efficiency of heat transfer by refinements of this type has made it possible for modern turbofans to operate at temperatures that are no longer dominated by the properties of the material.

But blade cooling has reached a limit: ducting still more cold air through the blades will begin to *reduce* the thermal efficiency by taking too much heat away from the combustion chamber. What do we do now?

23.5 Future developments: metals and metal–matrix composites

Driven by these impending limits to blade cooling emphasis has switched back to materials development. Because of the saturation in the development of nickel alloys, more revolutionary approaches are being explored.

One of these is to exploit a unique characteristic of some alloys — belonging to a class called *eutectics* — to form spontaneously an aligned reinforced structure when directionally solidified. Figure 23.9 shows how they are made. Table 23.3 lists some typical high-temperature composites under study for turbine-blade applications. The reinforcing phase is usually a compound with

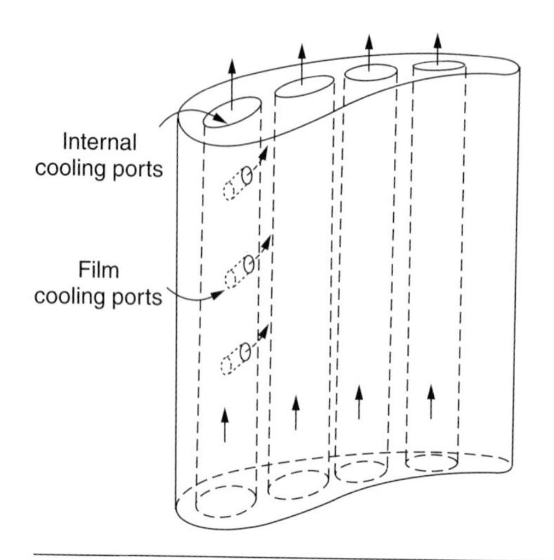

Figure 23.8 Air-cooled blades.

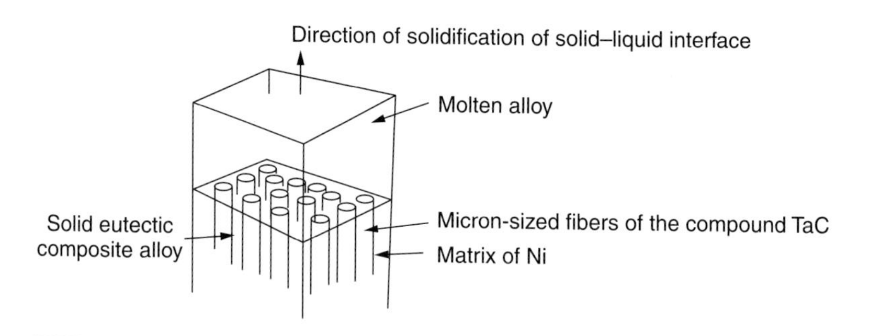

Figure 23.9 Directionally solidified eutectics for turbine blades.

Table 23.3 High-temperature composites

Matrix	*Reinforcing phase*	*Reinforcing phase geometry*
Ni	TaC	Fibers
Co	TaC	Fibers
Ni_3Al	Ni_3Nb	Plates
Co	Cr_7C_3	Fibers
Nb	Nb_2C	Fibers

a high melting temperature; this compound would be far too brittle on its own, but the surrounding metal matrix gives the composite the required toughness. The important feature of these alloys is that the high-melting-point fibers greatly improve the creep properties of the composite alloy, as one might anticipate. The bonding between fibers and matrix is atomic and therefore strong. The structure is very fine (the fibers are only microns in diameter) and the accidental breaking of a few of the brittle fibers in service would have little effect on the composite as a whole.

As Figure 23.7 shows, if DS eutectics (DSEs) prove successful, they will allow the metal temperature to be increased by ≈100°C above conventional DS nickel alloys, and the inlet temperature by ≈200°C (because of a temperature scaling effect caused by the blade cooling). Further improvements in alloy design are under way in which existing nickel alloys and DS eutectics are being blended to give a fiber-reinforced structure with precipitates in the matrix.

23.6 Future developments: high-temperature ceramics

The ceramics best suited for structural use at high temperatures ($\gtrsim$1000°C) are listed in Table 23.4 and compared with nickel-based super-alloys. The comparison shows that all the ceramics have attractively low densities, high moduli and high melting points (and thus excellent creep strength at 1000°C); and many are completely resistant to oxidation—they are oxides already. But several have poor thermal conductivity (leading to high thermal stresses) and all have very low toughness.

Table 23.4 Ceramics for high-temperature structures

Material	*Density* (Mg m^{-3})	*Melting or decomposition* (D) *temperature* (K)	*Modulus* (GN m^{-2})	*Expansion coefficient* $\times 10^{+6}$ (K^{-1})	*Thermal conductivity at* 1000K (Wm^{-1}K^{-1})	*Fracture toughness* K_c(MN m$^{-3/2}$)
Alumina, Al_2O_3	4.0	2320	360	6.9	7	≈5
Glass-ceramics (pyrocerams)	2.7	>1700	≈120	≈3	≈3	≈3
Hot-pressed silicon nitride, Si_3N_4	3.1	2173 (D)	310	3.1	16	≈5
Hot-pressed silicon carbide, SiC	3.2	3000 (D)	≈420	4.3	60	≈3.5
Nickel alloys (Nimonics)	8.0	1600	200	12.5	12	≈100

Alumina (Al_2O_3) was one of the first pure oxides to be produced in complex shapes, but its combination of high expansion coefficient, poor conductivity and low toughness gives it bad thermal-shock resistance.

Glass-ceramics are made by forming a complex silicate glass and then causing it to crystallize partly. They are widely used for ovenware and for heat exchangers for small engines. Their low thermal expansion gives them much better thermal shock resistance than most other ceramics, but the upper working temperature of 900 °C (when the glass phase softens) limits their use.

The covalently-bonded silicon carbide, silicon nitride, and sialons (alloys of Si_3N_4 and Al_2O_3) seem to be the best bet for high-temperature structural use. Their creep resistance is outstanding up to 1300°C, and their low expansion and high conductivity (better than nickel alloys!) makes them resist thermal shock well in spite of their typically low toughness. They can be formed by hot-pressing fine powders, by vapor deposition, or by nitriding silicon which is already pressed to shape: in this way, precise shapes (like turbine blades) can be formed without the need for machining (they are much too hard to machine). Their brittleness, however, creates major design problems. It might be over-come by creating ceramics with a fibrous structure, a bit like wood. In an effort to do this, ceramic-matrix components are under development: they combine strong, exceptionally perfect, fibers (such as silicon carbide or alumina, grown by special techniques) in a matrix of ceramic (silicon carbide or alumina again), made by conventional means to give a material that you might think of as "high-temperature wood".

These materials are now under intensive study for turbine use.

23.7 Cost effectiveness

Any major materials development programme, such as that on the eutectic super-alloys, can only be undertaken if a successful outcome would be cost effective. As Figure 23.10 shows, the costs of development can be colossal. Even before a new material is out of the laboratory, 25–50 million pounds (45–90 million dollars) can have been spent, and failure in an engine test can be expen-sive. Because the performance of a new alloy cannot finally be verified until it has been extensively flight-tested, at each stage of development risk decisions have to be taken whether to press ahead, or cut losses and abandon the programme.

One must consider, too, the cost of the materials themselves. Some of the metals used in conventional nickel alloys—such as hafnium—are hideously expensive and extremely scarce; and the use of greater and greater quantities of exotic materials in an attempt to improve the creep properties will drive the cost of blades up. But expensive though it is, the cost of the turbine blades is still only a small fraction of the cost of an engine, or of the fuel it will consume in its lifetime. Blade costs *are* important, but if new alloys offer improved life or inlet temperature, there is a strong incentive to pursue them.

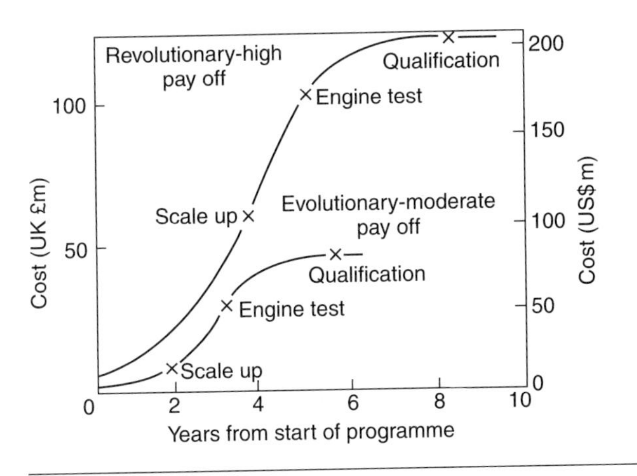

Figure 23.10 Development costs of new turbine-blade materials.

Examples

23.1 Why is there an incentive to maximize the turbine inlet temperature in aircraft engines?

23.2 Why are DS and single-crystal turbine blades better than conventional investment-cast blades?

23.3 What are the advantages and disadvantages of high-performance ceramics as turbine-blade materials?

23.4 How can blade cooling be used to increase engine performance without changing the blade material?

Part G

Oxidation and corrosion

Chapter 24

Oxidation of materials

Chapter contents

24.1 Introduction

In the last chapter we said that one of the requirements of a high-temperature material — in a turbine blade, or a super-heater tube, for example — was that it should resist attack by gases at high temperatures and, in particular, that it should resist oxidation. Turbine blades *do* oxidize in service, and react with H_2S, SO_2, and other combustion products. Excessive attack of this sort is obviously undesirable in such a highly stressed component. Which materials best resist oxidation, and how can the resistance to gas attack be improved?

Well, the earth's atmosphere is oxidizing. We can get some idea of oxidation-resistance by using the earth as a laboratory, and looking for materials which survive well in its atmosphere. All around us we see ceramics: the earth's crust (Chapter 2) is almost entirely made of oxides, silicates, aluminates, and other compounds of oxygen; and being oxides already, they are completely stable. Alkali halides, too, are stable: NaCl, KCl, NaBr — all are widely found in nature. By contrast, metals are not stable: only gold is found in "native" form under normal circumstances (it is completely resistant to oxidation at all temperatures); all the others in our data sheets will oxidize in contact with air. Polymers are not stable either: most will burn if ignited, meaning that they oxidize readily. Coal and oil (the raw materials for polymers), it is true, are found in nature, but that is only because geological accidents have sealed them off from all contact with air. A few polymers, among them PTFE (a polymer based on $-CF_2-$), are so stable that they survive long periods at high temperatures, but they are the exceptions. And polymer-based composites, of course, are just the same: wood is not noted for its high-temperature oxidation resistance.

How can we categorize in a more precise way the oxidation-resistance of materials? If we can do so for oxidation, we can obviously follow a similar method for sulphidation or nitrogenation.

24.2 The energy of oxidation

This tendency of many materials to react with oxygen can be quantified by laboratory tests which measure the energy needed for the reaction

$$\text{Material} + \text{Oxygen} + \text{Energy} \rightarrow \text{Oxide of material}$$

If this energy is *positive*, the material is stable; if *negative*, it will oxidize. The bar-chart of Figure 24.1 shows the energies of oxide formation for our four categories of materials; numerical values are given in Table 24.1.

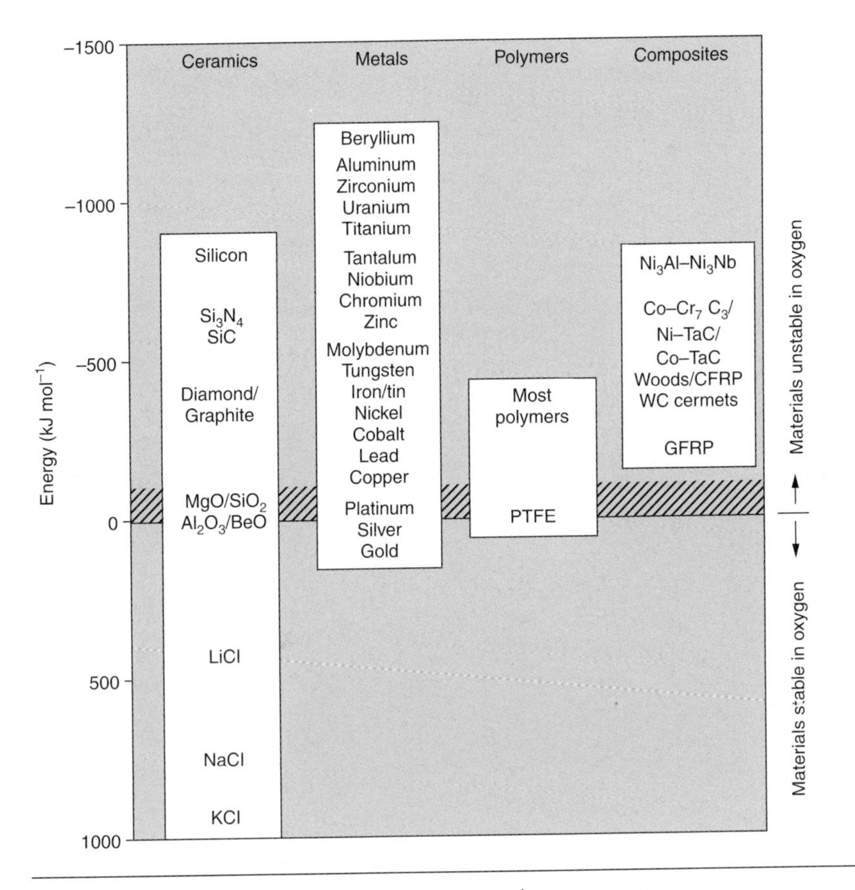

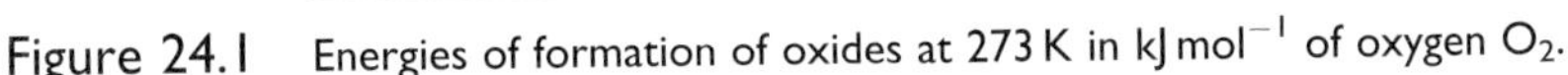

Figure 24.1 Energies of formation of oxides at 273 K in kJ mol^{-1} of oxygen O_2.

24.3 Rates of oxidation

When designing with oxidation-prone materials, it is obviously vital to know how *fast* the oxidation process is going to be. Intuitively one might expect that, the larger the energy released in the oxidation process, the faster the *rate* of oxidation. For example, one might expect aluminum to oxidize 2.5 times faster than iron from the energy data in Figure 24.1. In fact, aluminum oxidizes much more slowly than iron. Why should this happen?

If you heat a piece of bright iron in a gas flame, the oxygen in the air reacts with the iron at the surface of the metal where the oxygen and iron atoms can contact, creating a thin layer of iron oxide on the surface, and making the iron turn black. The layer grows in thickness, quickly at first, and then more slowly because iron atoms now have to diffuse through the film before they make contact and react with oxygen. If you plunge the piece of hot iron into a dish of water the shock of the quenching breaks off the iron oxide layer, and you can

Table 24.1 Energies of formation of oxides at 273 K

Material (oxide)		*Energy* (kJ mol^{-1} *of oxygen,* O_2)	*Material (oxide)*		*Energy* (kJ mol^{-1} *of oxygen,* O_2)
Beryllium	(BeO)	−1182	Woods, most polymers, CFRP		≈−400
Magnesium	(MgO)	−1162			
Aluminium	(Al_2O_3)	−1045	Diamond, graphite	(CO_2)	−389
Zirconium	(ZrO_2)	−1028	Tungsten carbide cermet (mainly WC)	(WO_3+ CO_2)	−349
Uranium	(U_3O_8)	≈−1000			
Titanium	(TiO)	−848	Lead	(Pb_3O_4)	−309
Silicon	(SiO_2)	−836	Copper	(CuO)	−254
Tantalum	(Ta_2O_5)	−764	GFRP		≈−200
Niobium	(Nb_2O_5	−757	Platinum	(PtO_2)	≈−160
Chromium	(Cr_2O_3)	−701	Silver	(Ag_2O)	−5
Zinc	(ZnO)	−636	PTFE		≈zero
Silicon nitride, Si_3N_4	($3SiO_2$+$2N_2$)	≈−629	Gold	(Au_2O_3)	+80
Silicon carbide, SiC	(SiO_2+CO_2)	≈−580	Alkali halides		≈+400 to ≈+1400
Molybdenum	(MoO_2)	−534			
Tungsten	(WO_3)	−510	Magnesia, MgO	Higher oxides	Large and positive
Iron	(Fe_3O_4)	−508	Silica, SiO_2		
Tin	(SnO)	−500	Alumina, Al_2O_3		
Nickel	(NiO)	−439	Beryllia, BeO		
Cobalt	(CoO)	−422			

see the pieces of layer in the dish. The iron surface now appears bright again, showing that the shock of the quenching has completely stripped the metal of the oxide layer which formed during the heating; if it were reheated, it would oxidize at the old rate.

The important thing about the oxide film is that it acts as a *barrier* which keeps the oxygen and iron atoms apart and cuts down the rate at which these atoms react to form more iron oxide. Aluminum, and most other materials, form oxide barrier layers in just the same sort of way—but the oxide layer on aluminum is a *much* more effective barrier than the oxide film on iron is.

How do we measure rates of oxidation in practice? Well, because oxidation proceeds by the addition of oxygen atoms to the surface of the material, the *weight* of the material usually goes up in proportion to the amount of material that has become oxidized. This weight increase, Δm, can be monitored continuously with time t in the way illustrated in Figure 24.2. Two types of behavior are usually observed at high temperature. The first is *linear oxidation*, with

$$\Delta m = k_{\mathrm{L}} t \tag{24.1}$$

where k_{L} is a *kinetic constant*. Naturally, k_{L} is usually positive. (In a few materials, however, the oxide evaporates away as soon as it has formed; the material then *loses* weight and k_{L} is then negative.)

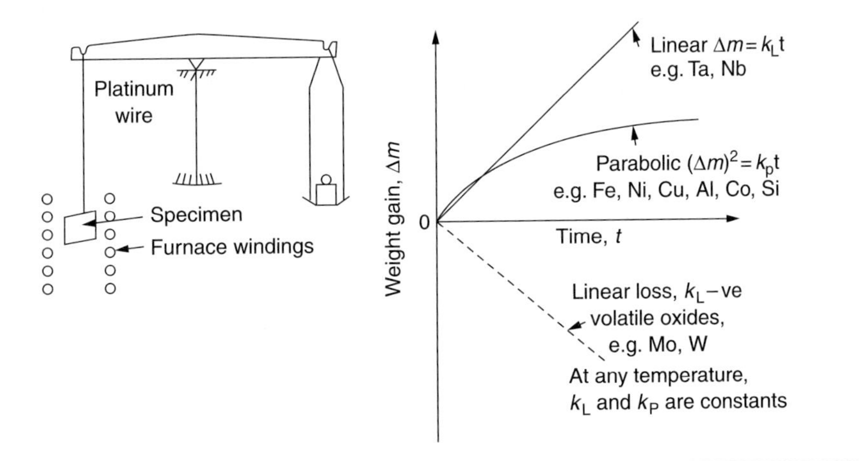

Figure 24.2 Measurement of oxidation rates.

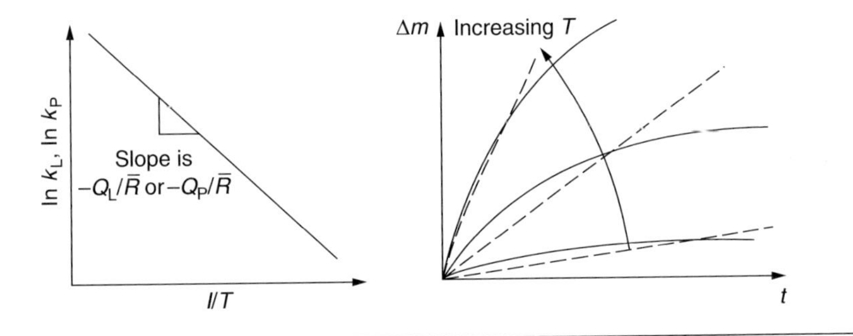

Figure 24.3 Oxidation rates increase with temperature according to Arrhenius's law.

The second type of oxidation behavior is *parabolic oxidation*, with

$$(\Delta m)^2 = k_p t \tag{24.2}$$

where k_p is another kinetic constant, this time always positive.

Oxidation rates follow Arrhenius's law (Chapter 21), that is, the kinetic constants k_L and k_p increase exponentially with temperature:

$$k_L = A_L e^{-Q_L/\overline{R}T} \text{ and } k_P = A_P e^{-Q_P/\overline{R}T} \tag{24.3}$$

where A_L and A_p, Q_L and Q_p are constants. Thus, as the temperature is increased, the rate of oxidation increases exponentially (Figure 24.3).

Finally, oxidation rates obviously increase with increasing partial pressure of oxygen, although rarely in a simple way. The partial pressure of oxygen in a gas turbine atmosphere, for example, may well be very different from that in air, and it is important to conduct oxidation tests on high-temperature components under the right conditions.

24.4 Data

Naturally, it is important from the design standpoint to know how much material is replaced by oxide. The mechanical properties of the oxide are usually grossly inferior to the properties of the material (e.g. oxides are comparatively brittle), and even if the layer is firmly attached to the material—which is certainly not always the case—the effective section of the component is reduced. The reduction in the section of a component can obviously be calculated from data for Δm.

Table 24.2 gives the times for a range of materials required to oxidize them to a depth of 0.1 mm from the surface when exposed to air at $0.7T_M$ (a typical figure for the operating temperature of a turbine blade or similar component): these times vary by many orders of magnitude, and clearly show that there is no correlation between oxidation rate and energy needed for the reaction (see Al, W for extremes: Al, *very* slow—energy = $-1045\,\text{kJ mol}^{-1}$ of O_2; W, very fast—energy = $-510\,\text{kJ mol}^{-1}$ of O_2).

24.5 Micromechanisms

Figure 24.4 illustrates the mechanism of parabolic oxidation. The reaction

$$M + O \rightarrow MO$$

(where M is the material which is oxidizing and O is oxygen) really goes in two steps. First M forms an ion, releasing electrons, *e*:

$$M \rightarrow M^{++} + 2e$$

These electrons are then absorbed by oxygen to give an oxygen ion:

$$O + 2e \rightarrow O^{--}$$

Either the M^{++} and the two *e*'s diffuse outward through the film to meet the O^{--} at the outer surface, *or* the oxygen diffuses inwards (with two electron holes) to meet the M^{++} at the inner surface. The concentration gradient of oxygen is simply the concentration in the gas, *c*, divided by the film thickness, *x*; and the rate of growth of the film dx/dt is obviously proportional to the flux of atoms diffusing through the film. So, from Fick's law (equation (21.1)):

$$\frac{dx}{dt} \propto D\frac{c}{x}$$

where *D* is the diffusion coefficient. Integrating with respect to time gives

$$x^2 = K_p t \tag{24.4}$$

Table 24.2 Time in hours for material to be oxidized to a depth of 0.1 mm at 0.7 T_M in air

Material	*Time*	*Melting point (K)*	*Material*	*Time*	*Melting point (K)*
Au	infinite?	1336	Ni	600	1726
Ag	very long	1234	Cu	25	1356
Al	very long	933	Fe	24	1809
Si_3N_4	very long	2173	Co	7	1765
SiC	very long	3110	Ti	<6	1943
Sn	very long	505	WC cermet	<5	1700
Si	2×10^6	1683	Ba	$\ll 0.5$	983
Be	10^6	1557	Zr	0.2	2125
Pt	1.8×10^5	2042	Ta	Very short	3250
Mg	$> 10^5$	923	Nb	Very short	2740
Zn	$> 10^4$	692	U	Very short	1405
Cr	1600	2148	Mo	Very short	2880
Na	> 1000	371	W	Very short	3680
K	> 1000	337			

Note: Data subject to considerable variability due to varying degrees of material purity, prior surface treatment and presence of atmospheric impurities like sulphur.

where

$$K_P \,\alpha\, c D_0 e^{-Q/\overline{R}T} \tag{24.5}$$

This growth law has exactly the form of equation (24.2) and the kinetic constant is analogous to* that of equation (24.3). This success lets us explain why some films are more protective than others: protective films are those with low diffusion coefficients — and thus high melting points. That is one reason why Al_2O_3 protects aluminum, Cr_2O_3 protects chromium and SiO_2 protects silicon so well, whereas Cu_2O and even FeO (which have lower melting points) are less protective. But there is an additional reason: electrons must also pass through the film and these films are insulators (the electrical resistivity of Al_2O_3 is 10^9 times greater than that of FeO).

Although our simple oxide film model explains most of the experimental observations we have mentioned, it does not explain the linear laws. How, for example, can a material *lose* weight linearly when it oxidizes as is sometimes observed (see Figure 24.2)? Well, some oxides (e.g. MoO_3, WO_3) are very volatile. During oxidation of Mo and W at high temperature, the oxides evaporate as soon as they are formed, and offer no barrier at all to oxidation. Oxidation, therefore, proceeds at a rate that is independent of time, and the material loses weight because the oxide is *lost*. This behavior explains the catastrophically rapid section loss of Mo and W shown in Table 24.2.

* It does not have the same value, however, because equation (24.5) refers to thickness gain and not *mass* gain; the two can be easily related if quantities like the density of the oxide are known.

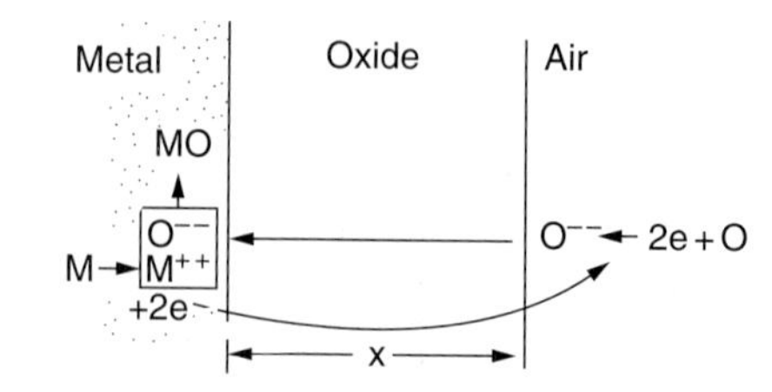

Case 1. M^{++} diffuses very slowly in oxide. Oxide grows at metal–oxide interface.
Examples: Ti, Zr, U.

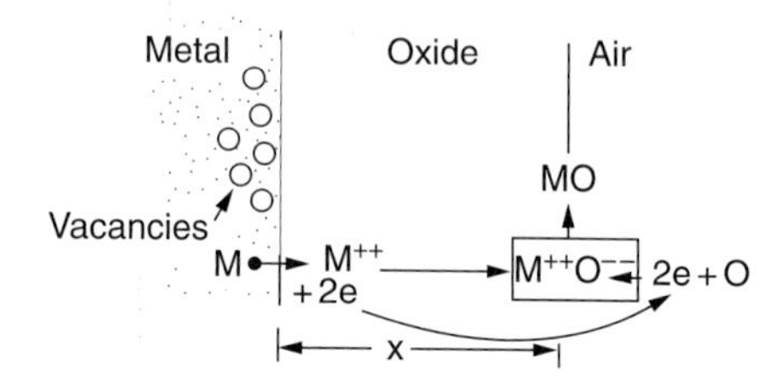

Case 2. O^{--} diffuses very slowly in oxide. Oxide grows at oxide–air interface.
Vacancies form between metal and oxide.
Examples: Cu, Fe, Cr, Co.

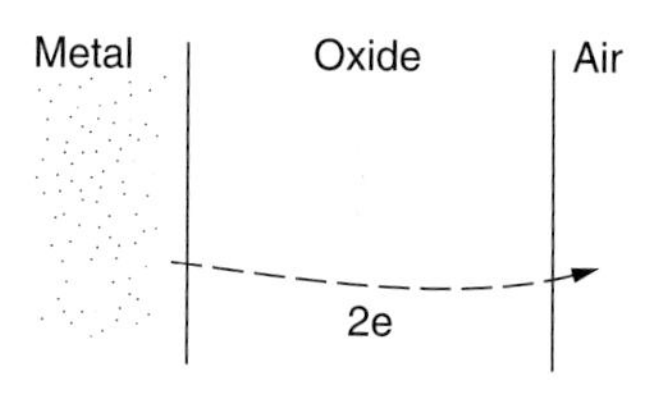

Case 3. Electrons move very slowly. Oxide can grow (slowly) at metal–oxide interface or oxide–air interface depending on whether M^{++} diffuses faster than O^{--} or not.
Example: Al.

Figure 24.4 How oxide layers grow to give parabolic oxidation behavior.

The explanation of a linear weight *gain* is more complex. Basically, as the oxide film thickens, it develops cracks, or partly lifts away from the material, so that the barrier between material and oxide does not become any more effective as oxidation proceeds. Figure 24.5 shows how this can happen. If the volume of the oxide is much less than that of the material from which it is formed, it will crack to relieve the strain (oxide films are usually brittle). If the volume of the oxide is much greater, on the other hand, the oxide will tend to release the strain energy by breaking the adhesion between material and oxide, and springing away. For protection, then, we need an oxide skin which is neither too small and splits open (like the bark on a fir tree) nor one which is too big and wrinkles up (like the skin of a rhinoceros), but one which is just right. Then, and only then, do we get protective parabolic growth.

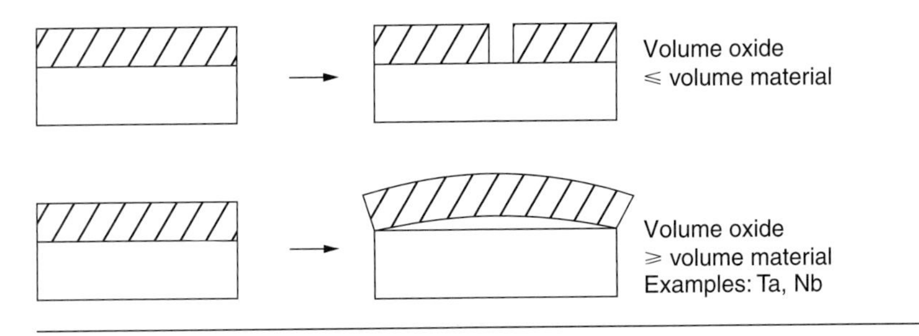

Figure 24.5 Breakdown of oxide films, leading to linear oxidation behavior.

In the next chapter we use this understanding to analyze the design of oxidation-resistant materials.

Examples

24.1 The oxidation of a particular metal in air is limited by the outward diffusion of metallic ions through an unbroken surface film of one species of oxide. Assume that the concentration of metallic ions in the film immediately next to the metal is c_1, and that the concentration of ions in the film immediately next to the air is c_2, where c_1 and c_2 are constants. Use Fick's law to show that the oxidation of the metal should satisfy parabolic kinetics, with weight gain Δm given by

$$(\Delta m)^2 = k_p t$$

24.2 The oxidation of another metal is limited by the outward flow of electrons through a uniform, unbroken oxide film. Assume that the electrical potential in the film immediately next to the metal is V_1, and the potential immediately next to the free surface is V_2, where V_1 and V_2 are constants. Use Ohm's law to show that parabolic kinetics should apply in this case also.

24.3 The kinetics of oxidation of mild steel at high temperature are parabolic, with

$$k_p\,(\text{kg}^2\,\text{m}^{-4}\,\text{s}^{-1}) = 37\,\exp\left\{-\frac{138\,\text{kJ}\,\text{mol}^{-1}}{\overline{R}T}\right\}$$

Find the depth of metal lost from the surface of a mild steel tie bar in a furnace at 500°C after 1 year. You may assume that the oxide scale is predominantly FeO. The atomic weight and density of iron are 55.9 kg kmol^{-1} and 7.87 Mg m^{-3}; the atomic weight of oxygen is 16 kg kmol^{-1}. What would be the loss at 600°C?

Answers

0.33 mm at 500°C; 1.13 mm at 600°C.

24.4 Of the many metals found in the earth's crust, why is gold the only one which is found "native", that is, as pieces of pure metal?

24.5 The energy released when silver reacts with oxygen at room temperature is small—only 5 kJ mol^{-1} of O_2. At 230°C, the energy is zero, and above 230°C the energy is positive (i.e. silver is stable and does not oxidize). Use this information to explain why small electrical contacts are usually plated with a thin layer of silver. Why is gold not used instead?

24.6 Why does the rate at which a metal oxidizes not necessarily correlate with the energy released in the oxidation reaction?

24.7 Explain why some metals show either a linear weight gain or a linear weight loss with time during oxidation.

Chapter 25

Case studies in dry oxidation

Chapter contents

25.1 Introduction

In this chapter we look first at an important class of alloys designed to resist corrosion: the stainless steels. We then examine a more complicated problem: that of protecting the most advanced gas turbine blades from gas attack. The basic principle applicable to both cases is to coat the steel or the blade with a stable ceramic: usually Cr_2O_3 or Al_2O_3. But the ways this is done differ widely. The most successful are those which produce a ceramic film which heals itself if damaged—as we shall now describe.

25.2 Case study 1: making stainless alloys

Mild steel is an excellent structural material—cheap, easily formed and strong mechanically. But at low temperatures it rusts, and at high, it oxidizes rapidly. There is a demand, for applications ranging from kitchen sinks via chemical reactors to superheater tubes, for a corrosion-resistant steel. In response to this demand, a range of stainless irons and steels has been developed. When mild steel is exposed to hot air, it oxidizes quickly to form FeO (or higher oxides). But if one of the elements near the top of Table 24.1 with a large energy of oxidation is dissolved in the steel, then this element oxidizes preferentially (because it is much more stable than FeO), forming a layer of its oxide on the surface. And if this oxide is a protective one, like Cr_2O_3, Al_2O_3, SiO_2 or BeO, it stifles further growth, and protects the steel.

A considerable quantity of this foreign element is needed to give adequate protection. The best is chromium, 18 percent of which gives a very protective oxide film: it cuts down the rate of attack at 900°C, for instance, by more than 100 times.

Other elements, when dissolved in steel, cut down the rate of oxidation, too. Al_2O_3 and SiO_2 both form in preference to FeO (Table 24.1) and form protective films (see Table 24.2). Thus 5 percent Al dissolved in steel decreases the oxidation rate by 30 times, and 5 percent Si by 20 times. The same principle can be used to impart corrosion resistance to other metals. We shall discuss nickel and cobalt in the next case study—they can be alloyed in this way. So, too, can copper; although it will not dissolve enough chromium to give a good Cr_2O_3 film, it *will* dissolve enough aluminum, giving a range of stainless alloys called "aluminum bronzes". Even silver can be prevented from tarnishing (reaction with sulfur) by alloying it with aluminum or silicon, giving protective Al_2O_3 or SiO_2 surface films. And archaeologists believe that the Delhi Pillar—an ornamental pillar of cast iron which has stood, uncorroded, for some hundreds of years in a *particularly* humid spot—survives because the iron has some 6 percent silicon in it.

Ceramics themselves are sometimes protected in this way. Silicon carbide, SiC, and silicon nitride, Si_3N_4 both have large negative energies of oxidation (meaning that they oxidize easily). But when they do, the silicon in them turns to SiO_2 which quickly forms a protective skin and prevents further attack.

This protection-by-alloying has one great advantage over protection by a surface coating (like chromium plating or gold plating): it repairs itself when damaged. If the protective film is scored or abraded, fresh metal is exposed, and the chromium (or aluminum or silicon) it contains immediately oxidizes, healing the break in the film.

25.3 Case study 2: protecting turbine blades

As we saw in Chapter 23, the materials at present used for turbine blades consist chiefly of nickel, with various foreign elements added to get the creep properties right. With the advent of DS blades, such alloys will normally operate around 950°C, which is close to $0.7T_M$ for Ni (1208 K, 935°C). If we look at Table 24.2 we can see that at this temperature, nickel loses 0.1 mm of metal from its surface by oxidation in 600 h. Now, the thickness of the metal between the outside of the blade and the integral cooling ports is about 1 mm, so that in 600 h a blade would lose about 10 percent of its cross section in service. This represents a serious loss in mechanical integrity and, moreover, makes no allowance for statistical variations in oxidation rate—which can be quite large—or for preferential oxidation (at grain boundaries, for example) leading to pitting. Because of the large cost of replacing a set of blades (≈UK£50,000 or US$90,000 per rotor) they are expected to last for more than 5000 h. Nickel oxidizes with parabolic kinetics (equation (24.4)) so that, after a time t_2, the loss in section x_2 is given by substituting our data into:

$$\frac{x_2}{x_1} = \left(\frac{t_2}{t_1}\right)^{1/2}$$

giving

$$x_2 = 0.1\left(\frac{5000}{600}\right)^{1/2} = 0.29\,\text{mm}$$

Obviously this sort of loss is not admissible, but how do we stop it?

Well, as we saw in Chapter 23, the alloys used for turbine blades contain large amounts of chromium, dissolved in solid solution in the nickel matrix. Now, if we look at our table of energies (Table 24.1) released when oxides are formed from materials, we see that the formation of Cr_2O_3 releases much more energy (701 kJ mol^{-1} of O_2) than NiO (439 kJ mol^{-1} of O_2). This means that Cr_2O_3 will form in *preference* to NiO on the surface of the alloy. Obviously, the more Cr there is in the alloy, the greater is the preference for Cr_2O_3. At the 20 percent level, enough Cr_2O_3 forms on the surface of the turbine blade to make the material act a bit as though it were chromium.

Suppose for a moment that our material *is* chromium. Table 24.2 shows that Cr would lose 0.1 mm in 1600 h at $0.7T_M$. Of course, we have forgotten about one thing. $0.7T_M$ for Cr is 1504 K (1231°C), whereas, as we have said, for Ni, it is 1208 K (935°C). We should, therefore, consider how Cr_2O_3 would act as a barrier to oxidation at 1208 K rather than at 1504 K (Figure 25.1). The oxidation of Cr follows parabolic kinetics with an activation energy of $330\,\text{kJ}\,\text{mol}^{-1}$. Then the ratio of the times required to remove 0.1 mm (from equation (24.3)) is

$$\frac{t_2}{t_1} = \frac{\exp -(Q/\bar{R}T_1)}{\exp -(Q/\bar{R}T_2)} = 0.65 \times 10^3$$

Thus the time at 1208 K is

$$\begin{aligned} t_2 &= 0.65 \times 10^3 \times 1600\,\text{h} \\ &= 1.04 \times 10^6\,\text{h} \end{aligned}$$

Now, as we have said, there is only at most 20 percent Cr in the alloy, and the alloy behaves only *partly* as if it were protected by Cr_2O_3. In fact, experimentally, we find that 20 percent Cr increases the time for a given metal

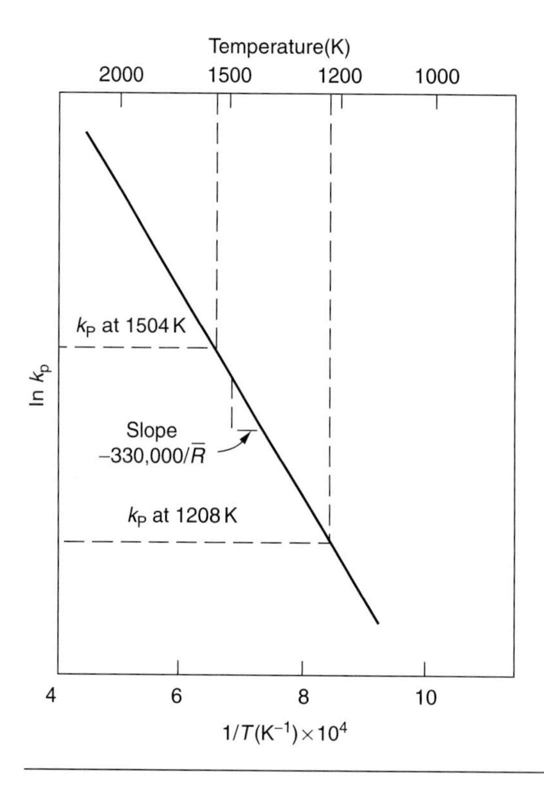

Figure 25.1 The way in which k_p varies with temperature.

loss by only about *ten* times, i.e. the time taken to lose 0.1 mm at blade working temperature becomes $600 \times 10\,h = 6000\,h$ rather than 10^6 hours.

Why this large difference? Well, whenever you consider an *alloy* rather than a pure material, the oxide layer—whatever its nature (NiO, Cr_2O_3, etc.)—has foreign elements contained in *it*, too. Some of these will greatly increase either the diffusion coefficients in, or electrical conductivity of, the layer, and make the rate of oxidation through the layer much more than it would be in the absence of foreign element contamination.

One therefore has to be very careful in transferring data on film protectiveness from a pure material to an alloyed one, but the approach does, nevertheless, give us an *idea* of what to expect. As in all oxidation work, however, experimental determinations on actual alloys are *essential* for working data.

This 0.1 mm loss in 6000 h from a 20 percent Cr alloy at 935°C, though better than pure nickel, is still not good enough. What is worse, we saw in Chapter 23 that, to improve the creep properties, the quantity of Cr has been reduced to 10 percent, and the resulting oxide film is even less protective. The obvious way out of this problem is to *coat* the blades with a protective layer (Fig. 25.2). This is usually done by spraying molten droplets of aluminum on to the blade surface to form a layer, some microns thick. The blade is then heated in a furnace to allow the Al to diffuse into the surface of the Ni. During this process, some of the Al forms compounds such as AlNi with the nickel—which are themselves good barriers to oxidation of the Ni, whilst the rest of the Al becomes oxidized up to give Al_2O_3—which, as we can see from our oxidation-rate data—should be a very good barrier to oxidation even allowing for the high temperature ($0.7\,T_M$ for Al = 653 K, 380°C). An incidental benefit of the relatively thick AlNi layer is its poor thermal conductivity—this helps insulate the metal of the cooled blade from the hot gases, and allows a slight extra increase in blade working temperature.

Other coatings, though more difficult to apply, are even more attractive. AlNi is brittle, so there is a risk that it may chip off the blade surface exposing the unprotected metal. It is possible to diffusion-bond a layer of a Ni–Cr–Al

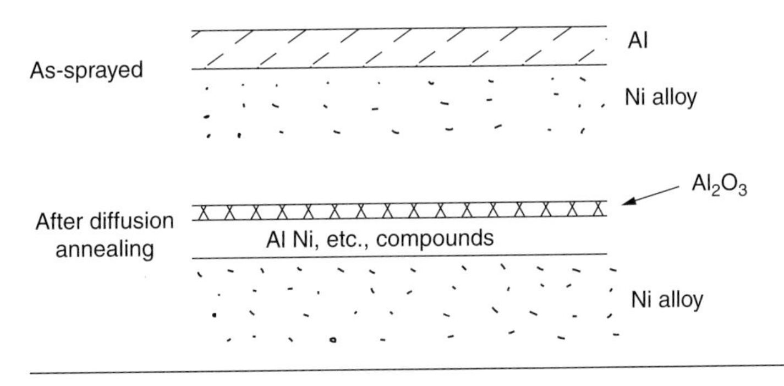

Figure 25.2 Protection of turbine blades by sprayed-on aluminum.

alloy to the blade surface (by spraying on a powder or pressing on a thin sheet and then heating it up) to give a ductile coating which still forms a very protective film of oxide.

Influence of coatings on mechanical properties

So far, we have been talking in our case study about the *advantage* of an oxide layer in reducing the rate of metal removal by oxidation. Oxide films do, however, have some *disadvantages*.

Because oxides are usually quite brittle at the temperatures encountered on a turbine blade surface, they can crack, especially when the temperature of the blade changes and differential thermal contraction and expansion stresses are set up between alloy and oxide. These can act as ideal nucleation centers for thermal fatigue cracks and, because oxide layers in nickel alloys are stuck well to the underlying alloy (they would be useless if they were not), the crack can spread into the alloy itself (Figure 25.3). The properties of the oxide film are thus very important in affecting the fatigue properties of the whole component.

Protecting future blade materials

What of the corrosion resistance of new turbine-blade alloys like DS eutectics? Well, an alloy like Ni_3Al–Ni_3Nb loses 0.05 mm of metal from its surface in 48 h at the anticipated operating temperature of 1155°C for such alloys. This is obviously not a good performance, and coatings will be required before these materials are suitable for application. At lower oxidation rates, a more insidious effect takes place — preferential attack of one of the phases, with penetration along interphase boundaries. Obviously this type of attack, occurring under a break in the coating, can easily lead to fatigue failure and raises another problem in the use of DS eutectics.

You may be wondering why we did not mention the pure "refractory" metals Nb, Ta, Mo, W in our chapter on turbine-blade materials (although we *did*

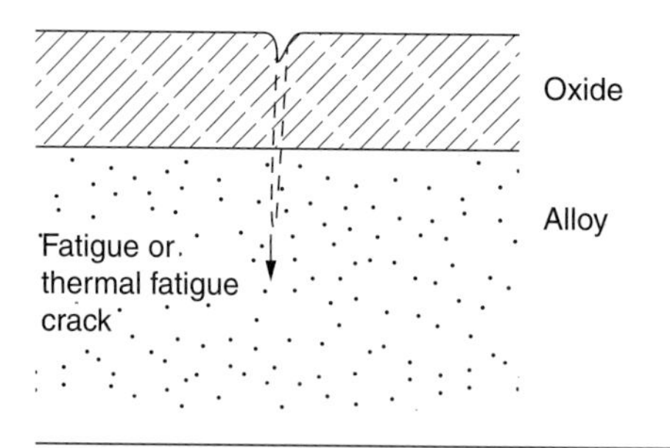

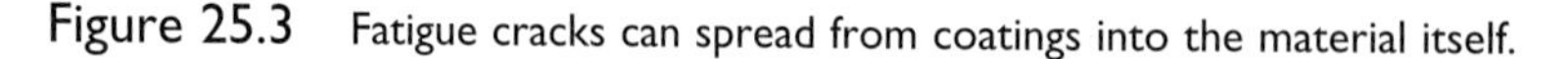

Figure 25.3 Fatigue cracks can spread from coatings into the material itself.

show one of them on Figure 23.7). These metals have very high melting temperatures, as shown, and should therefore have very good creep properties.

$$\left.\begin{array}{ll} \text{Nb} & 2740\,\text{K} \\ \text{Ta} & 3250\,\text{K} \\ \text{Mo} & 2880\,\text{K} \\ \text{W} & 3680\,\text{K} \end{array}\right\} T_M$$

But they all oxidize very rapidly indeed (see Table 24.2), and are utterly useless without coatings. The problem with coated refractory metals is, that if a break occurs in the coating (e.g. by thermal fatigue, or erosion by dust particles, etc.), catastrophic oxidation of the underlying metal will take place, leading to rapid failure. The "unsafeness" of this situation is a major problem that has to be solved before we can use these on-other-counts potentially excellent materials.

The ceramics SiC and Si_3N_4 do not share this problem. They oxidize readily (Table 24.1); but in doing so, a surface film of SiO_2 forms which gives adequate protection up to 1300°C. And because the film forms by oxidation of the material itself, it is self-healing.

25.4 Joining operations: a final note

One might imagine that it is always a good thing to have a protective oxide film on a material. Not always; if you wish to join materials by brazing or soldering, the protective oxide film can be a problem. It is this which makes stainless steel hard to braze and almost impossible to solder; even spot-welding and diffusion bonding become difficult. Protective films create poor electrical contacts; that is why aluminum is not more widely used as a conductor. And production of components by powder methods (which involve the compaction and sintering—really diffusion bonding—of the powdered material to the desired shape) is made difficult by protective surface films.

Examples

25.1 Diffusion of aluminum into the surface of a nickel super-alloy turbine blade reduced the rate of high-temperature oxidation. Explain.

25.2 Explain why many high-temperature components, e.g. furnace flues and heat exchanger shells, are made from stainless steel rather than mild steel.

25.3 Give examples of the use of alloying elements to reduce the rate at which metals oxidize.

25.4 Why are the refractory metals (Nb, Ta, Mo, W) of very limited use at high temperature? What special steps are taken to ensure that tungsten electric light filaments can operate for long times at white heat?

25.5 Why is oxidation a problem when soldering or brazing metal parts? Why are the connection tabs on small electrical components usually supplied pre-tinned with electrical solder?

25.6 Why is nichrome wire (Ni + 30Fe + 20 Cr), rather than mild steel, used for heating elements in electric fires?

25.7 Explain, giving examples, why ceramics are good at resisting high-temperature oxidation.

Chapter 26

Wet corrosion of materials

Chapter contents

26.1 Introduction

In the last two chapters we showed that most materials that are unstable in oxygen tend to oxidize. We were principally concerned with loss of material at high temperatures, in dry environments, and found that, under these conditions, oxidation was usually controlled by the diffusion of ions or the conduction of electrons through oxide films that formed on the material surface (Figure 26.1). Because of the thermally activated nature of the diffusion and reaction processes we saw that the rate of oxidation was much greater at high temperature than at low, although even at room temperature, very thin films of oxide *do* form on all unstable metals. This minute amount of oxidation is important: it protects, preventing further attack; it causes tarnishing; it makes joining difficult; and (as we shall see in Chapters 28 and 29) it helps keep sliding surfaces apart, and so influences the coefficient of friction. But the *loss* of material by oxidation at room temperature under these *dry* conditions is very slight.

Under *wet* conditions, the picture is dramatically changed. When mild steel is exposed to oxygen and water at room temperature, it rusts rapidly and the loss of metal quickly becomes appreciable. Unless special precautions are taken, the life of most structures, from bicycles to bridges, from buckets to battleships, is limited by wet corrosion. The annual bill in the United Kingdom for either replacing corroded components, or preventing corrosion (e.g. by painting the Forth Bridge), is around UK£4000 m or US$7000 m a year.

26.2 Wet corrosion

Why the dramatic effect of water on the rate of loss of material? As an example we shall look at *iron*, immersed in *aerated water* (Figure 26.2).

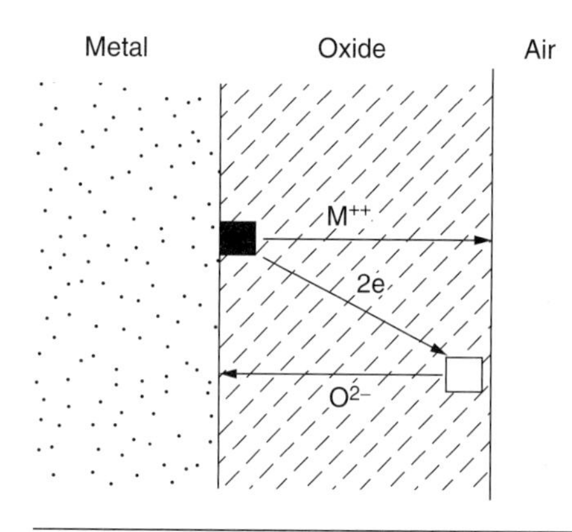

Figure 26.1 *Dry oxidation.*

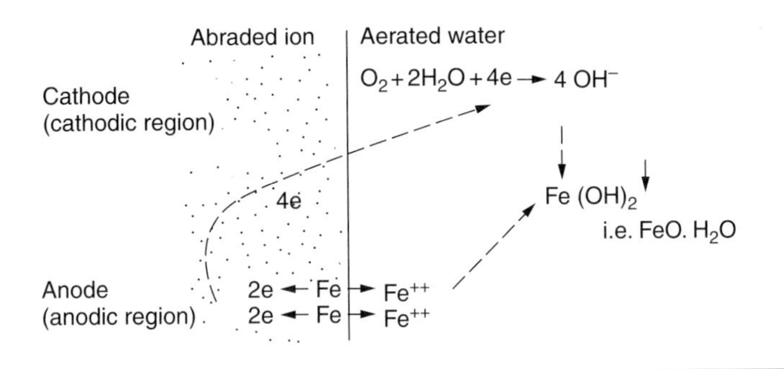

Figure 26.2 Wet corrosion.

Iron atoms pass into solution in the water as Fe^{++}, leaving behind two electrons each (the *anodic* reaction). These are conducted through the metal to a place where the "oxygen reduction" reaction can take place to consume the electrons (the *cathodic* reaction). This reaction generates OH^- ions which then combine with the Fe^{++} ions to form a *hydrated iron oxide* $Fe(OH)_2$ (really $FeO \cdot H_2O$); but instead of forming on the surface where it might give some protection, it often forms as a precipitate in the water itself. The reaction can be summarized by

$$\text{Material} + \text{Oxygen} \rightarrow \text{(Hydrated) Material Oxide}$$

just as in the case of dry oxidation.

Now the formation and solution of Fe^{++} is analogous to the formation and diffusion of M^{++} in an oxide film under dry oxidation; and the formation of OH^- is closely similar to the reduction of oxygen on the surface of an oxide film. However, the much faster attack found in wet corrosion is due to the following:

(a) The $Fe(OH)_2$ either deposits *away* from the corroding material; or, if it deposits on the surface, it does so as a loose deposit, giving little or no protection.
(b) Consequently M^{++} and OH^- usually diffuse in the *liquid* state, and therefore do so very rapidly.
(c) In *conducting* materials, the electrons can move very easily as well.

The result is that the oxidation of iron in aerated water (rusting) goes on at a rate which is millions of times faster than that in dry air. Because of the importance of (c), wet oxidation is a particular problem with metals.

26.3 Voltage differences as a driving force for wet oxidation

In dry oxidation we quantified the tendency for a material to oxidize in terms of the energy needed, in $kJ\,mol^{-1}$ of O_2, to manufacture the oxide from the material and oxygen. Because wet oxidation involves electron flow in conductors, which is easier to measure, the tendency of a metal to oxidize in solution is described by using a *voltage* scale rather than an *energy* one.

Figure 26.3 shows the voltage differences that would just stop various metals oxidizing in aerated water. As we should expect, the information in the figure is similar to that in our previous bar-chart (see Chapter 24) for the *energies* of oxidation. There are some differences in ranking, however, due to the differences between the detailed reactions that go on in dry and wet oxidation.

What do these voltages mean? Suppose we could separate the cathodic and the anodic regions of a piece of iron, as shown in Figure 26.4. Then at the cathode, oxygen is reduced to OH^-, absorbing electrons, and the metal therefore becomes positively charged. The reaction continues until the potential rises to +0.401 V. Then the coulombic attraction between the positive charged metal and the negative charged OH^- ion becomes so large that the OH^- is pulled back to the surface, and reconverted to H_2O and O_2; in other words, the reaction stops. At the anode, Fe^{++} forms, leaving electrons behind in the metal which acquires a negative charge. When its potential falls to

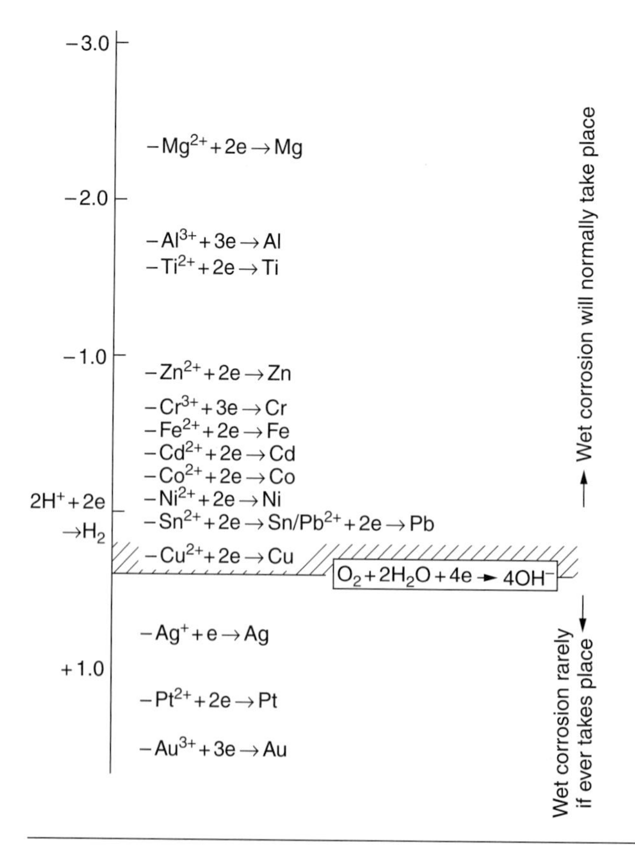

Figure 26.3 Wet corrosion voltages (at 300 K).

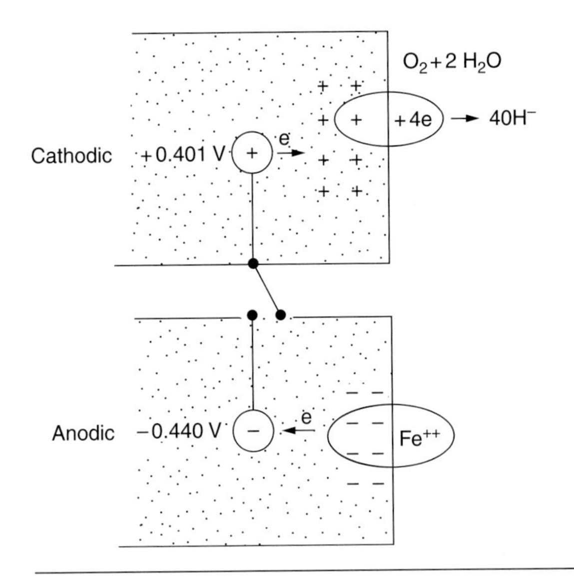

Figure 26.4 The voltages that drive wet corrosion.

− 0.440 V, that reaction, too, stops (for the same reason as before). If the anode and cathode are now *connected*, electrons flow from the one to the other, the potentials fall, and both reactions start up again. The difference in voltage of 0.841 V is the driving potential for the oxidation reaction. The bigger it is, the bigger the tendency to oxidize.

Now a note of caution about how to interpret the voltages. For convenience, the voltages given in reference books always relate to ions having certain specific concentrations (called "unit activity" concentrations). These concentrations are high—and make it rather hard for the metals to dissolve (Figure 26.5). In dilute solutions, metals can corrode more easily, and this sort of effect tends to move the voltage values around by up to 0.1 V or more for some metals. The important thing about the voltage figures given therefore is that they are only a *guide* to the driving forces for wet oxidation.

Obviously, it is not very easy to measure voltage variations *inside* a piece of iron, but we can artificially transport the "oxygen-reduction reaction" away from the metal by using a piece of metal that does not normally undergo wet oxidation (e.g. platinum) and which serves merely as a *cathode* for the oxygen-reduction reaction.

The corrosion voltages of Figure 26.3 also tell you what will happen when two *dissimilar* metals are joined together and immersed in water. If copper is joined to zinc, for instance, the zinc has a larger corrosion voltage than the copper. The zinc therefore becomes the anode, and is attacked; the copper becomes the cathode, where the oxygen reaction takes place, and it is unattacked. Such *couples* of dissimilar metals can be dangerous: the attack at the anode is sometimes very rapid, as we shall see in the next chapter.

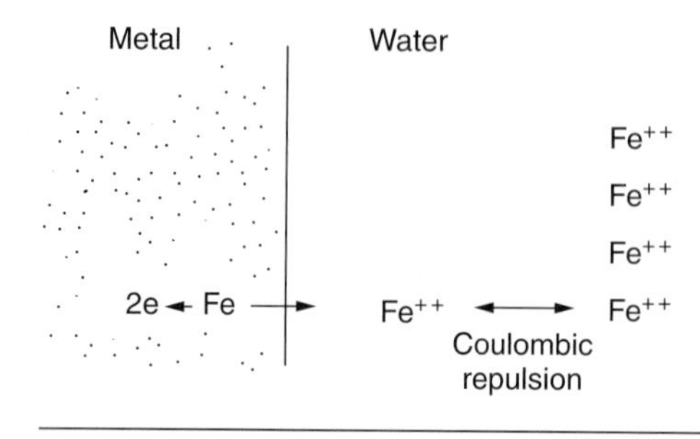

Figure 26.5 Corrosion takes place less easily in concentrated solution.

26.4 Rates of wet oxidation

As one might expect on the basis of what we said in the chapters on dry oxidation, the rates of wet oxidation found in practice bear little relationship to the voltage driving forces for wet oxidation, provided these are such that the metal is prone to corrosion in the first place. To take some examples, the approximate surface losses of some metals in mm per year in clean water are shown in Figure 26.6. They are almost the reverse of the order expected in terms of the voltage driving forces for wet oxidation. The slow rate of wet oxidation for Al, for example, arises because it is very difficult to prevent a thin, dry oxidation film of Al_2O_3 forming on the metal surface. In *sea* water, on the other hand, Al corrodes very rapidly because the chloride ions tend to break down the protective Al_2O_3 film. Because of the effect of "foreign" ions like this in most practical environments, corrosion rates vary very widely indeed for most materials. Materials handbooks often list rough figures of the wet oxidation resistance of metals and alloys in various environments (ranging from beer to sewage!).

As we saw in Chapter 21, rates of reaction (including corrosion) follow Arrhenius's law. This means that, with a given metal/environment combination, the rate of corrosion increases rapidly with increasing temperature. This behavior is shown in Figure 26.7, which plots the relative corrosion rate of mild steel in aerated mains water as a function of temperature. With this metal/environment combination, the corrosion rate increases by a factor of 6 going from 0 to 100°C.

26.5 Localized attack

It is often found that wet corrosion attacks metals *selectively* as well as, or instead of, uniformly, and this can lead to component failure *much* more rapidly and insidiously than one might infer from average corrosion rates

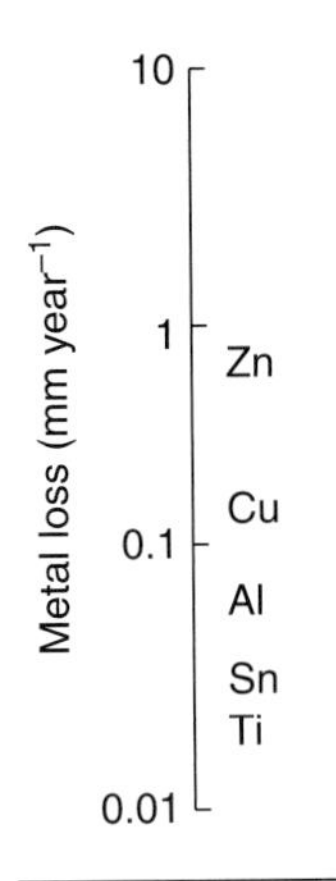

Figure 26.6 Corrosion rates of some metals in clean water.

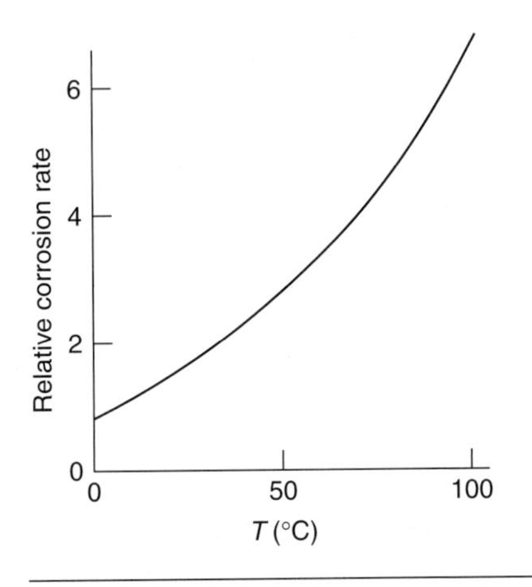

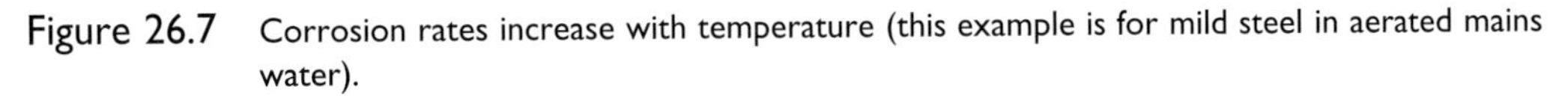

Figure 26.7 Corrosion rates increase with temperature (this example is for mild steel in aerated mains water).

(Figure 26.8). Stress and corrosion acting together can be particularly bad, giving cracks which propagate rapidly and unexpectedly. Four types of localized attack commonly lead to unplanned failures. These are:

(a) Stress corrosion cracking

In some materials and environments, cracks grow steadily under a constant stress intensity K which is much less than K_c (Figure 26.9). This is obviously dangerous: a structure which is safe when built can become unsafe with time. Examples are brass in ammonia, mild steel in caustic soda, austenitic stainless steel in hot chloride solutions, and some Al and Ti alloys in salt water.

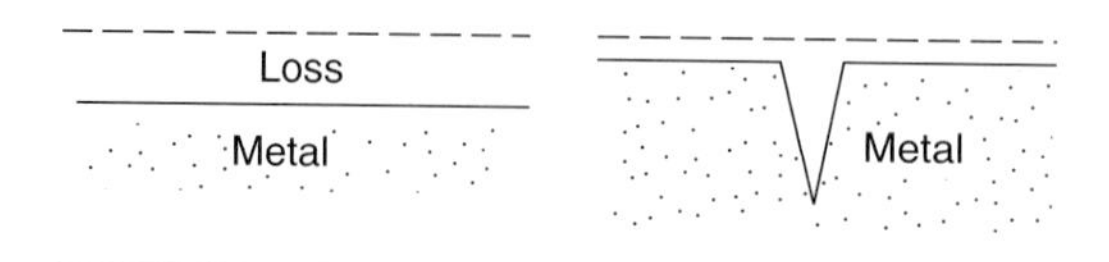

Figure 26.8 Localized attack.

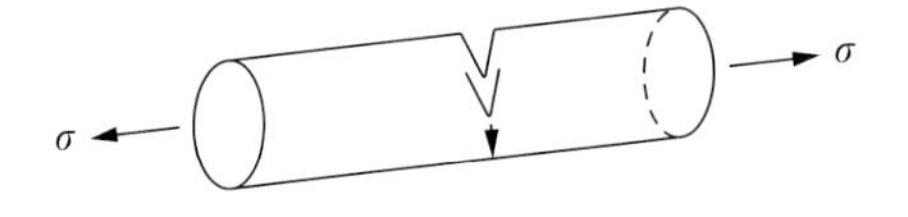

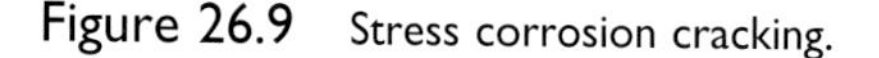

Figure 26.9 Stress corrosion cracking.

(b) Corrosion fatigue

Corrosion increases the rate of growth of fatigue cracks in most metals and alloys, for example, the stress to give $N_f = 5 \times 10^7$ cycles decreases by four times in salt water for many steels (Figure 26.10). The crack growth rate is larger—often much larger—than the sum of the rates of corrosion and fatigue, each acting alone.

(c) Intergranular attack

Grain boundaries have different corrosion properties from the grain and may corrode preferentially, giving cracks that then propagate by stress corrosion or corrosion fatigue (Figure 26.11).

(d) Pitting

Preferential attack can also occur at breaks in the oxide film (caused by abrasion), or at precipitated compounds in certain alloys (Figure 26.12).

To summarize, unexpected corrosion failures are much more likely to occur by localized attack than by uniform attack (which can easily be detected); and although corrosion handbooks are useful for making initial choices of materials for applications where corrosion is important, critical components must be checked for life-to-fracture in closely controlled experiments resembling the actual environment as nearly as possible.

In the next chapter we shall look at some case studies in corrosion-resistant designs which are based on the ideas we have just discussed.

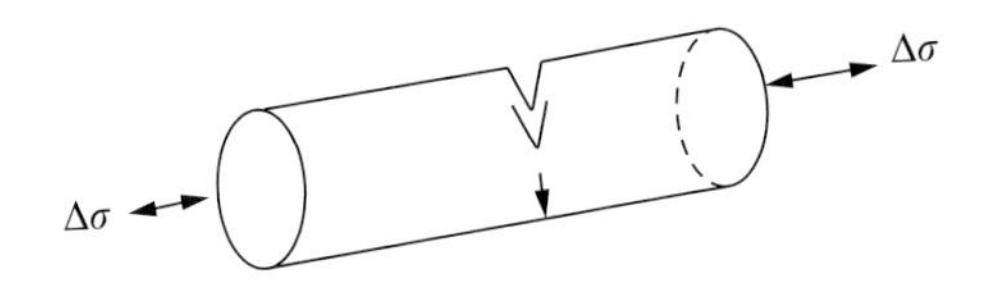

Figure 26.10 Corrosion fatigue.

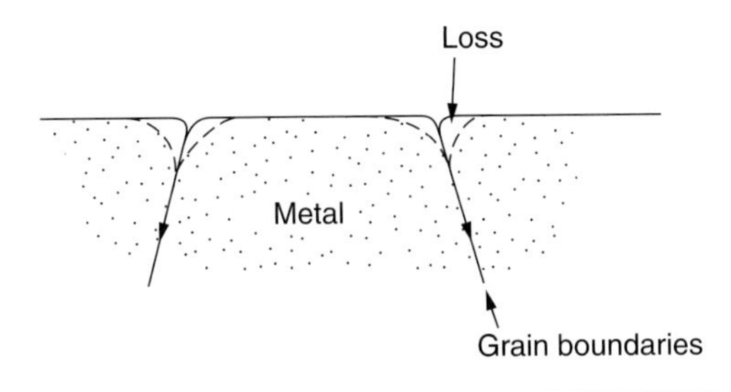

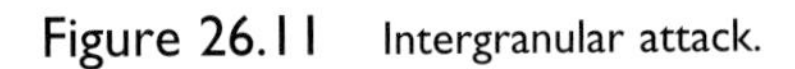
Figure 26.11 Intergranular attack.

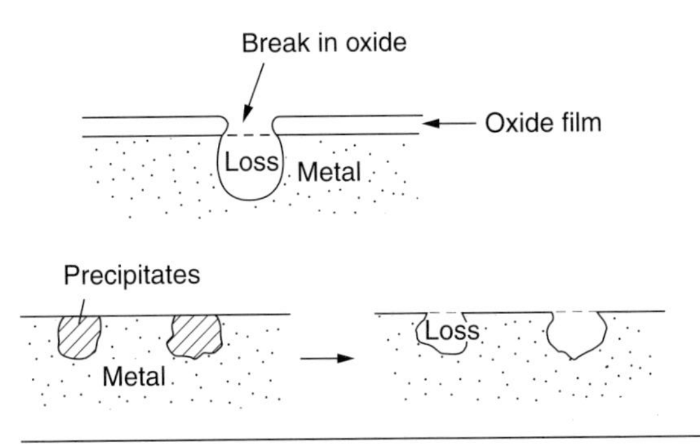

Figure 26.12 Pitting corrosion.

Examples

26.1 Measurements of the rate of crack growth in brass exposed to ammonium sulphate solution and subjected to a constant tensile stress gave the following data:

Nominal stress σ (MN m^{-2})	*Crack depth* a (mm)	*Crack growth rate* da/dt (mm year^{-1})
4	0.25	0.3
4	0.50	0.6
8	0.25	1.2

Show that these data are consistent with a relationship of form

$$\frac{da}{dt} = AK^n$$

where $K = \sigma\sqrt{\pi a}$ is the stress intensity factor. Find the values of the integer n and the constant A.

Answers

$n = 2, \quad A = 0.0239\,\text{m}^4\ \text{MN}^{-2}\ \text{year}^{-1}$

26.2 The diagram shows a mild steel pipe which carried water around a sealed central heating system. The pipes were laid in the concrete floor slabs of the building at construction, and thermally insulated with blocks of expanded polystyrene foam. After 15 years, the pipes started to leak. When the floor was dug up, it was found that the pipe had rusted through from the *outside*. It was known that rainwater had been finding its way into the building along joints between the concrete slabs, and consequently the foam in the pipe ducts would have been wet. Explain why the corrosion occurred, and also why the pipe did not rust from the inside.

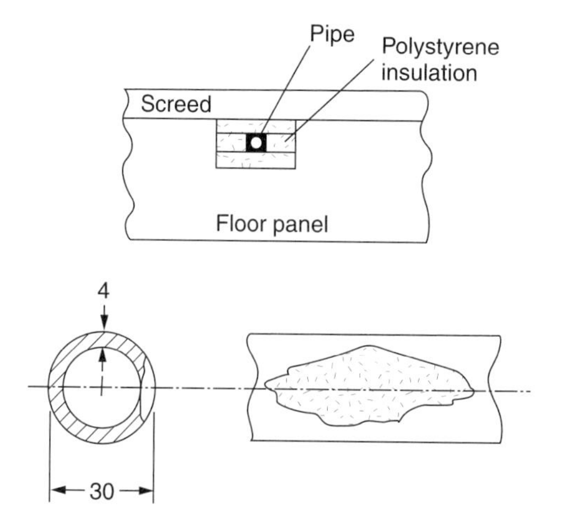

26.3 The photograph shows the inside of a centrifugal pump used for pumping industrial effluent. The material is cast iron. Identify the type of corrosion attack.

26.4 Why is resistance to stress corrosion cracking sometimes more important than resistance to uniform corrosion?

26.5 Because polymer molecules are covalently bonded (so do not ionize in water), plastics are essentially immune to wet corrosion (although they can suffer environmental degradation from UV light, ozone and some chemicals). Give examples from your own experience where polymers are used in applications where corrosion resistance is one of the prime factors in the choice of material.

26.6 The photograph shows the surface of an expansion bellows made from austenitic stainless steel, which carried wet steam at 184°C. The stress in the steel was approximately 50 percent of the yield stress. The system had been treated with zinc chloride corrosion inhibitor. Identify the type of corrosion attack.

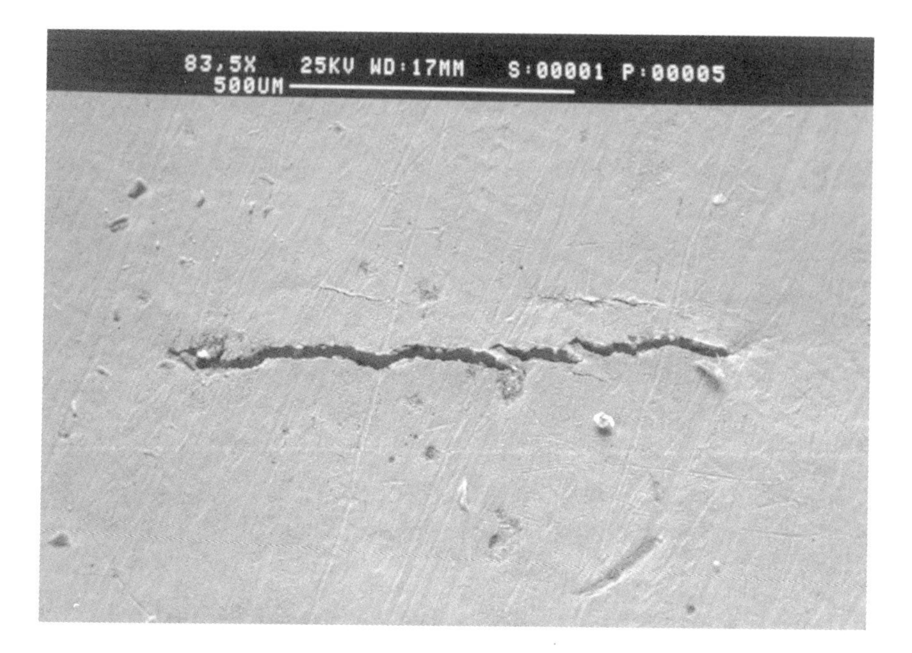
83,5X 25KV WD:17MM S:00001 P:00005
500UM

Chapter 27

Case studies in wet corrosion

Chapter contents

27.1 Introduction

We now examine three real corrosion problems: the protection of pipelines, the selection of a material for a factory roof, and materials for car exhaust systems. The rusting of iron appears in all three case studies, but the best way of overcoming it differs in each. Sometimes the best thing is to change to a new material which does not rust; but often economics prevent this, and ways must be found to slow down or stop the rusting reaction.

27.2 Case study 1: the protection of underground pipes

Many thousands of miles of steel pipeline have been laid under, or in contact with, the ground for the long-distance transport of oil, natural gas, etc. Obviously corrosion is a problem if the ground is at all damp, as it usually will be, and if the depth of soil is not so great that oxygen is effectively excluded. Then the oxygen reduction reaction

$$O_2 + 2H_2O + 4e \rightarrow 4OH^-$$

and the metal-corroding reaction

$$Fe \rightarrow Fe^{++} + 2e$$

can take place, causing the pipe to corrode. Because of the capital cost of pipelines, their inaccessibility if buried, the disruption to supplies caused by renewal, and the potentially catastrophic consequences of undetected corrosion failure, it is obviously very important to make sure that pipelines do not corrode. How is this done?

One obvious way of protecting the pipe is by covering it with some inert material to keep water and oxygen out: thick polyethylene sheet stuck in position with a butyl glue, for example. The end sections of the pipes are left uncovered ready for welding—and the welds are subsequently covered on site. However, such coverings rarely provide complete protection—rough handling on site frequently leads to breakages of the film, and careless wrapping of welds leaves metal exposed. What can we do to prevent localized attack at such points?

Sacrificial protection

If the pipe is connected to a slab of material which has a more negative corrosion voltage (Figure 27.1), then the couple forms an electrolytic cell. As explained in Chapter 26, the more electronegative material becomes the anode (and dissolves), and the pipe becomes the cathode (and is protected).

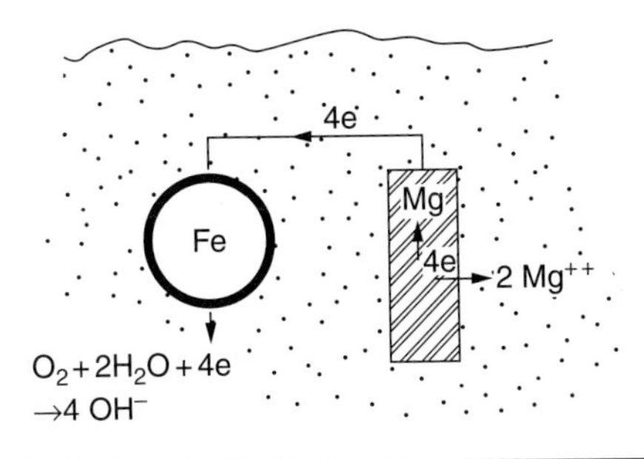

Figure 27.1 Sacrificial protection of pipelines. Typical materials used are Mg (with 6% Al, 3% Zn, 0.2% Mn), Al (with 5% Zn) and Zn.

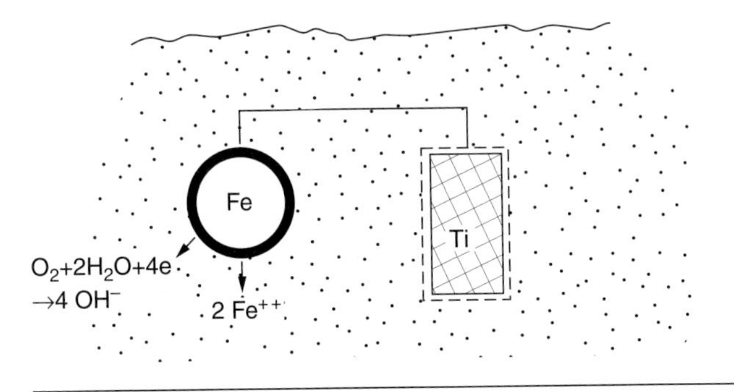

Figure 27.2 Some sacrificial materials do not work because they carry a "passivating" oxide layer.

As Figure 27.1 shows, pipelines are protected from corrosion by being wired to anodes in just this way. Magnesium alloy is often used because its corrosion voltage is very low (much lower than that of zinc) and this attracts Fe^{++} to the steel very strongly; but aluminum alloys and zinc are used widely too. The alloying additions help prevent the formation of a protective oxide on the anode—which might make it become cathodic. With some metals in particular environments (e.g. titanium in sea water) the nature of the oxide film is such that it effectively prevents metal passing through the film into solution. Then, although titanium is very negative with respect to iron (see Figure 26.3), it fails to protect the pipeline (Figure 27.2). Complications like this can also affect other metals (e.g. Al, Cd, Zn) although generally to a much smaller extent. *This sort of behavior is another reason for our earlier warning that corrosion voltage are only general guides to corrosion behavior*—again, experimental work is usually a necessary prelude to design against corrosion.

Naturally, because the protection depends on the dissolution of the anodes, these require replacement from time to time (hence the term "sacrificial"

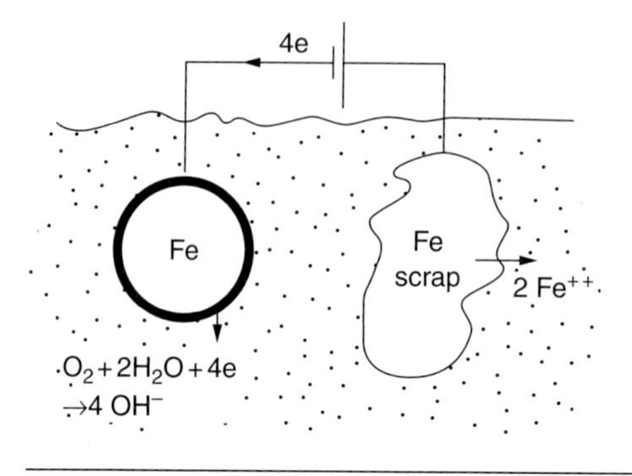

Figure 27.3 Protection of pipelines by imposed potential.

anodes). In order to minimize the loss of anode metal, it is important to have as good a barrier layer around the pipe as possible, even though the pipe would still be protected with no barrier layer at all.

Protection by imposing a potential

An alternative way of protecting the pipe is shown in Figure 27.3. Scrap steel is buried near the pipe and connected to it through a battery or d.c. power supply, which maintains a sufficient potential difference between them to make sure that the scrap is always the anode and the pipe the cathode (it takes roughly the corrosion potential of iron—a little under 1 volt). This alone will protect the pipe, but unless the pipe is coated, a large current will be needed to maintain this potential difference.

Alternative materials

Cost rules out almost all alternative materials for long-distance pipelines: it is much cheaper to build and protect a mild steel pipe than to use stainless steel instead—even though no protection is then needed. The only competing material is a polymer, which is completely immune to wet corrosion of this kind. City gas mains are now being replaced by polymeric ones; but for large diameter transmission lines, the mechanical strength of steel makes it the preferred choice.

27.3 Case study 2: materials for a lightweight factory roof

Let us now look at the corrosion problems that are involved in selecting a material for the lightweight roof of a small factory. Nine out of 10 people

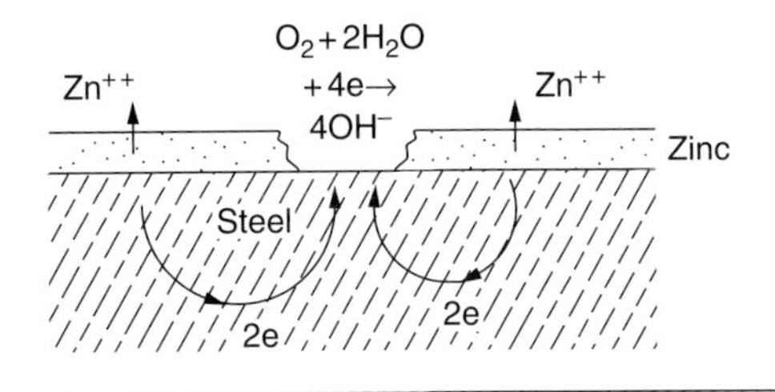

Figure 27.4 Galvanized steel is protected by a sacrificial layer of zinc.

asked to make a selection would think first of corrugated, *galvanized steel*. This is strong, light, cheap, and easy to install. Where's the catch? Well, fairly new galvanized steel is rust-free, but after 20–30 years, rusting sets in and the roof eventually fails.

How does galvanizing work? As Figure 27.4 shows, the galvanizing process leaves a thin layer of zinc on the surface of the steel. This acts as a barrier between the steel and the atmosphere; and although the driving voltage for the corrosion of zinc is greater than that for steel (see Figure 26.3) in fact zinc corrodes quite slowly in a normal urban atmosphere because of the barrier effect of its oxide film. The loss in thickness is typically 0.1 mm in 20 years.

If scratches and breaks occur in the zinc layers by accidental damage—which is certain to occur when the sheets are erected—then the zinc will cathodically protect the iron (see Figure 27.4) in exactly the way that pipelines are protected using zinc anodes. This explains the long postponement of rusting. But the coating is only about 0.15 mm thick, so after about 30 years most of the zinc has gone, rusting suddenly becomes chronic, and the roof fails.

At first sight, the answer would seem to be to increase the thickness of the zinc layer. This is not easily done, however, because the hot dipping process used for galvanizing is not sufficiently adjustable; and electroplating the zinc onto the steel sheet increases the production cost considerably. Painting the sheet (e.g. with a bituminous paint) helps to reduce the loss of zinc considerably, but at the same time should vastly decrease the area available for the cathodic protection of the steel; and if a scratch penetrates both the paint and the zinc, the exposed steel may corrode through much more quickly than before.

Alternative materials

A relatively recent innovation has been the architectural use of *anodized aluminum*. Although the driving force for the wet oxidation of aluminum is very large, aluminum corrodes very slowly in fresh-water environments because it carries a very adherent film of the poorly conducting Al_2O_3. In anodized aluminum, the Al_2O_3 film is artificially thickened in order to make this barrier to corrosion extremely effective. In the anodizing process, the aluminum part is put into water containing various additives to promote compact film growth (e.g. boric acid). It is then made positive electrically which attracts the oxygen atoms in the polar water molecules (see Chapter 4). The attached oxygen atoms react continuously with the metal to give a thickened oxide film as shown on Figure 27.5. The film can be coloured for aesthetic purposes by adding colouring agents towards the end of the process and changing the composition of the bath to allow the colouring agents to be incorporated.

Finally, what of polymeric materials? Corrugated plastic sheet is commonly used for roofing small sheds, car ports, and similar buildings; but although polymers do not generally corrode — they are often used in nasty environments

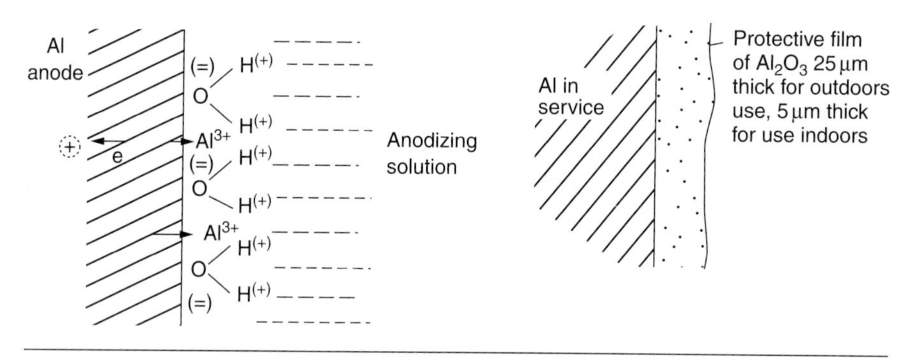

Figure 27.5 Protecting aluminum by anodizing it.

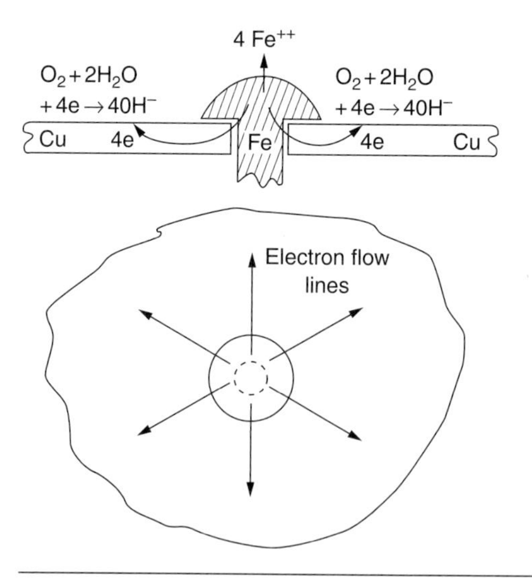

Figure 27.6 Large cathodes can lead to very rapid corrosion.

like chemical plant — they *are* prone to damage by the ultraviolet wavelengths of the sun's radiation. These high-energy photons, acting over a period of time, gradually break up the molecular chains in the polymer and degrade its mechanical properties.

A note of caution about roof fasteners. A common mistake is to fix a galvanized or aluminum roof in place with nails or screws of a *different* metal: copper or brass, for instance. The copper acts as cathode, and the zinc or aluminum corrodes away rapidly near to the fastening. A similar sort of goof has been known to occur when copper roofing sheet has been secured with steel nails. As Figure 27.6 shows, this sort of situation leads to catastrophically rapid corrosion not only because the iron is anodic, but because it is so easy for the electrons generated by the anodic corrosion to get away to the large copper cathode.

27.4 Case study 3: automobile exhaust systems

The lifetime of a conventional exhaust system on an average family car is only 3 years or so. This is hardly surprising — mild steel is the usual material and, as we have shown, it is not noted for its corrosion resistance. The interior of the system is not painted and begins to corrode immediately in the damp exhaust gases from the engine. The single coat of cheap cosmetic paint soon falls off the outside and rusting starts there, too, aided by the chloride ions from road salt, which help break down the iron oxide film.

The lifetime of the exhaust system could be improved by galvanizing the steel to begin with. But there are problems in using platings where steel has to be joined by *welding*. Zinc, for example, melts at 420°C and would be burnt off the welds; and breaks would still occur if plating metals of higher melting point (e.g. Ni, 1455°C) were used. Occasionally manufacturers fit chromium-plated exhaust systems but this is for appearance only: if the plating is done before welding, the welds are unprotected and will corrode quickly; and if it is done after welding, the interior of the system is unplated and will corrode.

Alternative materials

The most successful way of combating exhaust-system corrosion is, in fact, stainless steel. This is a good example of how — just as with dry oxidation — the addition of foreign atoms to a metal can produce stable oxide films that act as barriers to corrosion. In the case of stainless steel, Cr is dissolved in the steel in solid solution, and Cr_2O_3 forms on the surface of the steel to act as a corrosion barrier.

There is one major pitfall which must be avoided in using stainless-steel components joined by welding: it is known as *weld decay*. It is sometimes found that the *heat-affected zone* — the metal next to the weld which got hot but did not melt — corrodes badly.

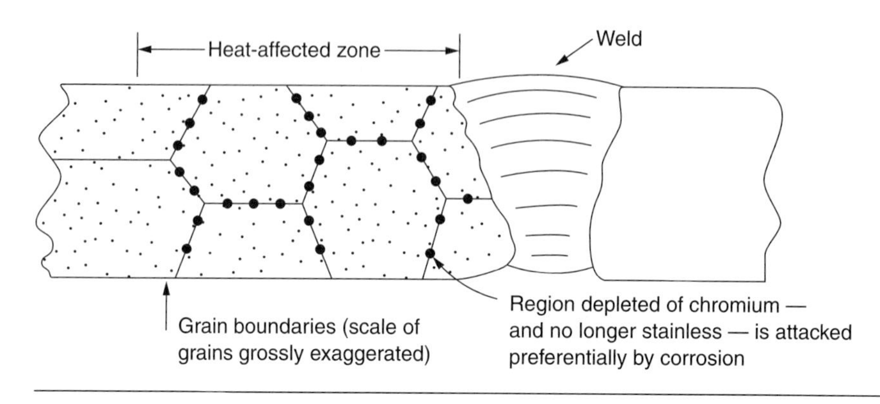

Figure 27.7 Weld decay in stainless steel.

Figure 27.7 explains why. All steels contain carbon — for their mechanical properties — and this carbon can "soak up" chromium (at grain boundaries in particular) to form precipitates of the compound *chromium carbide*. Because the regions near the grain boundaries lose most of their chromium in this way, they are no longer protected by Cr_2O_3, and corrode badly. The cure is to *stabilize* the stainless steel (by adding Ti or Nb which soaks up the carbon near the grain boundaries) or use low-carbon grades of stainless steel.

Examples

27.1 Mild-steel radiators in a sealed (oxygen free) central-heating system were found to have undergone little corrosion after several years service. Explain.

27.2 In order to prevent the corrosion of a mild-steel structure immersed in sea water, a newly qualified engineer suggested the attachment of titanium plates in the expectation of powerful cathodic action. He later found that the structure had corroded badly. Explain.

27.3 The corrosion of an underground steel pipeline was greatly reduced when the pipeline was connected to a buried bar of magnesium alloy. Explain.

27.4 Under aggressive corrosion conditions it is estimated that the maximum corrosion current density in a galvanized steel sheet will be $6 \times 10^{-3}\,A\,m^{-2}$. Estimate the thickness of the galvanized layer needed to give a rust-free life of at least 5 years. The density of zinc is $7.13\,Mg\,m^{-3}$, and its atomic weight is 65.4. Assume that the zinc corrodes to give Zn^{2+} ions.

Answer

0.045 mm.

27.5 A sheet of steel of thickness 0.50 mm is tinplated on both sides and subjected to a corrosive environment. During service, the tinplate becomes scratched, so that steel is exposed over 0.5 percent of the area of the sheet. Under these conditions it is estimated that the current consumed at the tinned surface by the oxygen-reduction reaction is $2 \times 10^{-3}\,A\,m^{-2}$. Will the sheet rust through within 5 years in the scratched condition? The density of steel is $7.87\,Mg\,m^{-3}$. Assume that the steel corrodes to give Fe^{2+} ions. The atomic weight of iron is 55.9.

Answer

Yes.

27.6 Steel nails used to hold copper roofing sheet in position failed rapidly by wet corrosion. Explain.

27.7 A reaction vessel for a chemical plant was fabricated by welding together stainless-steel plates (containing 18% chromium, 8% nickel and 0.1% carbon by weight). During service the vessel corroded badly at the grain boundaries near the welds. Explain.

Part H

Friction, abrasion and wear

Chapter 28

Friction and wear

Chapter contents

28.1 Introduction

We now come to the final properties that we shall be looking at in this book on engineering materials: the frictional properties of materials in contact, and the wear that results when such contacts slide. This is of considerable importance in mechanical design. Frictional forces are undesirable in bearings because of the power they waste; and wear is bad because it leads to poor working tolerances, and ultimately to failure. On the other hand, when selecting materials for clutch and brake linings—or even for the soles of shoes—we aim to maximize friction but still to minimize wear, for obvious reasons. But wear is not always bad: in operations such as grinding and polishing, we try to achieve maximum wear with the minimum of energy expended in friction; and without wear you could not write with chalk on a blackboard, or with a pencil on paper. In this chapter and the next we shall examine the origins of friction and wear and then explore case studies which illustrate the influence of friction and wear on component design.

28.2 Friction between materials

As you know, when two materials are placed in contact, any attempt to cause one of the materials to slide over the other is resisted by a *friction force* (Figure 28.1). The force that will just cause sliding to start, F_s, is related to the force P acting normal to the contact surface by

$$F_s = \mu_s P \tag{28.1}$$

Where μ_s is the *coefficient of static friction.* Once sliding starts, the limiting frictional force decreases slightly and we can write

$$F_k = \mu_k P \tag{28.2}$$

where μ_k $(<\mu_s)$ is the *coefficient of kinetic friction* (Figure 28.1). The work done in sliding against kinetic friction appears as *heat.*

These results at first sight run counter to our intuition—how is it that the friction between two surfaces can depend only on the *force P* pressing them together and not on their area? In order to understand this behavior, we must first look at the geometry of a typical surface.

If the surface of a fine-turned bar of copper is examined by making an oblique slice through it (a "taper section" which magnifies the height of any asperities), or if its profile is measured with a "Talysurf" (a device like a gramophone pick-up which, when run across a surface, plots out the hills and valleys), it is found that the surface looks like Figure 28.2. The figure shows a large number of projections or *asperities*—it looks rather like a cross-section through Switzerland. If the metal is abraded with the finest abrasive paper, the

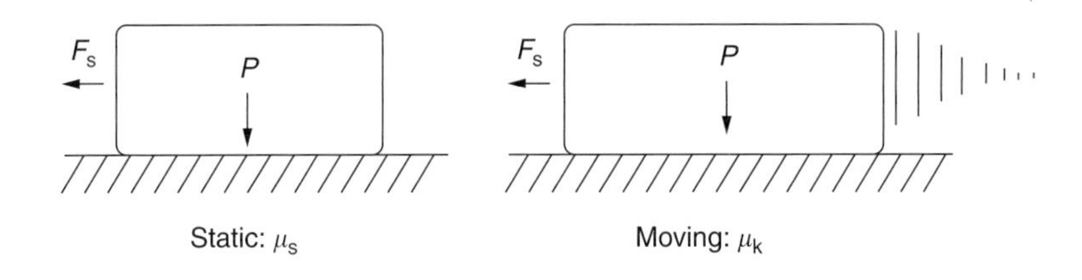

Figure 28.1 Static and kinetic coefficients of friction.

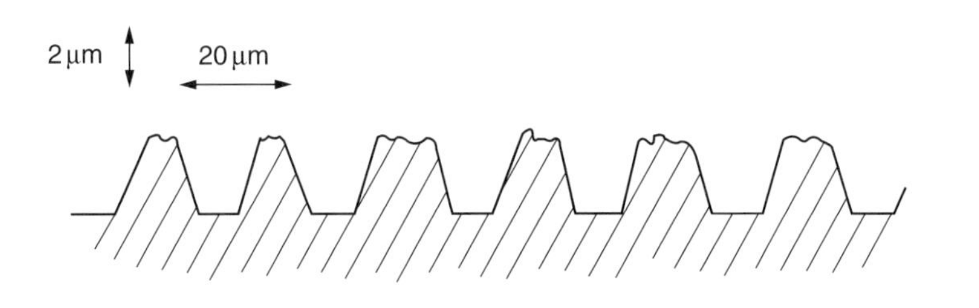

Figure 28.2 What a finely machined metal surface looks like at high magnification (the heights of the asperities are plotted on a much more exaggerated scale than the lateral distances between asperities).

scale of the asperities decreases but they are still there—just smaller. Even if the surface is polished for a long time using the finest type of metal polish, micro-asperities still survive.

So it follows that, if two surfaces are placed in contact, no matter how carefully they have been machined and polished, they will contact only at the occasional points where one set of asperities meets the other. It is rather like turning Austria upside down and putting it on top of Switzerland. The load pressing the surfaces together is supported solely by the contacting asperities. The *real* area of contact, a, is very small and because of this the stress P/a (load/area) on each asperity is very large.

Initially, at very low loads, the asperities deform elastically where they touch. However, for realistic loads, the high stress causes extensive *plastic* deformation at the tips of asperities. If each asperity yields, forming a junction with its partner, the total load transmitted across the surface (Figure 28.3) is

$$P \approx a\sigma_y \tag{28.3}$$

where σ_y is the compressive yield stress. In other words, the real area of contact is given by

$$a \approx \frac{P}{\sigma_y} \tag{28.4}$$

Obviously, if we double P we double the real area of contact, a.

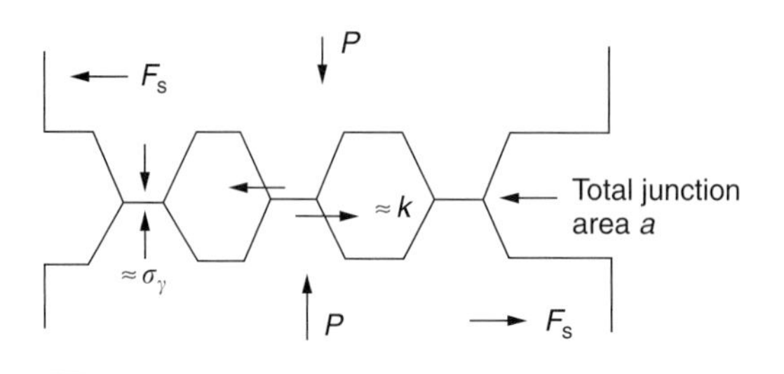

Figure 28.3 The real contact area between surfaces is less than it appears to be, because the surfaces touch only where asperities meet.

Let us now look at how this contact geometry influences friction. If you attempt to slide one of the surfaces over the other, a shear stress F_s/a appears at the asperities. The shear stress is greatest where the cross-sectional area of asperities is least, that is, at or very near the contact plane. Now, the intense plastic deformation in the regions of contact presses the asperity tips together so well that there is atom-to-atom contact across the junction. The junction, therefore, can withstand a shear stress as large as k approximately, where k is the shear-yield strength of the material (Chapter 11).

The asperities will give way, allowing sliding, when

$$\frac{F_s}{a} \geq k$$

or, since $k \approx \sigma_y/2$, when

$$F_s \approx ak \approx a\sigma_y/2 \tag{28.5}$$

Combining this with equation (28.3), we have

$$F_s \approx \frac{P}{2} \tag{28.6}$$

This is just the empirical equation (28.1) we started with, with $\mu_s \approx 1/2$, but this time it is not empirical—we derived it from a model of the sliding process. The value $\mu_s \approx 1/2$ is close to the value of coefficients of static friction between unlubricated metal, ceramic and glass surfaces—a considerable success.

How do we explain the lower value of μ_k? Well, once the surfaces are sliding, there is less *time* available for atom-to-atom bonding at the asperity junctions than when the surfaces are in static contact, and the contact area over which shearing needs to take place is correspondingly reduced. As soon as sliding stops, creep allows the contacts to grow a little, and diffusion allows the bond there to become stronger, and μ rises again to μ_s.

28.3 Data for coefficients of friction

If *metal* surfaces are thoroughly cleaned in vacuum it is almost impossible to slide them over each other. Any shearing force causes further plasticity at the junctions, which quickly grow, leading to complete seizure ($\mu > 5$). This is a problem in outer space, and in atmospheres (e.g. H_2) which remove any surface films from the metal. A little oxygen or H_2O greatly reduces μ by creating an oxide film which prevents these large metallic junctions forming.

We said in Chapter 24 that all metals except gold have a layer, no matter how thin, of metal oxide on their surfaces. Experimentally, it is found that for some metals the junction between the oxide films formed at asperity tips is weaker in shear than the metal on which it grew (Figure 28.4). In this case, sliding of the surfaces will take place in the thin oxide layer, at a stress less than in the metal itself, and lead to a corresponding reduction in μ to a value between 0.5 and 1.5.

When soft metals slide over each other (e.g., lead on lead, Figure 28.5) the junctions are weak but their area is large so μ is large. When hard metals slide (e.g. steel on steel) the junctions are small, but they are strong, and again friction is large (Figure 28.5). Many bearings are made of a thin film of a soft metal between two hard ones, giving weak junctions of small area. *White metal* bearings, for example, consist of soft alloys of lead or tin supported in a matrix of stronger phases; *bearing bronzes* consist of soft lead particles (which smear out to form the lubricating film) supported by a bronze matrix; and polymer-impregnated *porous bearings* are made by partly sintering copper with a polymer (usually PTFE) forced into its pores. Bearings like these are not designed to run dry—but if lubrication does break down, the soft component gives a coefficient of friction of 0.1–0.2 which may be low enough to prevent catastrophic overheating and seizure.

When ceramics slide on ceramics (Figure 28.5), friction is lower. Most ceramics are very hard—good for resisting wear—and, because they are stable in air and water (metals, except gold, are not genuinely stable, even if they appear so)—they have less tendency to bond, and shear more easily.

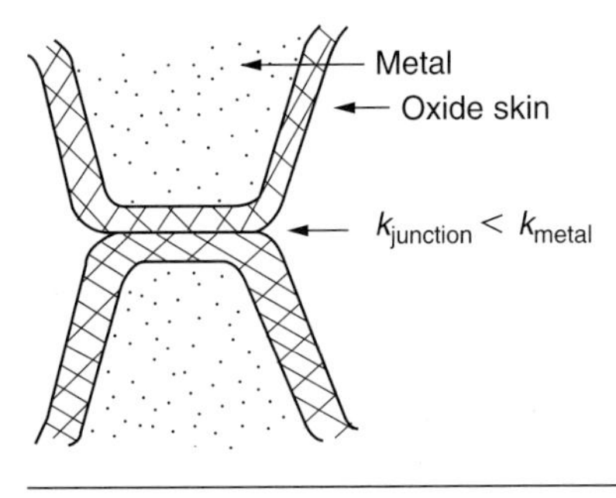

Figure 28.4 Oxide-coated junctions can often slide more easily than ones which are clean.

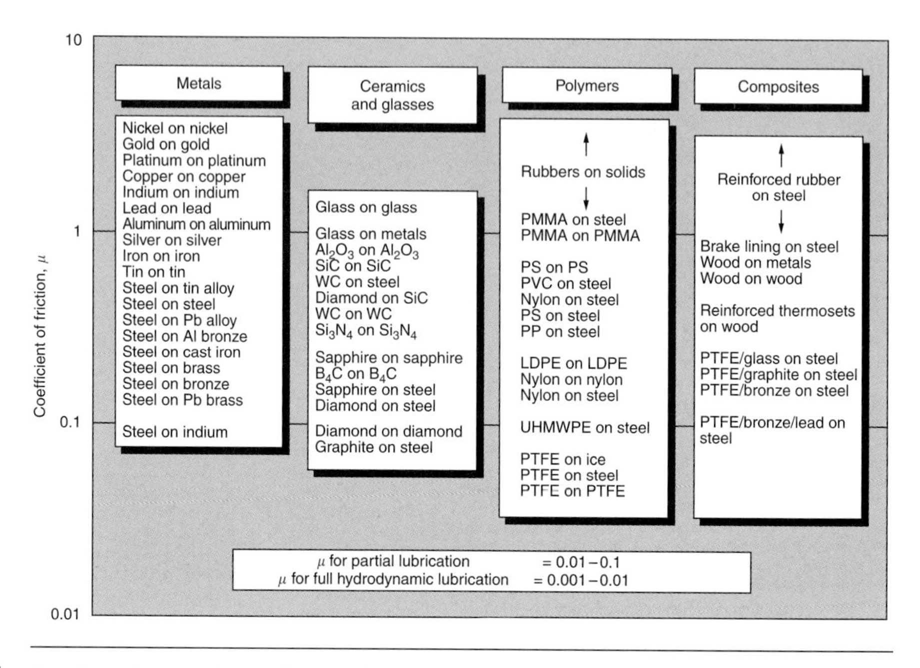

Figure 28.5 Bar chart showing the coefficient of static friction for various material combinations.

When metals slide on bulk polymers, friction is still caused by adhesive junctions, transferring a film of polymer to the metal. And any plastic flow tends to orient the polymer chains parallel to the sliding surface, and in this orientation they shear easily, so μ is low—0.05 to 0.5 (Figure 28.5). Polymers make attractive low-friction bearings, although they have some drawbacks: polymer molecules peel easily off the sliding surface, so wear is heavy; and because creep allows junction growth when the slider is stationary, the coefficient of static friction, μ_s, is sometimes much larger than that for sliding friction, μ_k.

Composites can be designed to have high friction (brake linings) or low friction (PTFE/bronze/lead bearings), as shown in Figure 28.5. More of this presently.

28.4 Lubrication

As we said in the introduction, friction absorbs a lot of work in machinery and as well as wasting power, this work is mainly converted to heat at the sliding surfaces, which can damage and even melt the bearing. In order to minimize frictional forces we need to make it as easy as possible for surfaces to slide over one another. The obvious way to try to do this is to contaminate the asperity tips with something that: (a) can stand the pressure at the bearing surface and so prevent atom-to-atom contact between asperities; (b) can itself shear easily.

Polymers and soft metal, as we have said, can do this; but we would like a much larger reduction in μ than these can give, and then we must use *lubricants*. The standard lubricants are oils, greases and fatty materials such as soap and animal fats. These "contaminate" the surfaces, preventing adhesion, and the thin layer of oil or grease shears easily, obviously lowering the coefficient of friction. What is *not* so obvious is why the very fluid oil is not squeezed out from between the asperities by the enormous pressures generated there. One reason is that oils nowadays have added to them small amounts (≈ 1 percent) of active organic molecules. One end of each molecule reacts with the metal oxide surface and sticks to it, while the other ends attract one another to form an oriented "forest" of molecules (Figure 28.6), rather like mould on cheese. These forests can resist very large forces normal to the surface (and hence separate the asperity tips very effectively) while the two layers of molecules can shear over each other quite easily. This type of lubrication is termed *partial or boundary lubrication*, and is capable of reducing μ by a factor of 10 (Figure 28.5). *Hydrodynamic lubrication* is even more effective: we shall discuss it in the next chapter.

Even the best boundary lubricants cease to work above about 200°C. Soft metal bearings like those described above can cope with *local* hot spots: the soft metal melts and provides a local lubricating film. But when the entire bearing is designed to run hot, special lubricants are needed. The best are a suspension of PTFE in special oils (good to 320°C); graphite (good to 600°C); and molybdenum disulphide (good to 800°C).

28.5 Wear of materials

Even when solid surfaces are protected by oxide films and boundary lubricants, some solid-to-solid contact occurs at regions where the oxide film breaks down under mechanical loading, and adsorption of active boundary lubricants is poor. This intimate contact will generally lead to *wear*. Wear is normally divided into two main types: *adhesive* and *abrasive wear*.

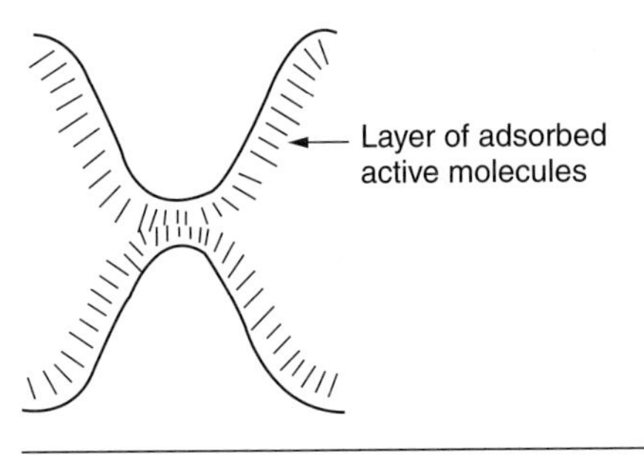

Figure 28.6 Boundary lubrication.

Adhesive wear

Figure 28.7 shows that, if the adhesion between A and B atoms is good enough, wear fragments will be removed from the softer material A. If materials A and B are the same, wear takes place from *both* surfaces—the wear bits fall off and are lost or get trapped between the surfaces and cause further trouble (see below). The size of the bits depends on how far away from the junction the shearing takes place: if work-hardening extends well into the asperity, the tendency will be to produce large pieces. In order to minimize the *rate of wear* we obviously need to minimize the size of each piece removed. The obvious way to do this is to minimize the area of contact a. Since $a \approx P/\sigma_y$ reducing the loading on the surfaces will reduce the wear, as would seem intuitively obvious. Try it with chalk on a blackboard: the higher the pressure, the stronger the line (a wear track). The second way to reduce a is to increase σ_y, i.e. the *hardness*. This is why hard pencils write with a lighter line than soft pencils.

Abrasive wear

Wear fragments produced by adhesive wear often become detached from their asperities during further sliding of the surfaces. Because oxygen is desirable in lubricants (to help maintain the oxide-film barrier between the sliding metals) these detached wear fragments can become oxidized to give hard oxide particles which *abrade* the surfaces in the way that sandpaper might.

Figure 28.8 shows how a hard material can "plough" wear fragments from a softer material, producing severe abrasive wear. Abrasive wear is not, of course, confined to indigenous wear fragments, but can be caused by dirt particles (e.g. sand) making their way into the system, or—in an engine—by combustion products: that is why it is important to filter the oil.

Obviously, the rate of abrasive wear can be reduced by reducing the load—just as in a hardness test. The particle will dig less deeply into the metal, and

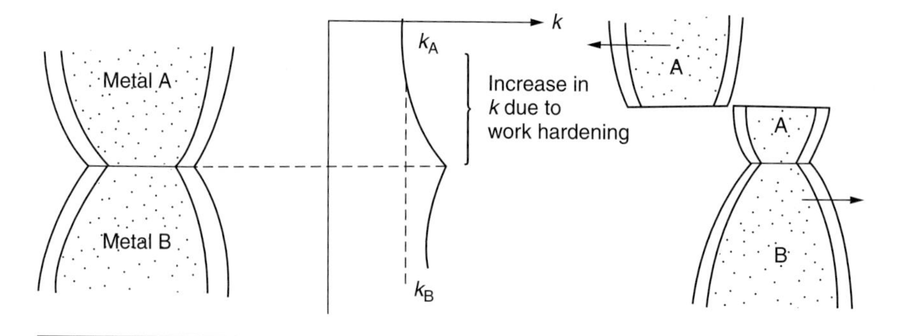

Figure 28.7 Adhesive wear.

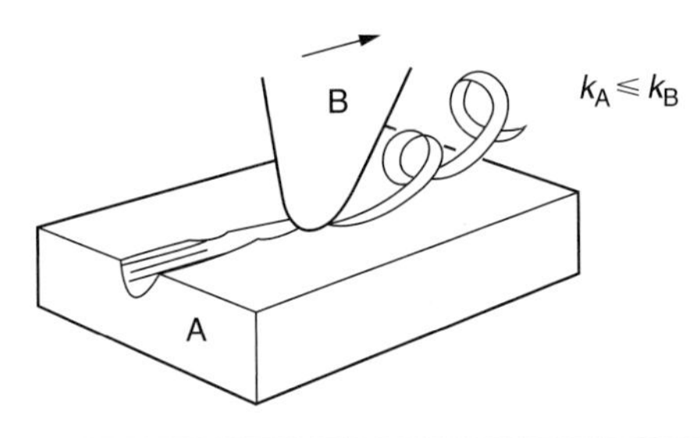

Figure 28.8 Abrasive wear.

plough a smaller furrow. Increasing the hardness of the metal will have the same effect. Again, although abrasive wear is usually bad — as in machinery — we would find it difficult to sharpen lathe tools, or polish brass ornaments, or drill rock, without it.

28.6 Surface and bulk properties

Many considerations enter the choice of material for a bearing. It must have bulk properties which meet the need to support loads and transmit heat fluxes. It must be processable: that is, capable of being shaped, finished, and joined. It must meet certain economic criteria: limits on cost, availability, and suchlike. If it can do all these things it must further have — or be given — necessary surface properties to minimize wear, and, when necessary, resist corrosion.

So, bearing materials are not chosen for their wear or friction properties (their "tribological" properties) alone; they have to be considered in the framework of the overall design. One way forward is to choose a material with good bulk properties, and then customize the surface with exotic treatments or coatings. For the most part, it is the properties of the surface which determine tribological response, although the immediate subsurface region is obviously important because it supports the surface itself.

There are two general ways of tailoring surfaces. The aim of both is to increase the surface hardeness, or to reduce friction, or all of these. The first is *surface treatment* involving only small changes to the chemistry of the surface. They exploit the increase in the hardness given by embedding foreign atoms in a thin surface layer: in carburizing (carbon), nitriding (nitrogen) or boriding (boron) the surface is hardened by diffusing these elements into it from a gas, liquid or solid powder at high temperatures. Steels, which already contain carbon, can be surface-hardened by rapidly heating and then cooling their surfaces with a flame, an electron beam, or a laser. Elaborate though these processes sound, they are standard procedures, widely used, and to very good effect.

The second approach, that of *surface coating*, is more difficult, and that means more expensive. But it is often worth it. Hard, corrosion resistant layers of alloys rich in tungsten, cobalt, chromium, or nickel can be sprayed onto surfaces, but a refinishing process is almost always necessary to restore the dimensional tolerances. Hard ceramic coatings such as Al_2O_3, Cr_2O_3, TiC, or TiN can be deposited by plasma methods and these not only give wear resistance but resistance to oxidation and other sorts of chemical attack as well. And—most exotic of all—it is now possible to deposit diamond (or something very like it) on to surfaces to protect them from almost anything.

Enough of this here. Surfaces resurface (as you might say) in the next chapter.

Examples

28.1 Explain the origins of friction between solid surfaces in contact.

28.2 The diagram shows a compression joint for fixing copper water pipe to plumbing fittings. When assembling the joint the gland nut is first passed over the pipe followed by a circular olive made from soft copper. The nut is then screwed onto the end of the fitting and the backlash is taken up. Finally the nut is turned through a specified angle which compresses the olive on to the surface of the pipe. The angle is chosen so that it is just sufficient to make the cross-section of the pipe yield in compression over the length which is in contact with the olive. Show that the water pressure required to make the pipe shoot out of the fitting is given approximately by

$$p_{\mathrm{w}} = 2\mu\sigma_{\mathrm{y}}\left(\frac{t}{r}\right)\left(\frac{l}{r}\right)$$

where μ is the coefficient of friction between the olive and the outside of the pipe.

Calculate p_{w} given the following information: $t = 0.65$ mm, $l = 7.5$ mm, $r = 7.5$ mm, $\mu = 0.15$, $\sigma_{\mathrm{y}} = 120$ MPa. Comment on your answer in relation to typical hydrostatic pressures in water systems. [Hints: the axial load on the

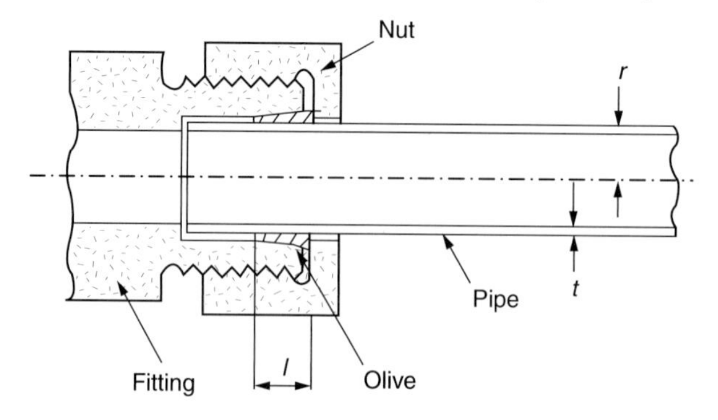

joint is $p_w \pi r^2$; the radial pressure applied to the outside of the pipe by the olive is $P = \sigma_y t/r$.]

Answer

3.1 MPa, or 31 bar.

28.3 Give examples, from your own experience, of situations where *friction* is (a) desirable, and (b) undesirable.

28.4 Give examples, from your own experience, of situations where *wear* is (a) desirable, and (b) undesirable.

Chapter 29

Case studies in friction and wear

Chapter contents

29.1 Introduction

In this chapter we examine three quite different problems involving friction and wear. The first involves most of the factors that appeared in Chapter 28: it is that of a round shaft or journal rotating in a cylindrical bearing. This type of *journal bearing* is common in all types of rotating or reciprocating machinery: the crankshaft bearings of an automobile are good examples. The second is quite different: it involves the frictional properties of ice in the design of skis and sledge runners. The third case study introduces us to some of the frictional properties of polymers: the selection of rubbers for anti-skid tyres.

29.2 Case study 1: the design of journal bearings

In the proper functioning of a well-lubricated journal bearing, the frictional and wear properties of the materials are, surprisingly, irrelevant. This is because the mating surfaces never touch: they are kept apart by a thin pressurized film of oil formed under conditions of *hydrodynamic lubrication.* Figure 29.1 shows a cross section of a bearing operating hydrodynamically. The load on the journal pushes the shaft to one side of the bearing, so that the working clearance is almost all concentrated on one side. Because oil is viscous, the revolving shaft drags oil around with it. The convergence of the oil stream towards the region of nearest approach of the mating surfaces causes an increase in the pressure of the oil film, and this pressure lifts the shaft away from the bearing surface. Pressures of 10 to 100 atmospheres are common under such conditions. Provided the oil is sufficiently viscous, the film at its thinnest region is still thick enough to cause complete separation of the mating surfaces. Under *ideal* hydrodynamic conditions there is no asperity contact and no wear. Sliding of the mating surfaces takes place by shear in the liquid oil itself, giving coefficients of friction in the range 0.001 to 0.005.

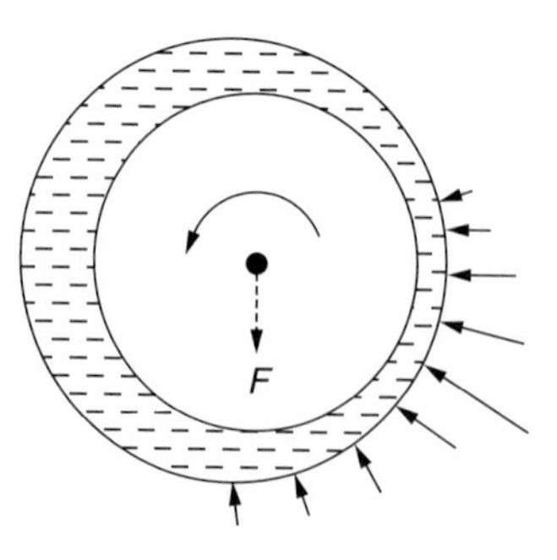

Figure 29.1 Hydrodynamic lubrication.

Hydrodynamic lubrication is all very well when it functions properly. But real bearings contain dirt—hard particles of silica, usually—and new automobile engines are notorious for containing hard cast-iron dust from machining operations on the engine block. Then, if the particles are thicker than the oil film at its thinnest, abrasive wear will take place. There are two ways of solving this problem. One is to make the mating surfaces harder than the dirt particles. Crankshaft journals are "case-hardened" by special chemical and heat treatments (Chapter 28) to increase the surface hardness to the level at which the dirt is abraded by the journal. (It is important not to harden the *whole* shaft because this will make it brittle and it might then break under shock loading.) However, the bearing surfaces are not hardened in this way; there are benefits in keeping them soft. First, if the bearing metal is soft enough, dirt particles will be pushed into the surface of the bearing and will be taken largely out of harm's way. This property of bearing material is called *embeddability*. And, second, a bearing only operates under conditions of hydrodynamic lubrication when the rotational speed of the journal is high enough. When starting an engine up, or running slowly under high load, hydrodynamic lubrication is not present, and we have to fall back on *boundary lubrication* (see Chapter 28). Under these conditions some contact and wear of the mating surfaces will occur (this is why car engines last less well when used for short runs rather than long ones). Now crankshafts are difficult and expensive to replace when worn, whereas bearings can be designed to be cheap and easy to replace as shown in Figure 29.2. It is thus good practice to concentrate as much of the wear as possible on the *bearing*—and, as we showed in our section on *adhesive wear* in the previous chapter, this is done by having a soft bearing material: lead, tin, zinc or alloys of these metals.

Now for the snag of a soft bearing material—will it not fail to support the normal operating forces imposed on it by the crankshaft? All bearing materials have a certain 'p–v' envelope within which they function safely (Figure 29.3). The maximum pressure, p, that the bearing can accept is determined by the hardness of the surface; the maximum velocity, v, is determined by heating,

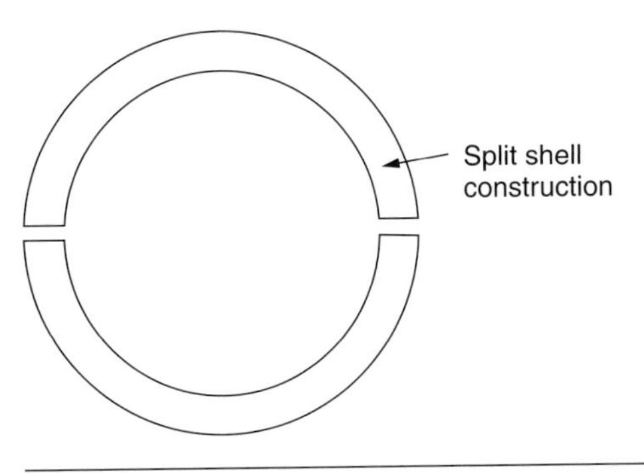

Figure 29.2 Easily replaceable bearing shells.

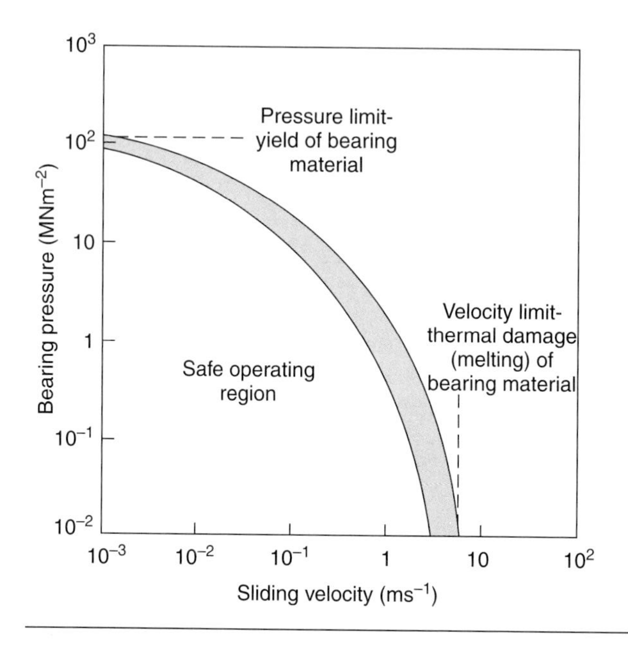

Figure 29.3 The pressure–velocity envelope for a bearing material.

and thus by the thermal conductivity of the material of which the bearing is made. So, if nothing special were done to prevent it, a soft bearing would deform under the imposed pressure like putty. In practice, by making the layer of soft material *thin* and backing it with something much harder, this difficulty can be avoided.

The way it works is this. If you squeeze a slug of plasticine between two blocks of wood, the slug deforms easily at first, but as the plasticine layer gets thinner and thinner, more and more lateral flow is needed to make it spread, and the pressure required to cause this flow gets bigger and bigger. The plasticine is constrained by the blocks so that it can never be squeezed out altogether — that would take an infinite pressure. This principle of *plastic constraint* is used in bearing design by depositing a very thin layer (about 0.03 mm thick) of soft alloy on to the bearing shell. This is thick enough to embed most dirt particles, but thin enough to support the journal forces.

This soft bearing material also has an important role to play if there is a failure in the oil supply to the bearing. In this case, frictional heating will rapidly increase the bearing temperature, and would normally lead to pronounced metal-to-metal contact, gross atomic bonding between journal and bearing, and seizure. The soft bearing material of low melting point will be able to shear in response to the applied forces, and may also melt locally. This helps protect the journal from more severe surface damage, and also helps to avoid component breakages that might result from sudden locking of mating surfaces.

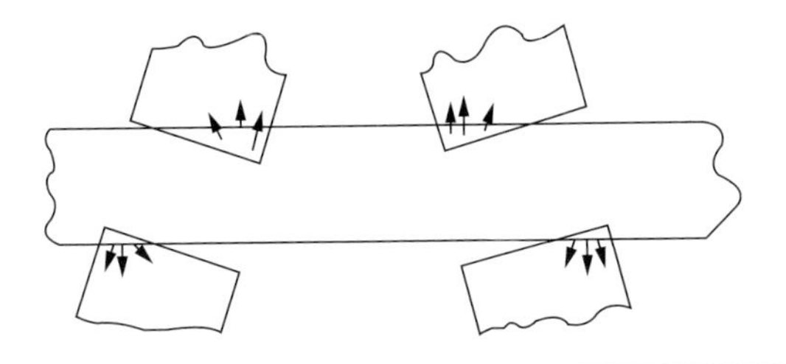

Figure 29.4 Conformability of bearings; a conformable bearing material will flow to adjust to minor misalignments.

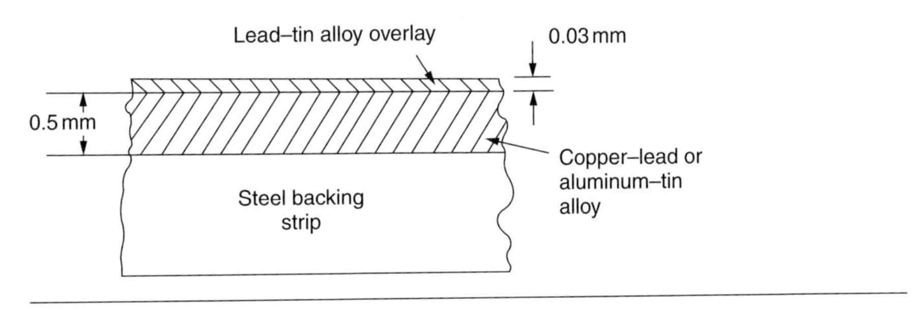

Figure 29.5 A schematic cross section through a typical layered bearing shell.

The third advantage of a soft bearing material is *conformability*. Slight misalignments of bearings can be self-correcting if plastic flow occurs easily in the bearing metal (Figure 29.4). Clearly there is a compromise between load-bearing ability and conformability.

Because our thin overlay of lead–tin can get worn away under severe operating conditions before the end of the normal life of the bearing, it is customary to put a second thicker, and therefore harder, layer between the overlay and the steel backing strip (Figure 29.5). The alloys normally used are copper–lead, or aluminum–tin. In the event of the wearing through of the overlay they are still soft enough to act as bearing materials without immediate damage to the journal.

In the end it is through experience as much as by science that bearing materials have evolved. Table 29.1 lists some of these.

29.3 Case study 2: materials for skis and sledge runners

Skis, both for people and for aircraft, used to be made of waxed wood. Down to about -10°C, the friction of waxed wood on snow is very low—μ is about 0.02—and if this were not so, planes equipped with skis could not take off

Table 29.1 Materials for oil lubricated bearings

Material	*Application*
Tin-based white metal 88% Sn–8% Sb–4% Cu	Camshaft, cross-head bearings of i.c. engines
Lead-based white metal 75% Pb–12% Sn–13% Sb– 1% Cu	General machinery operating at low bearing pressures
Copper–lead 70% Cu–30% Pb	Crankshaft and camshaft bearings of i.c. engines; turbocharger bearings
Lead–bronze 75% Cu–20% Pb–5% Sn	High performance i.c. engines, camshafts, gearboxes
Phosphor–bronze 88% Cu–11% Sn–1% P	High-load bearings for gearboxes, rolling mills, gudgeon pin bushes, etc., often used with overlay of softer material
Aluminum–tin 60% Al–40% Sn	Heavily loaded crankshaft bearings for diesel engines. Can be used with overlay of softer material
Acetal–steel	General machinery where low start-up friction is required

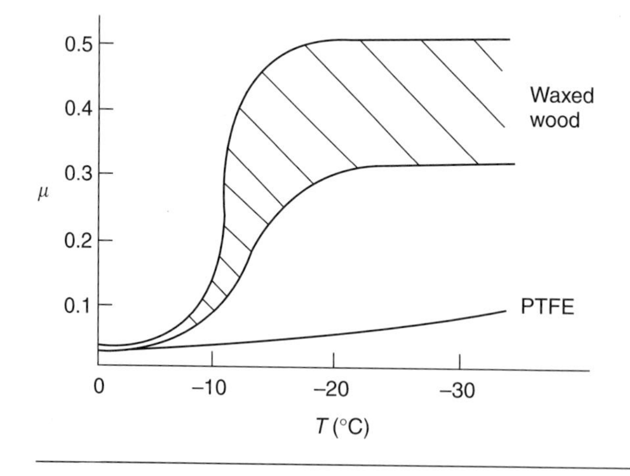

Figure 29.6 Friction of materials on ice at various temperatures.

from packed-snow runways, and the winter tourist traffic to Switzerland would drop sharply. Below $-10°C$, bad things start to happen (Figure 29.6): μ rises sharply to about 0.4. Polar explorers have observed this repeatedly. Wright, a member of the 1911–13 Scott expedition, writes: "Below 0°F ($-18°C$) the friction (on the sledge runners) seemed to increase progressively as the temperature fell"; it caused the expedition considerable hardship. What determines the friction of skis on snow?

Ice differs from most materials in that its melting point *drops* if you compress it. It is widely held that pressure from the skis causes the snow beneath to melt, but this is nonsense: the pressure of a large person distributed over a ski lowers the melting point of ice by about 0.0001°C and, even if the weight is carried by asperities which touch the ski over only 10^{-3} of its nominal area, the depression of the melting point is still only 0.1°C. Pressure melting, then, cannot account for the low friction shown in Figure 29.6. But as the large person starts to descend the ski slope, work is done against the frictional forces, heat is generated at the sliding surface, and the "velocity limit" of Figure 29.3 is exceeded. The heat melts a layer of ice, producing a thin film of water, at points where asperities touch the ski: the person hydroplanes along on a layer of water generated by friction. The principle is exactly like that of the lead–bronze bearing, in which local hot spots melt the lead, producing a lubricating film of liquid which lowers μ and saves the bearing.

Below −10°C, heat is conducted away too quickly to allow this melting—and because their thermal conductivity is high, skis with exposed metal (aluminum or steel edges) are slower at low temperatures than those without. At these low temperatures, the mechanism of friction is the same as that of metals: ice asperities adhere to the ski and must be sheared when it slides. The value of μ (0.4) is close to that calculated from the shearing model in Chapter 28. This is a large value of the coefficient of friction—enough to make it very difficult for a plane to take off, and increasing by a factor of more than 10 the work required to pull a loaded sledge. What can be done to reduce it?

This is a standard friction problem. A glance at Figure 28.5 shows that, when polymers slide on metals and ceramics, μ can be as low as 0.04. Among the polymers with the lowest coefficients are PTFE ("Teflon") and polyethylene. By coating the ski or sledge runners with these materials, the coefficient of friction stays low, *even* when the temperature is so low that frictional heating is unable to produce a boundary layer of water. Aircraft and sports skis now have polyethylene or Teflon undersurfaces; the Olympic Committee has banned their use on bob-sleds, which already, some think, go fast enough.

29.4 Case study 3: high-friction rubber

So far we have talked of ways of reducing friction. But for many applications—brake pads, clutch linings, climbing boots, and above all, car tyres—we want as much friction as we can get.

The frictional behavior of rubber is quite different from that of metals. In Chapter 28 we showed that when metallic surfaces were pressed together, the bulk of the deformation at the points of contact was plastic; and that the friction between the surfaces arose from the forces needed to shear the junctions at the areas of contact.

But rubber deforms *elastically* up to very large strains. When we bring rubber into contact with a surface, therefore, the deformation at the contact points is *elastic*. These elastic forces still squeeze the atoms together at the areas of contact, of course; adhesion will still take place there, and shearing will still be necessary if the surfaces are to slide. This is why car tyres grip well in dry conditions. In *wet* conditions, the situation is different; a thin lubricating film of water and mud forms between rubber and road, and this will shear at a stress a good deal lower than previously, with dangerous consequences. Under these circumstances, another mechanism of friction operates to help prevent a skid.

It is illustrated in Figure 29.7. All roads have a fairly rough surface. The high spots push into the tyre, causing a considerable local elastic deformation. As the tyre skids, it slips forward over the rough spots. The region of rubber that was elastically deformed by the rough spot now relaxes, while the rubber just behind this region becomes compressed as it reaches the rough spot. Now, all rubbers exhibit some *anelasticity* (Chapter 8); the stress–strain curve looks like Figure 29.8. As the rubber is compressed, work is done on it equal to the area under the upper curve; but if the stress is removed we do not get all this work back. Part of it is dissipated as heat—the part shown as the shaded area between the loading and the unloading curve. So to make the tyre slide on a rough road we have to do work, even when the tyre is well lubricated, and if we have to do work, there is friction. Special rubbers have been developed for high-loss characteristics (called "high-loss" or "high-hysteresis" rubbers) and these have excellent skid resistance even in wet conditions.

There is one obvious drawback of high-hysteresis rubber. In normal rolling operation, considerable elastic deformations still take place in the tyre wall, and high-loss tyres will consume fuel and generate considerable heat. The way

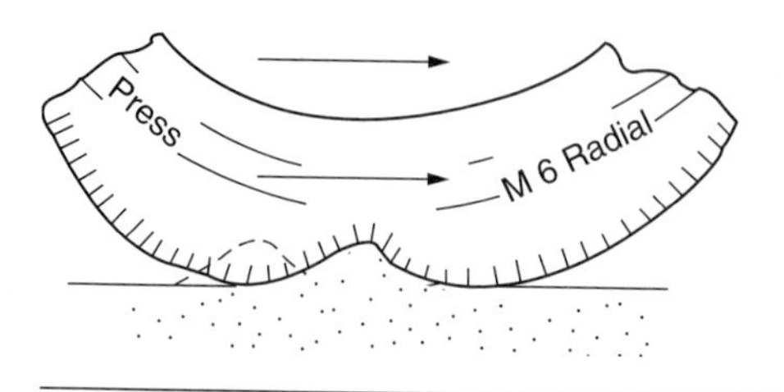

Figure 29.7 Skidding on a rough road surface deforms the tyre material elastically.

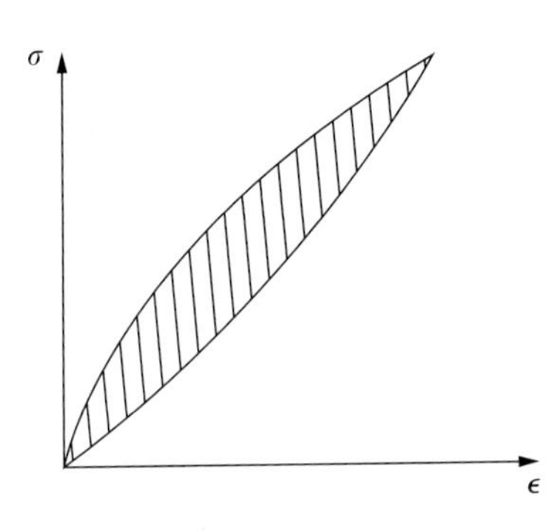

Figure 29.8 Work is needed to cycle rubber elastically.

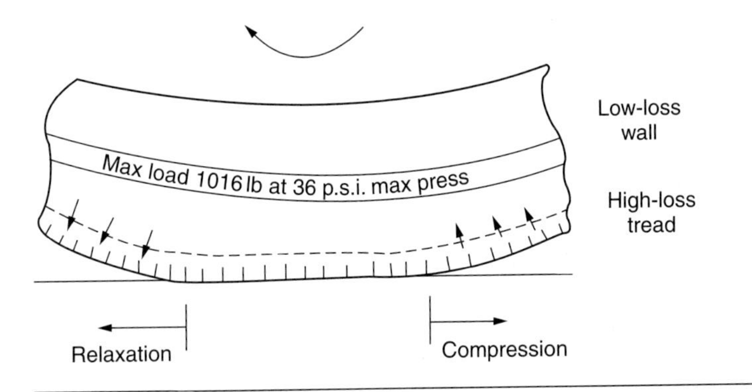

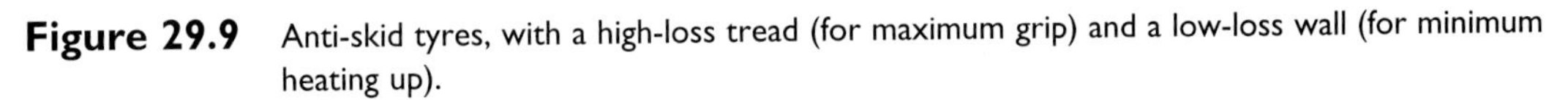

Figure 29.9 Anti-skid tyres, with a high-loss tread (for maximum grip) and a low-loss wall (for minimum heating up).

out is to use a low-loss tyre covered with a high-loss tread—another example of design using composite materials (Figure 29.9).

Examples

29.1 How does lubrication reduce friction?

29.2 It is observed that snow lies stably on roofs with a slope of less than 24°, but that it slides off roofs with a greater slope. Skiers, on the other hand, slide on a snow-covered mountainside with a slope of only 2°. Why is this?

29.3 A person of weight 100 kg standing on skis 2 m long and 0.10 m wide slides on the 2° mountain slope, at 0°C. Calculate the loss of potential energy when the ski slides a distance equal to its own length. Hence calculate the average thickness of the water film beneath each ski. (The latent heat of fusion of ice is 330 MJ m^{-3}.)

Answers

Work done 68 J; average film thickness = 0.5 μm.

29.4 How can friction between road and tyre be maintained even under conditions of appreciable lubrication?

29.5 In countries which have cold winters (e.g. Sweden) it is usual for road vehicles to be fitted with special tyres in the winter months. These tyres have hard metal studs set into the tread. Why is this necessary?

29.6 Explain the principle behind hydrodynamic lubrication. Under what conditions is hydrodynamic lubrication likely to break down? What then saves the bearing?

29.7 Why do plain bearings usually consists of a hardened shaft running in a soft shell?

Part I

Designing with metals, ceramics, polymers and composites

Chapter 30

Design with materials

Chapter contents

30.1 Introduction

Design is an iterative process. You start with the definition of a function (a pen, a hairdryer, a fuel pin for a nuclear reactor) and draw on your knowledge (the contents of this book, for instance) and experience (your successes and failures) in formulating a tentative design. You then refine this by a systematic process that we shall look at later.

Materials selection is an integral part of design. And because the principles of mechanics, dynamics, and so forth are all well established and not changing much, whereas new materials are appearing all the time, innovation in design is frequently made possible by the use of new materials. Designers have at their disposal the range of materials that we have discussed in this book: metals, ceramics, polymers, and combinations of them to form composites. Each class of material has its own strengths and limitations, which the designer must be fully aware of. Table 30.1 summarizes these.

At and near room temperature, metals have well-defined, almost constant, moduli, and yield strengths (in contrast to polymers, which do not). And most metallic alloys have a ductility of 20 percent or better. Certain high-strength alloys (spring steel, for instance) and components made by powder methods, have less—as little as 2 percent. But even this is enough to ensure that an unnotched component yields before it fractures, and that fracture, when it occurs, is of a tough, ductile, type. But—partly because of their ductility—metals are prey to cyclic fatigue and, of all the classes of materials, they are the least resistant to corrosion and oxidation.

Historically, design with ceramics has been empirical. The great gothic cathedrals, still the most impressive of all ceramic designs, have an aura of stable permanence. But many collapsed during construction; the designs we know evolved from these failures. Most ceramic design is like that. Only recently, and because of more demanding structural applications, have design methods evolved.

In designing with ductile materials, a *safety-factor* approach is used. Metals can be used under static loads within a small margin of their ultimate strength with confidence that they will not fail prematurely. Ceramics cannot. As we saw earlier, brittle materials always have a wide scatter in strength, and the strength itself depends on the volume of material under stress. The use of a single, constant, safety factor is no longer adequate, and the statistical approach of Chapter 16 must be used instead.

We have seen that the "strength" of a ceramic means, almost always, the fracture or crushing strength. Then (unlike metals) the compressive strength is 10 to 20 times larger than the tensile strength. And because ceramics have no ductility, they have a low tolerance for stress concentrations (such as holes and flaws) or for high contact stresses (at clamping or loading points, for instance). If the pin of a pin-jointed frame, made of metal, fits poorly, then the metal deforms locally, and the pin beds down, redistributing the load. But if the pin

Table 30.1 Design-limiting properties of materials

Material	*Good*	*Poor*
Metals High E, K_{Ic} Low σ_y	Stiff ($E \approx 100\,\text{GN m}^{-2}$) Ductile ($\varepsilon_f \approx 20\%$) – formable Tough ($K_c$) $> 50\,\text{MN m}^{-3/2}$ High MP ($T_m \approx 1000°\text{C}$)	Yield (pure, $\sigma_y \approx 1\,\text{MN m}^{-2}$) $\rightarrow$ alloy Hardness ($H \approx 3\sigma_y$) $\rightarrow$ alloy Fatigue strength ($\sigma_e = \frac{1}{2}\sigma_y$) Corrosion resistance $\rightarrow$ coatings
Ceramics High E, σ_y Low K_{Ic}	Stiff ($E \approx 200\,\text{GN m}^{-2}$) Very high yield, hardness ($\sigma_y > 3\,\text{GN m}^{-2}$) High MP ($T_m \approx 2000°\text{C}$) Corrosion resistant Moderate density	Very low toughness ($K_c \approx 2\,\text{MN m}^{-3/2}$) Formability $\rightarrow$ powder methods
Polymers Adequate σ_y, K_{Ic} Low E	Ductile and formable Corrosion resistant Low density	Low stiffness ($E \approx 2\,\text{GN m}^{-2}$) Yield ($\sigma_y = 2\text{–}100\,\text{MN m}^{-2}$) Low glass temp ($T_G \approx 100°\text{C}$) $\rightarrow$ creep Toughness often low ($K_c \approx 1\,\text{MN m}^{-3/2}$)
Composites High E, σ_y, K_{Ic} but cost	Stiff ($E > 50\,\text{GN m}^{-2}$) Strong ($\sigma_y \approx 200\,\text{MN m}^{-2}$) Tough ($K_c > 20\,\text{MN m}^{-3/2}$) Fatigue resistant Corrosion resistant Low density	Formability Cost Creep (polymer matrices)

and frame are made of a brittle material, the local contact stresses nucleate cracks which then propagate, causing sudden collapse. Obviously, the process of design with ceramics differs in detail from that of design with metals.

That for polymers is different again. When polymers first became available to the engineer, it was common to find them misused. A "cheap plastic" product was one which, more than likely, would break the first time you picked it up. Almost always this happened because the designer used a polymer to replace a metal component, without redesign to allow for the totally different properties of the polymer. Briefly, there are three:

(a) Polymers have much lower moduli than metals—roughly 100 times lower. So elastic deflections may be large.
(b) The deflection of polymers depends on the time of loading: they creep at room temperature. A polymer component under load may, with time, acquire a permanent set.
(c) The strengths of polymers change rapidly with temperature near room temperature. A polymer which is tough and flexible at 20°C may be brittle at the temperature of a household refrigerator, 4°C.

With all these problems, why use polymers at all? Well, complicated parts performing several functions can be molded in a single operation. Polymer components can be designed to snap together, making assembly fast and cheap. And by accurately sizing the mold, and using pre-coloured polymer, no finishing operations are necessary. So great economies of manufacture are possible: polymer parts really can be cheap. But are they inferior? Not necessarily. Polymer densities are low (all are near $1\,\mathrm{Mg\,m^{-3}}$); they are corrosion-resistant; they have abnormally low coefficients of friction; and the low modulus and high strength allows very large elastic deformations. Because of these special properties, polymer parts may be distinctly superior.

Composites overcome many of the remaining deficiencies. They are stiff, strong, and tough. Their problem lies in their cost: composite components are usually expensive, and they are difficult and expensive to form and join. So, despite their attractive properties, the designer will use them only when the added performance offsets the added expense.

New materials are appearing all the time. New polymers with greater stiffness and toughness appear every year; composites are becoming cheaper as the volume of their production increases. Ceramics with enough toughness to be used in conventional design are becoming available, and even in the metals field, which is a slowly developing one, better quality control, and better understanding of alloying, leads to materials with reliably better properties. All of these offer new opportunities to the designer who can frequently redesign an established product, making use of the properties of new materials, to reduce its cost or its size and improve its performance and appearance.

30.2 Design methodology

Books on design often strike the reader as vague and qualitative; there is an implication that the ability to design is like the ability to write music: a gift given to few. And it is true that there is an element of creative thinking (as opposed to logical reasoning or analysis) in good design. But a design methodology can be formulated, and when followed, it will lead to a practical solution to the design problem.

Figure 30.1 summarizes the methodology for designing a component which must carry load. At the start there are two parallel streams: materials selection and component design. A tentative material is chosen and data for it are assembled from data sheets like the ones given in this book or from data books (referred to in the References). At the same time, a tentative component design is drawn up, able to fill the function (which must be carefully defined at the start); and an approximate stress analysis is carried out to assess the stresses, moments, and stress concentrations to which it will be subjected.

The two streams merge in an assessment of the material performance in the tentative design. If the material can bear the loads, moments, concentrated

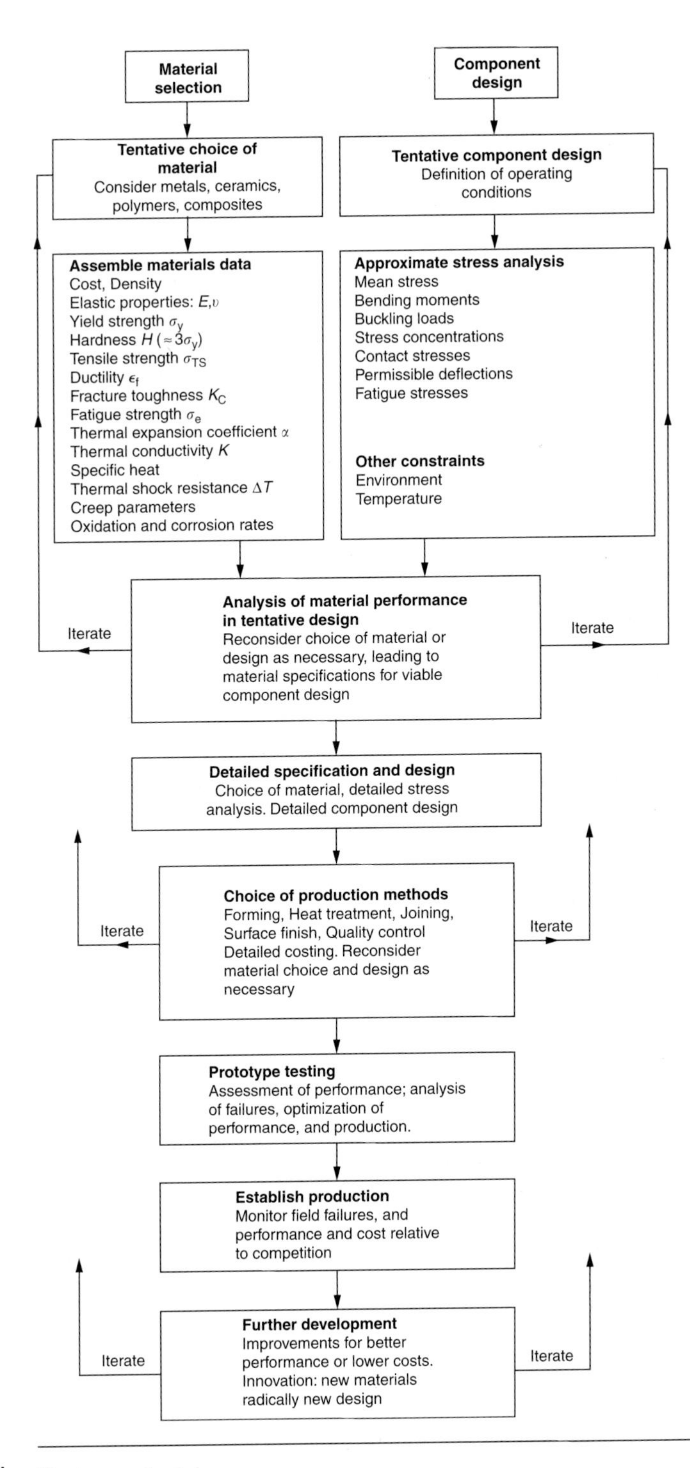

Figure 30.1 Design methodology.

stresses (etc.) without deflecting too much, collapsing or failing in some other way, then the design can proceed. If the material cannot perform adequately, the first iteration takes place: either a new material is chosen, or the component design is changed (or both) to overcome the failing.

The next step is a detailed specification of the design and of the material. This may require a detailed stress analysis, analysis of the dynamics of the system, its response to temperature and environment, and a detailed consideration of the appearance and feel (the aesthetics of the product). And it will require better material data: at this point it may be necessary to get detailed material properties from possible suppliers, or to conduct tests yourself.

The design is viable only if it can be produced economically. The choice of production and fabrication method is largely determined by the choice of material. But the production route will also be influenced by the size of the production run, and how the component will be finished and joined to other components. The choice of material and production route will, ultimately, determine the price of the product, so a second major iteration may be required if the costing shows the price to be too high. Then a new choice of material or component design, allowing an alternative production path, may have to be considered.

At this stage a prototype product is produced, and its performance in the market is assessed. If this is satisfactory, full-scale production is established. But the designer's role does not end at this point. Continuous analysis of the performance of a component usually reveals weaknesses or ways in which it could be improved or made more cheaply. And there is always scope for further innovation: for a radically new design, or for a radical change in the material which the component is made from. Successful designs evolve continuously, and only in this way does the product retain a competitive position in the marketplace.

Chapter 31

Final case study: materials and energy in car design

Chapter contents

31.1 Introduction

The status of steel as the raw material of choice for the manufacture of car bodies rests principally on its *price*. It has always been the cheapest material which meets the necessary strength, stiffness, formability, and weldability requirements of large-scale car body production. Until recently, the fact that steel has a density two-thirds that of lead was accorded little significance. But increasing environmental awareness and tightening legislative requirements concerning damaging emissions and fuel efficiency are now changing that view. Car makers are looking hard at alternatives to steel.

31.2 Energy and cars

Energy is used to build a car, and energy is used to run it. Rising oil prices mean that, since about 1980, the cost of the petrol consumed during the average life of a car is comparable with the cost of the car itself. Consumers, therefore, now want more fuel-efficient cars, and more fuel-efficient cars pollute less.

A different perspective on the problem is shown in Table 31.1: 15 percent of all the energy used is consumed by private cars. The dependence of most countries on imported oil is such a liability that they are seeking ways of reducing it. Private transport is an attractive target because to trim its energy consumption does not necessarily depress the economy. In the United States, for example, legislation is in place requiring that the *average* fleet-mileage of a manufacturer increase from 22.5 to 34.5 miles per gallon; and in certain cities (e.g. Los Angeles) a target of *zero* emission has been set. How can this be achieved?

31.3 Ways of achieving energy economy

It is clear from Table 31.1 that the energy content of the car itself—that is of the steel, rubber, glass, and of the manufacturing process itself—is small: less than one-tenth of that required to move the car. This means that there is little point in trying to save energy here; indeed (as we shall see) it may pay to use more energy to make the car (using, for instance, aluminum instead of steel) if this reduces the fuel consumption.

We must focus, then, on reducing the energy used to move the car. There are two routes.

(a) *Improve engine efficiency*. Engines are already remarkably efficient; there is a limit to the economy that can be achieved here, though it can help.

(b) *Reduce the weight of the car.* Figure 31.1 shows how the fuel consumption (g.p.m.) and the mileage (m.p.g.) vary with car weight. There is a linear correlation: halving the weight halves the g.p.m. This is why small cars are more economical than big ones: engine size and performance have some influence, but it is mainly the weight that determines the fuel consumption.

We can, then reduce the size of cars, but the consumer does not like that. Or we can reduce the *weight* of the car by substituting lighter materials for those used now. Lighter cars not only use less fuel, they also emit less gaseous pollution—hence the interest in producing lighter vehicles, reversing what historically has been a consistent trend in the opposite direction.

Table 31.1 Energy in manufacture and use of cars

Energy to produce cars, per year	=	0.8 percent to 1.5 percent of total energy consumed by nation
Energy to move cars, per year	=	15 percent of total energy consumed by nation
(Transportation of people and goods, total)	=	24 percent of total energy consumed by nation

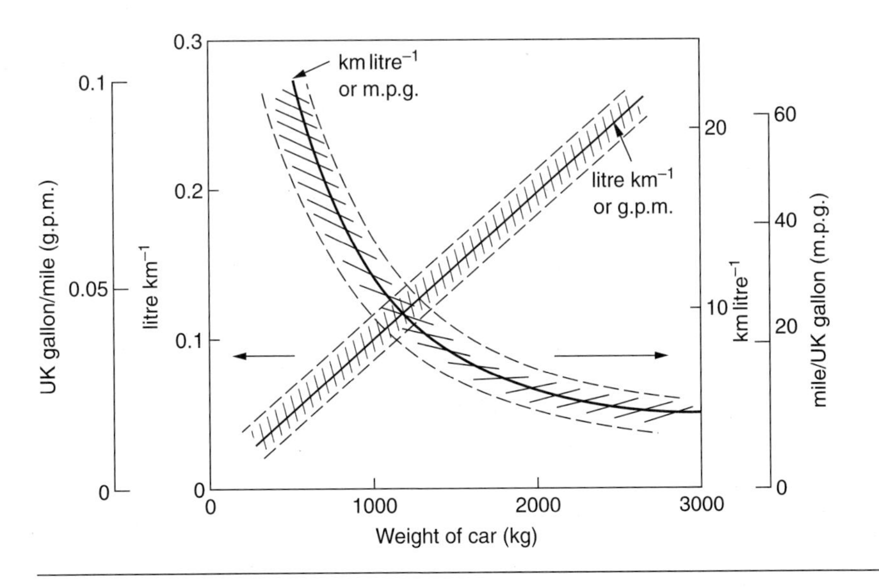

Figure 31.1 Fuel consumption of production cars.

31.4 Material content of a car

As Figure 31.1 suggests, most cars weigh between 400 kg and 2500 kg. In a typical modern production car (Figure 31.2) this is made up as shown in Table 31.2.

31.5 Alternative materials

Primary mechanical properties

Candidate materials for substitutes must be lighter than steel, but structurally equivalent. For the engine block, the choice is an obvious one: aluminum (density 2.7 Mg m^{-3}) or possibly magnesium (density 1.8 Mg m^{-3}) replace an *equal volume* of cast iron (density 7.7 Mg m^{-3}) with an immediate weight reduction on this component of 2.8–4.3 times. The production methods remain almost unchanged. Many manufacturers have made this change already.

The biggest potential weight saving, however, is in the body panels, which make up 60 percent of the weight of the vehicle. Here the choice is more difficult. Candidate materials are given in Table 31.3.

How do we assess the possible weight saving? Just replacing a steel panel with an aluminum-alloy or fiberglass one of equal thickness (giving a weight

Figure 31.2 The Volkswagen Passat—a typical modern pressed-steel body with no separate chassis. For a *given material* this "monocoque" construction gives a minimum weight-to-strength ratio.

Table 31.2 Contributors to the weight of car

71% Steel	Body shell, panels
15% Cast iron	Engine block; gear box; rear axle
4% Rubber	Tyres; hoses
Balance	Glass, zinc, copper, aluminum, polymers

Table 31.3 Properties of candidate materials for car bodies

Material	*Density* ρ ($Mg\,m^{-3}$)	*Young's modulus,* E ($GN\,m^{-2}$)	*Yield strength,* σ_y ($MN\,m^{-2}$)	($\rho/E^{1/3}$)	($\rho/\sigma_y^{1/2}$)
Mild steel	7.8	207	220	1.32	0.53
High-strength steel			up to 500		0.35
Aluminum alloy	2.7	69	193	0.66	0.19
GFRP (chopped fiber, molding grade)	1.8	15	75	0.73	0.21

Figure 31.3 The Morris Traveller—a classic of the 1950s with wooden members used as an integral part of the monocoque body shell.

saving which scales as the density) is unrealistic: both the possible substitutes have much lower moduli, and thus will deflect (under given loads) far more; and one of them has a much lower yield strength, and might undergo plastic flow. We need an analysis like that of Chapters 7 and 12.

If (as with body panels) *elastic* deflection is what counts, the logical comparison is for a panel of equal *stiffness*. And if, instead, it is resistance to *plastic* flow which counts (as with bumpers) then the proper thing to do is to compare sections with equal resistance to plastic flow.

The *elastic* deflection δ of a panel under a force F (Figure 31.4) is given by

$$\delta = \frac{Cl^3F}{Ebt^3} \tag{31.1}$$

where t is the sheet thickness, l the length and b the width of the panel. The constant C depends on how the panel is held: at two edges, at four edges, and so on—it does not affect our choice. The mass of the panel is

$$M = \rho btl \tag{31.2}$$

The quantities b and l are determined by the design (they are the dimensions of the door, trunk lid, etc.). The only variable in controlling the stiffness is t. Now from equation (31.1)

$$t = \left(\frac{Cl^3F}{\delta Eb}\right)^{1/3} \tag{31.3}$$

Substituting this expression for t in equation (31.2) gives, for the mass of the panel,

$$M = \left\{C\left(\frac{F}{\delta}\right)l^6b^2\right\}^{1/3}\left(\frac{\rho}{E^{1/3}}\right) \tag{31.4}$$

Thus, for a given *stiffness* (F/δ), *panel dimensions* (l, b) and edge-constraints (C) the lightest panel is the one with the smallest value of $\rho/E^{1/3}$.

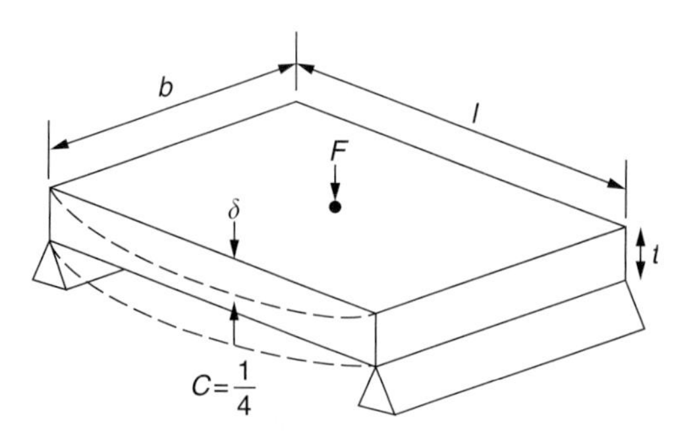

Figure 31.4 Elastic deflection of a car-body panel.

We can perform a similar analysis for *plastic* yielding. A panel with the section shown in Figure 31.5 yields at a load

$$F = \left(\frac{Cbt^2}{l}\right)\sigma_y \tag{31.5}$$

The panel mass is given, as before, by equation (31.2). The only variable is t given by

$$t = \left(\frac{Fl}{Cb\sigma_y}\right)^{1/2} \tag{31.6}$$

from equation (31.5). Substituting for t in equation (31.2) gives

$$M = \left(\frac{Fbl^3}{C}\right)^{1/2}\left(\frac{\rho}{\sigma_y^{1/2}}\right) \tag{31.7}$$

The panel with the smallest value of $\rho/\sigma_y^{1/2}$ is the one we want.

We can now assess candidate materials more sensibly on the basis of the data given in Table 31.3. The values of the two important property groups are shown in the last two columns. For the majority of body panels (for which *elastic* deflection determines design) high-strength steel offers no advantage over mild steel: it has the same value of $\rho/E^{1/3}$. GFRP is better (a lower $\rho/E^{1/3}$ and thus weight), and aluminum alloy is better still — that is one of the reasons aircraft are made of aluminum. But notice that the weight saving in going from a steel to an aluminum panel is not a factor of 3 (the ratio of the densities) but only 2 (the ratio of their $\rho/E^{1/3}$) because the aluminum panel has to be thicker to compensate for its lower E.

High strength steel *does* offer a weight saving for strength-limited components: bumpers, front, and rear header panels, engine mounts, bulkheads, and so forth; the weight saving ($\rho/\sigma_y^{1/2}$) is a factor of 1.5. Both aluminum alloy and

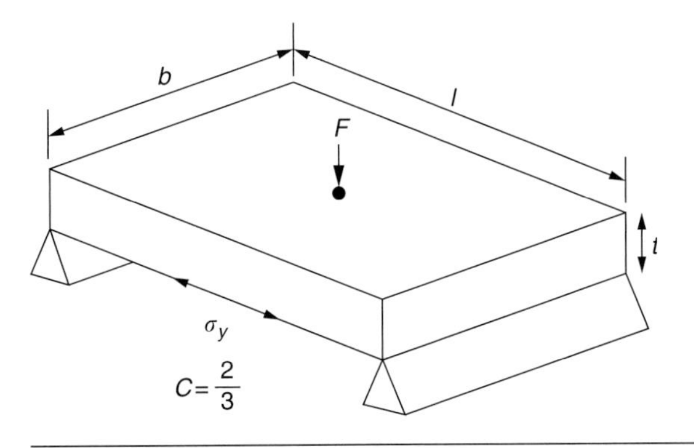

Figure 31.5 Yielding of a car-body panel.

fiberglass offer potential weight savings of up to 3 times ($\rho/\sigma_y^{1/2}$) on these components. This makes possible a saving of at least 30 percent on the weight of the vehicle; if, in addition, an aluminum engine block is used, the overall weight saving is larger still. These are very substantial savings—sufficient to achieve the increase in mileage per gallon from 22.5 to 34.5 without any decrease in the size of the car, or increase in engine efficiency. So they are obviously worth examining more closely. What, then, of the *other* properties required of the substitute materials?

Secondary properties

Although resistance to deflection and plastic yielding are obviously of first importance in choosing alternative materials, other properties enter into the selection. Let us look at these briefly. Table 31.4 lists the conditions imposed by the service environment.

Consider these in turn. *Elastic* and *plastic* deflection we have dealt with already. The toughness of steel is so high that *fracture* of a steel panel is seldom a problem. But what about the other materials? The data for toughness are given in Table 31.5.

But what is the proper way to use toughness values? The most sensible thing to do is ask: suppose the panel is loaded up to its yield load (above this load we *know* it will begin to fail—by plastic flow—so it does not matter whether other failure mechanisms also appear); what is the maximum crack size that is still stable? If this is large enough that it should not appear in service, we are satisfied; if not, we must increase the section. This crack size is given (Chapter 13) by

$$\sigma_y\sqrt{\pi a} = K_c = \sqrt{EG_c}$$

Table 31.4 The service environment of the average car

Loading	Static →	Elastic or plastic deflection	
	Impact →	Elastic or plastic deflection	
	Impact →	Fracture	
	Fatigue →	Fatigue fracture	
	Long-term static →	Creep	
Physical environment	$-40°C <$	$T < 120°C$	
	55% <	relative humidity < 100%	
Chemical environment		Water	Petrol
		Oil	Antifreeze
		Brake fluid	Salt
		Transmission fluid	

Table 31.5 Properties of body-panel materials: toughness, fatigue, and creep

Material	*Toughness* G_c (kJ m^{-2})	*Tolerable crack length* (mm)	*Fatigue*	*Creep*
Mild steel	≈100	≈140	OK	OK
High-strength steel	≈100	≈26		
Aluminum alloy	≈20	≈12	OK	OK
GFRP (chopped fiber, molding grade)	≈37	≈30	OK	Creep above 60°C

from which

$$a_{\max} = \frac{EG_c}{\pi\sigma_y^2}$$

The resulting crack lengths are given in Table 31.5. A panel with a crack longer than this will fail by 'tearing'; one with a short crack will simply fail by general yield, that is it will bend permanently. Although the tolerable crack lengths are shorter in replacement materials than in steel, they are still large enough to permit the replacement materials to be used.

Fatigue (Chapter 17) is always a potential problem with any structure subject to varying loads: anything from the loading due to closing the door to that caused by engine vibration can, potentially, lead to failure. The fatigue strength of all these materials is adequate.

Creep (Chapter 20) is not normally a problem a designer considers when designing a car body with metals: the maximum service temperature reached is 120°C (panels near the engine, under extreme conditions), and neither steel nor aluminum alloys creep significantly at these temperatures. But GFRP does. Above 60°C creep rates are significant. GFRP shows a classic three-stage creep curve, ending in failure; so that extra reinforcement or heavier sections will be necessary where temperatures exceed this value.

More important than either creep or fatigue in current car design is the *effect of environment* (Chapter 26). An appreciable part of the cost of a new car is contributed by the manufacturing processes designed to prevent rusting; and these processes only partly work—it is bodyrust that ultimately kills a car, since the mechanical parts (engine, etc.) can be replaced quite easily, as often as you like.

Steel is particularly bad in this regard. In ordinary circumstances, aluminum is much better as we showed in the chapters on corrosion. Although the effect of salt on aluminum is bad, heavy anodizing will slow down even that form of attack to tolerable levels (the masts of modern yachts are made of anodized aluminum alloy, for example).

So aluminum alloy is good: it resists all the fluids likely to come in contact with it. What about GFRP? The strength of GFRP is reduced by up to 20 percent by continuous immersion in most of the fluids—even salt-water — with which it is likely to come into contact; but (as we know from fiberglass boats) this drop in strength is not critical, and it occurs without visible corrosion, or loss of section. In fact, GFRP is much more corrosion-resistant, in the normal sense of "loss-of-section", than steel.

31.6 Production methods

The biggest penalty one has to pay in switching materials is likely to be the higher material and production costs. *High-strength steel,* of course, presents almost no problem. The yield strength is higher, but the section is thinner, so that only slight changes in punches, dies and presses are necessary, and once these are paid for, the extra cost is merely that of the material.

At first sight, the same is true of *aluminum alloys*. But because they are heavily alloyed (to give a high yield strength) their *ductility* is low. If expense is unimportant, this does not matter; some early Rolls-Royce cars (Figure 31.6) had aluminum bodies which were formed into intricate shapes by laborious hand-beating methods, with frequent annealing of the aluminum to restore its ductility. But in mass production we should like to deep draw body panels in one operation — and then low ductility is much more serious. The result is a loss of design flexibility: there are more constraints on the use of aluminum alloys than on steel; and it is this, rather than the cost, which is the greatest obstacle to the wholesale use of aluminum in cars.

Figure 31.6 A 1932 Rolls-Royce. Mounted on a separate steel chassis is an all-aluminum hand-beaten body by the famous coach building firm of James Mulliner. Any weight advantage due to the use of aluminum is totally outweighed by the poor weight-to-strength ratio of separate-chassis construction; but the bodywork remains immaculate after 73 years of continuous use!

Figure 31.7 A Lotus Esprit, with a GFRP body (but still mounted on a steel chassis—which does not give anything like the weight saving expected with an all-GFRP monocoque structure).

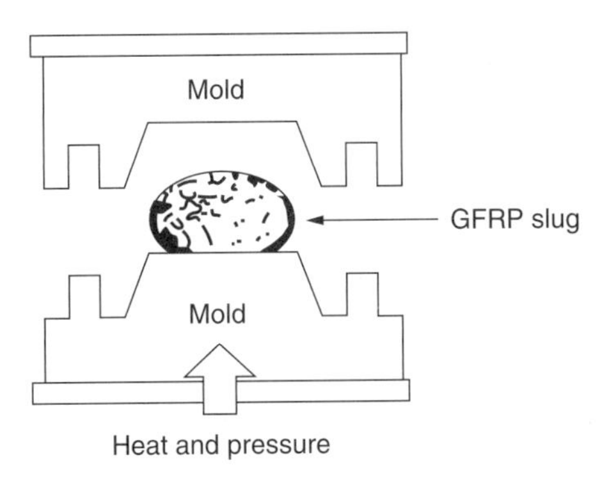

Figure 31.8 Compression molding of car-body components.

GFRP looks as if it would present production problems: you may be familiar with the tedious hand layup process required to make a fiberglass boat or canoe. But mass-production methods have now been developed to handle GFRP. Most modern cars have GFRP components (bumpers, facia panels, internal panels) and a few have GFRP bodies (Figure 31.7), usually mounted on a steel chassis; the full weight savings will only be realized if the *whole* load-bearing structure is made from GFRP. In producing GFRP car panels, a slug of polyester resin, with *chopped* glass fibers mixed in with it, is dropped into a heated split mold (Figure 31.8). As the polyester used is a *thermoset* it will "go off" in the hot mold, after which the solid molding can be ejected. Modern methods allow a press like this to produce one molding per minute— still slower than steel pressing, but practical. Molding (as this is called) brings certain advantages. It offers great design flexibility—particularly in change of section, and sharp detail—which cannot be achieved with steel. And GFRP moldings often result in consolidation of components, reducing assembly costs.

31.7 Conclusions

The conclusions are set out in the table below.

A. High-strength steel

For	*Against*
Retains all existing technology	Weight saving only appreciable in designing against *plastic* flow

Use in selected applications, for example bumpers.

B. Aluminum alloy

For	*Against*
Large weight saving in both body shell and engine block	Unit cost higher
Retains much existing technology	Deep drawing properties poor—loss in design flexibility
Corrosion resistance excellent	

Aluminum alloy offers saving of up to 40 percent in total car weight. The increased unit cost is offset by the lower running cost of the lighter vehicle, and the greater recycling potential of the aluminum.

C. GFRP

For	*Against*
Large weight saving in body shell	Unit cost higher
Corrosion resistance excellent	Massive changes in manufacturing technology
Great gain in design flexibility and some parts consolidation	Designer must cope with some creep

GFRP offers savings of up to 30 percent in total car weight, at some increase in unit cost and considerable capital investment in new equipment. Recycling problems still have to be overcome.

Appendix 1

Symbols and formulae

Chapter contents

List of principal symbols

Symbol	Meaning (units)
Note:	Multiples or sub-multiples of basic units indicate the unit suffixes typically used with materials data.
a	side of cubic unit cell (nm)
a	crack length (mm)
a	constant in Basquin's law (dimensionless)
A	constant in fatigue crack-growth law
A	constant in creep law $\dot{\epsilon}_{ss} = A\sigma^n e^{-Q/\bar{R}T}$
b	Burgers vector (nm)
b	constant in Coffin–Manson law (dimensionless)
c	concentration (m^{-3})
C_1	constant in Basquin's law ($MN\,m^{-2}$)
C_2	constant in Coffin–Manson law (dimensionless)
D	diffusion coefficient ($m^2\,s^{-1}$)
D_0	pre-exponential constant in diffusion coefficient ($m^2\,s^{-1}$)
E	Young's modulus of elasticity ($GN\,m^{-2}$)
f	force acting on unit length of dislocation line ($N\,m^{-1}$)
F	force (N)
g	acceleration due to gravity on the Earth's surface ($m\,s^{-2}$)
G	shear modulus ($GN\,m^{-2}$)
G_c	toughness (or critical strain energy release rate) ($kJ\,m^{-2}$)
H	hardness ($kg\,mm^{-2}$)
J	diffusion flux ($m^{-2}\,s^{-1}$)
k	shear yield strength ($MN\,m^{-2}$)
k	Boltzmann's constant $\bar{R}/N_A$ ($J\,K^{-1}$)
K	bulk modulus ($GN\,m^{-2}$)
K	stress intensity factor ($MN\,m^{-3/2}$)
K_c	fracture toughness (critical stress intensity factor) ($MN\,m^{-3/2}$)
ΔK	K range in fatigue cycle ($MN\,m^{-3/2}$)
m	constant in fatigue crack growth law (dimensionless)
n	creep exponent in $\dot{\epsilon}_{ss} = A\sigma^n e^{-Q/\bar{R}T}$
N	number of fatigue cycles
N_A	Avogadro's number (mol^{-1})
N_f	number of fatigue cycles leading to failure (dimensionless)
$\tilde{p}$	price of material (UK£ or US$ $tonne^{-1}$)
Q	activation energy per mole ($kJ\,mol^{-1}$)
r_0	equilibrium interatomic distance (nm)
$\bar{R}$	universal gas constant ($J\,K^{-1}\,mol^{-1}$)
S_0	bond stiffness ($N\,m^{-1}$)
t_f	time-to-failure (s)
T	line tension of dislocation (N)
T	absolute temperature (K)
T_M	absolute melting temperature (K)

Symbol	Meaning (units)
U^{el}	elastic strain energy (J)
γ	(true) engineering shear strain (dimensionless)
Δ	dilatation (dimensionless)
ϵ	true (logarithmic) strain (dimensionless)
ϵ_f	(nominal) strain after fracture; tensile ductility (dimensionless)
ϵ_n	nominal (linear) strain (dimensionless)
ϵ_0	permittivity of free space ($F\,m^{-1}$)
$\dot{\epsilon}_{ss}$	steady-state tensile strain-rate in creep (s^{-1})
$\Delta\epsilon^{pl}$	plastic strain range in fatigue (dimensionless)
μ_k	coefficient of kinetic friction (dimensionless)
μ_s	coefficient of static friction (dimensionless)
ν	Poisson's ratio (dimensionless)
ρ	density ($Mg\,m^{-3}$)
σ	true stress ($MN\,m^{-2}$)
σ_n	nominal stress ($MN\,m^{-2}$)
σ_{TS}	(nominal) tensile strength ($MN\,m^{-2}$)
σ_y	(nominal) yield strength ($MN\,m^{-2}$)
$\tilde{\sigma}$	ideal strength ($GN\,m^{-2}$)
$\Delta\sigma$	stress range in fatigue ($MN\,m^{-2}$)
τ	shear stress ($MN\,m^{-2}$)

Summary of principal formulae

Chapter 2

Exponential growth

$$\frac{dC}{dt} = \frac{rC}{100}$$

C = consumption rate (tonne per year); r = fractional growth rate (percent per year); t = time.

Chapter 3

Stress, strain, Poisson's ratio, elastic moduli

$$\sigma = \frac{F}{A}, \quad \tau = \frac{F_s}{A}, \quad p = -\frac{F}{A}, \quad \nu = -\frac{\text{lateral strain}}{\text{tensile strain}}$$

$$\epsilon_n = \frac{u}{l}, \quad \gamma = \frac{w}{l}, \quad \Delta = \frac{\Delta V}{V}$$

$$\sigma = E\epsilon_n, \quad \tau = G\gamma, \quad p = -K\Delta$$

$F(F_s)$ = normal (shear) component of force; A = area; $u(w)$ = normal (shear) component of displacement; $\sigma(\epsilon_n)$ = true tensile stress (nominal tensile strain); $\tau(\gamma)$ = true shear stress (true engineering shear strain); $p(\Delta)$ = external pressure (dilatation); ν = Poisson's ratio; E = Young's modulus; G = shear modulus; K = bulk modulus.

Chapter 8

Nominal/true stress/strain

$$\sigma_n = \frac{F}{A_0}, \quad \sigma = \frac{F}{A}, \quad \epsilon_n = \frac{u}{l_0} = \frac{l - l_0}{l_0}, \quad \epsilon = \int_{l_0}^{l} \frac{dl}{l} = \ln\left(\frac{l}{l_0}\right)$$

$A_0 l_0 = Al$ for *plastic* deformation; or for elastic or elastic/plastic deformation when $\nu = 0.5$. Hence

$$\sigma = \sigma_n(1 + \epsilon_n)$$

Also

$$\epsilon = \ln(1 + \epsilon_n)$$

Work of deformation, per unit volume

$$U = \int_{\epsilon_{n_1}}^{\epsilon_{n_2}} \sigma_n \, d\epsilon_n = \int_{\epsilon_1}^{\epsilon_2} \sigma \, d\epsilon$$

For linear-elastic deformation *only*

$$U = \frac{\sigma_n^2}{2E}$$

Hardness

$$H = F/A$$

σ_n = nominal stress; $A_0(l_0)$ = initial area (length); $A(l)$ = current area (length); ϵ = true strain.

Chapters 9 and 10

The dislocation yield-strength

$$\tau_y = \frac{2T}{bL}$$

$$\sigma_y = 3\tau_y$$

Grain-size effect

$$\tau_y = \beta d^{-1/2}$$

T = line tension (about $Gb^2/2$); b = Burgers vector; L = obstacle spacing; σ_y = yield strength; d = grain size; β = constant.

Chapter 11

Shear yield stress

$$k = \sigma_y/2$$

Hardness

$$H \approx 3\sigma_y$$

Necking starts when

$$\frac{d\sigma}{d\epsilon} = \sigma$$

Chapters 13 and 14

The stress intensity

$$K = Y\sigma\sqrt{\pi a}; \quad Y \approx 1$$

Fast fracture occurs when

$$K = K_c = \sqrt{EG_c}$$

a = crack length; Y = dimensionless constant; K_c = critical stress intensity or fracture toughness; G_c = critical strain energy release rate or toughness.

Chapter 16

Tensile strength of brittle material

$$\sigma_{TS} = \frac{K_c}{\sqrt{\pi a_m}}$$

Modulus of rupture

$$\sigma_r = \frac{6M_r}{bd^2}$$

Compressive strength of brittle material

$$\sigma_C \approx 15\sigma_{TS}$$

$$\sigma_C \approx \frac{15K_c}{\sqrt{\pi \tilde{a}}}$$

Weibull equation

$$P_s(V) = \exp\left\{-\frac{V}{V_0}\left(\frac{\sigma}{\sigma_0}\right)^m\right\} \quad \text{(constant stress)}$$

$$P_s(V) = \exp\left\{-\frac{1}{\sigma_0^m V_0}\int_V \sigma^m \mathrm{d}V\right\} \quad \text{(varying stress)}$$

$$P_s = 1 - P_f$$

K_c = fracture toughness; a_m = size of longest microcrack (crack depth for surface crack, crack half-length for buried crack); M_r = bending moment to cause fracture; b, d = width and depth of beam; $\tilde{a}$ = average crack length; P_s = survival probability of component; P_f = failure probability of component; V = volume of component; V_0 = volume of test specimen; σ = tensile stress in component; σ_0 = normalizing stress; m = Weibull modulus.

Chapters 17 and 18

Basquin's law (high cycle)

$$\Delta\sigma N_f^a = C_1$$

Coffin–Manson law (low cycle)

$$\Delta\epsilon^{pl} N_f^b = C_2$$

Goodman's rule

$$\Delta\sigma \text{ (for } \sigma_m = \sigma_m) = \Delta\sigma(\text{for } \sigma_m = 0)\left\{1 - \frac{\sigma_m}{\sigma_{TS}}\right\}$$

Miner's rule for cumulative damage

$$\sum_i \frac{N_i}{N_{fi}} = 1$$

Crack growth law

$$\frac{\mathrm{d}a}{\mathrm{d}N} = A\Delta K^m$$

Failure by crack growth

$$N_f = \int_{a_0}^{a_f} \frac{\mathrm{d}a}{A\Delta K^m}$$

$$SCF_{eff} = S(SCF - 1) + 1$$

$\Delta\sigma$ = tensile stress range; $\Delta\epsilon^{pl}$ = plastic strain range; ΔK = stress intensity range; N = cycles; N_f = cycles to failure; C_1, C_2, a, b, A, m = constants;

σ_m = tensile mean stress; σ_{TS} = tensile strength; a = crack length; SCF_{eff} = effective stress concentration factor; SCF = stress concentration factor; S = notch sensitivity factor.

Chapter 20

Creep rate

$$\dot{\epsilon}_{ss} = A\sigma^{n}e^{-Q/\bar{R}T}$$

$\dot{\epsilon}_{ss}$ = steady-state tensile strain-rate; Q = activation energy; $\bar{R}$ = universal gas constant; T = absolute temperature; A, n = constants.

Chapter 21

Fick's law

$$J = -D\frac{dc}{dx}$$

Arrhenius's law

$$\text{Rate} \propto e^{-Q/\bar{R}T}$$

Diffusion coefficient

$$D = D_0e^{-Q/\bar{R}T}$$

Diffusion distance

$$x \approx \sqrt{Dt}$$

J = diffusive flux; D = diffusion coefficient; c = concentration; x = distance; D_0 = pre-exponential exponential factor; t = time.

Chapter 24

Linear growth law for oxidation

$$\Delta m = k_L t; \quad k_L = A_L e^{-Q/\bar{R}T}$$

Parabolic growth law for oxidation

$$\Delta m^2 = k_p t; \quad k_p = A_p e^{-Q/\bar{R}T}$$

Δm = mass gain per unit area; k_L, k_P, A_L, A_P = constants.

Chapter 28

True contact area $a \approx P/\sigma_y$
P = contact force.

Magnitudes of properties

The listed properties lie, for most structural materials, in the range shown.

Moduli of elasticity, E	2–200 GN m^{-2}
Densities, ρ	1–10 Mg m^{-3}
Yield strengths, σ_y	20–200 MN m^{-2}
Toughnesses, G_c	0.2–200 kJ m^{-2}
Fracture toughnesses, K_c	0.2–200 MN m$^{-3/2}$

References

Ashby, M.F. and Cebon, D. *Case studies in materials selection.* Granta Design, 1996.

Ashby, M.F. and Johnson, K. *Materials and design – the art and science of material selection in product design.* Elsevier, 2002.

Ashby, M.F. *Materials selection in mechanical design*, 3rd edition. Elsevier, 2005.

ASM. *Metals handbook*, 2nd desk-top edition. ASM, 1999.

Bowden, F.P. and Tabor, D. *The friction and lubrication of solids, Part 1.* Oxford University Press, 1950.

Bowden, F.P. and Tabor, D. *The friction and lubrication of solids, Part 2.* Oxford University Press, 1965.

British Standards Institution. BS 7608: Code of practice for fatigue design and assessment of steel structures, 1993.

Broek, D. *The practical use of fracture mechanics.* Kluwer, 1989.

Calladine, C.R. *Plasticity for engineers.* Ellis Horwood, 1985.

Chapman, P.F. and Roberts, F. *Metal resources and energy.* Butterworths, 1983.

Charles, J.A., Crane, F.A.A., and Furness, J.A.G. *Selection and use of engineering materials.* 3rd edition. Butterworth-Heinemann, 1997.

Cottrell, A.H. *The mechanical properties of matter.* Wiley, 1964.

Cottrell, A.H. *Environmental economics.* Edward Arnold, 1977.

Crawford, R.J. *Plastics engineering*, 3rd edition. Butterworth-Heinemann, 1998.

Davidge, R.W. *Mechanical behaviour of ceramics.* Cambridge University Press, 1980.

Davies, G. *Materials for automotive bodies.* Elsevier, 2004.

Duggan, T.V. and Byrne, J. *Fatigue as a design criterion.* Macmillan, 1977.

Easterling, K.E. *Tomorrow's materials.* Institute of Materials, 1987.

Finnie, I. and Heller, W.R. *Creep of engineering materials.* McGraw-Hill, 1959.

Fontana, M.G. *Corrosion engineering.* 3rd edition. McGraw-Hill, 1987.

Frost, H.J. and Ashby, M.F. *Deformation mechanism maps.* Pergamon, 1982.

Gale, W. and Totemeier, T. *Smithells reference book*, 8th edition. Elsevier, 2003.

Gordon, J.E. *The new science of strong materials, or why you don't fall through the floor*, 2nd edition. Princeton University Press, 1988.

Gordon, J.E. *Structures - or why things don't fall down.* Da Capo Press, 2003.

Hertzberg, R.W. *Deformation and fracture of engineering materials*, 4th edition. Wiley, 1996.

Honeycombe, R.W.K. *The plastic deformation of metals.* Edward Arnold, 1968.

Hull, D. and Clyne, T.W. *An introduction to composite materials*, 2nd edition. Cambridge University Press, 1996.

Hull, D. *Introduction to dislocations*, 4th edition. Butterworth-Heinemann, 2001.

Hutchings, I.M. *Tribology: functions and wear of engineering materials.* Butterworth-Heinemann, 1992.

Kittel, C. *Introduction to solid state physics*, 7th edition. Wiley, 1996.

Kubaschewski, O. and Hopkins, B.E. *Oxidation of metals and alloys*, 2nd edition. Butterworths, 1962.

Lawn, B.R. *Fracture of brittle solids*, 2nd edition. Cambridge University Press, 1993.

Lewis, P.R., Reynolds, K., and Gagg, C. *Forensic materials engineering – case studies*. CRC Press, 2004.

Llewellyn, D.T. and Hudd, R.C. *Steels – metallurgy and applications*, 3rd edition. Butterworth-Heinemann, 1998.

Marx, S. and Pfau, W. *Observatories of the world*. Blandford Press, 1982.

McEvily, A.J. *Metal failures*. Wiley, 2002.

Murakami, Y. *Stress intensity factors handbook*. Pergamon, 1987.

Polmear, I.J. *Light alloys*. 3rd edition. Butterworth-Heinemann, 1995.

Powell, P.C. and Ingen Housz, A.J. *Engineering with polymers*. 2nd edition. Stanley Thornes, 1998.

Schijve, J. *Fatigue of structures and materials*. Kluwer, 2001.

Seymour, R.B. *Polymers for engineering applications*. ASM International, 1987.

Shewmon, P.G. *Diffusion in solids*, 2nd edition. TMS Publishers, 1989.

Ward, I.M. *Mechanical properties of solid polymers*, 2nd edition. Wiley, 1983.

Waterman, N.A. and Ashby, M.F. *Elsevier materials selector*. Elsevier, 1991.

Young, W.C. and Budynas, R.G. *Roark's formulas for stress and strain*, 7th edition. McGraw-Hill, 2001.

Index